AN INTRODUCTION TO
PROCESS DYNAMICS AND CONTROL

An Introduction to Process Dynamics and Control

THOMAS W. WEBER

*State University of New York
at Buffalo*

Krieger Publishing Company
Malabar, Florida

Original Edition 1973
Reprint Edition 1988 w/corrections

Printed and Published by
ROBERT E. KRIEGER PUBLISHING COMPANY, INC.
KRIEGER DRIVE
MALABAR, FLORIDA 32950

Library of Congress Cataloging-in-Publication Data

Weber, Thomas W., 1930-
    An introduction to process dynamics and control / Thomas W. Weber.
  -- Reprint ed. w/corrections.
      p.   cm.
    Reprint. Originally published: New York : Wiley, 1973.
    Bibliography: p.
    Includes index.
    ISBN 0-89464-301-0
    1. Process control. I. Title
TS156.8.W43 1988
660.2'81--dc19
                                      88-6410
                                        CIP

10   9   8   7   6   5   4   3

To

My Parents
My Family
and
My Teachers

Preface to Reprint Edition

On the occasion of the reprinting, I can write from the perspective of having used this textbook roughly twenty times over the past fifteen years. I should like to say a little bit about how the book is structured.

The book presents a balanced picture of control from the standpoints of theory and practical application. At the same time, about equal emphasis is given to control theory and process dynamics. Thus, about half of the material deals with the behavior of control systems, first qualitatively, and finally quantitatively; the other half considers the modeling and dynamic behavior of processes.

The content and style were developed with undergraduates in chemical engineering in mind. However, the "processes" analyzed are ones that are nearly all familiar to students in mechanical engineering. As a result, the book has had some adoptions over the years in courses in mechanical engineering.

Some may consider this book to be outdated. The controllers are assumed to be pneumatic, for example. The trend these days is toward computer control, but there are still many pneumatic controllers in industry. Moreover, whether a controller signal is sent in pneumatic form, or electrical form, the principles are basically the same. *This book does not consider digital or computer control.* This in no way compromises the material considered in this text, but rather is an extension of it that could be considered, perhaps even more naturally, in a more advanced or second course.

Process control is a subject that can be rather mysterious to someone who has never been exposed to it. The goal of this book is to remove the mystery, to prepare students in a rudimentary way for what they will confront in the industrial world, and to form the basis for further study along a number of specialized pathways. It is not intended to duplicate what a student might learn about control in an industrial environment, but rather to supplement or complement the industrial perspective. After having studied this book, the reader will have both a mathematical and a physical understanding of the dynamics of processes, why they behave the way they do under feedback control, and familiarity with two common methods for tuning controllers.

The pedagogical method of the book is mainly a series of examples, although in most cases, they are not labeled as such. It has been my experience that undergraduates grasp the material more rapidly this way than if it were presented more abstractly. The problems at the end of each chapter are designed to extend the concepts to other examples.

The book begins with a few introductory chapters to provide a broad, qualitative overview of some control methods such as feedback and feedforward. Familiar, easy-to-understand examples are used for these illustrations. Gradually some quantitative illustrations are introduced, first using only algebra, and then some calculus and simple differential equations. After a discussion of the Laplace transform, the process dynamics of a variety of simple processes are developed, beginning with first-order processes and advancing through second-order processes to some distributed ones. Finally, a quantitative overview is presented from the standpoints of transient response and frequency response.

I like to think of the book as being in the shape of an hourglass. The top and bottom view control systems as a whole, and the middle is devoted to a study of some of the details. The overview at the beginning provides a preview of the subject and points toward the areas that must be studied in order to predict the dynamic behavior of control systems in a quantitative way.

Because of this hourglass structure, the text has been divided into three parts. Part I begins with an introductory chapter that presents some historical background and a series of familiar examples. Chapter 2 discusses the four most common methods of control — open loop, environmental, feedback, and feedforward. Chapter 3 focuses on the two major reasons for control: servo operation and regulator operation. These chapters are so completely nonmathematical that almost anyone could read them!

The quantitative treatment of control commences with Chapter 4. Pure and simple algebra is used to discover the effect of proportional controller action on the steady state behavior of processes. Chapter 5 considers the unsteady state behavior of a simple process under proportional control using elementary calculus. The standard controller actions are taken up in Chapter 6 and the effects of some of them in simple systems are illustrated. Finally, Chapter 7 deals with the Laplace transformation and introduces the concept of a transfer function.

Having developed a broad picture of control systems and controller actions,

Part II examines the modeling and dynamics of the processes themselves. There are two chapters on first-order processes, a chapter on second-order processes, and finally a fairly long chapter on distributed processes. A unique feature of these chapters is the extensive development of the electrical analogies for nearly all of the processes that are examined. I believe that these analogies are a useful "second way" of looking at processes and moreover, elucidate the similarities between seemingly unrelated physical examples. A good understanding of these analogies can eliminate needless "reinventing the wheel" in process model development. Some students find the mastering of the electrical analogues to be "more work than they are worth." Fortunately, these analogues are not an essential element to understanding process control!

Part III returns to an examination of control system behavior but from a quantitative viewpoint. Chapter 12 provides an introduction to this area by means of an example drawn, to a considerable extent, from some experimental studies that I made. This development relies heavily on material in the previous four chapters and points the way toward the quantitative treatments of system behavior in the last two chapters. Chapter 13 considers the transient response of feedback systems to changes in set point and in load disturbances and leads quite naturally into a presentation of the reaction curve method for tuning controllers. Finally, Chapter 14 deals with frequency response analysis. The emphasis is on the Bode stability criterion, the use of frequency response as design tool, and a discussion of the continuous cycling method for tuning controllers.

Now to the matter of using the book in a course. Our semesters are fifteen weeks long and my three-credit course has three 50-minute lectures per week. Moreover, there is a one-credit laboratory which consists of four experiments. I feel that some laboratory component is essential for a course in control.

I cover the first three chapters in the first two lectures. The pace slows after that. We then go through Chapters 4 through 10 at the rate of about four lectures per chapter. Chapter 11 on distributed processes is completely skipped. While I don't believe that partial differential equations and the application of the Laplace transforms to them are beyond our students, it simply is not worth the effort and there is not enough time. I never cover Chapter 12 in detail, although I recommend that the students skim through the chapter, or at least take a look at it. I cover Chapter 13 completely in 4 lectures. That usually leaves me with five lectures to cover about two-thirds of Chapter 14 on frequency response. In particular, I cover about the first half of the chapter, and the last few pages on the continuous cycling method for controller tuning. Sometimes I give a lecture on pulsation damping as an example of a practical use of frequency response. I would like to cover more of this chapter, but about the only way would be to give up something else; one might consider a less complete coverage of Chapters 8 through 10 on modeling.

Problem solving is an essential to learning control. The problems in this book do not generally lend themselves to simply plugging numbers into a formula. I nearly always find it necessary to provide some hints to the students. A complete *Solutions Manual* is available from me.

My acknowledgments for this book remain the same as they were in the first printings: Professors Fred H. Rhodes and Julian C. Smith during my undergraduate days at Cornell University for their tireless pursuit of clear writing; and Professor Peter Harriott, who during my graduate studies at Cornell, whetted my thirst for the subject of process control. I should mention that my book is a companion to Professor Harriott's book, PROCESS CONTROL, also now published by the Robert E. Krieger Publishing Co., Inc. I am particularly grateful for the advice and help of my graduate student, Mohanlal A. Bhalodia. In addition to his technical advice and his preparation of the majority of the drawings for this book, I greatly appreciated his frankness and honesty.

Finally, I am indebted to my family who endured me during the final stages of the manuscript preparation. The writing of this book was the single most personally satisfying accomplishment of my professional career.

Contents

AN INTRODUCTION TO
PROCESS DYNAMICS AND CONTROL

PART I

Introductory Concepts of Control

This book is written in the shape of an hourglass. Part I is designed to convey an understanding of control systems without becoming overly mathematical. This is achieved by using simple models for processes and simple controller actions. Part II focuses on the various types of processes that are controlled. The section commences with relatively simple "lumped parameter" models and ends with an examination of some "distributed parameter" processes. Part III broadens the perspective again to see how the models of the middle section are combined to form a complete control system. The effects of various controller actions on system behavior are studied and some conventional methods for tuning controllers are discussed and compared.

The first three chapters of Part I constitute a qualitative introduction to control methods and systems. The development is based upon intuitive notions and past experience. The main objectives of control systems are discussed.

The quantitative treatment begins in Chapter 4. An understanding of the steady-state behavior of a system under proportional control can be obtained by just using algebra. The development of Chapter 4 is extended to unsteady-state behavior in Chapter 5. System complexity is minimized so that a first-order differential equation describes its behavior.

Chapter 6 examines the most common controller actions and their effects in controlling simple processes. The section concludes with an introduction to the Laplace transformation in Chapter 7. This tool is needed not only for the development of process models in the middle part, but also for the synthesis of complex systems in the last part.

1

CHAPTER I

Introduction

I. HOW OLD IS THE CONCEPT OF CONTROL?

Process control, of interest to chemical engineers, is a relatively new and specialized topic. However, the need for some background in process control is attested to by the very wide application of instruments and controllers in all types of chemical plants. Hundreds of controllers are used in a modern oil refinery, for example.

Within the past several decades there has been an enormous growth in the application of controllers, not only in chemical plants, but generally throughout technology. This has largely resulted from the increased research in this area. At the same time, there has been a gradual evolution from "classical control theory" to "modern control theory."

However, the concept of control and the use of control devices surely dates far back in history. Man himself was undoubtedly the first control "instrument." From his first appearance on earth, he has found it necessary to exercise disciplinary control over his offspring. For that matter, the idea of parental authority is readily observable in countless members of the animal kingdom. In recorded history, we find that the Romans invented a water-level control device similar to that used in many toilets.

Of course, some dividing lines can be drawn. There is a clear distinction between manual control and automatic control. Manual control implies that there is a man involved in the control system. Parental discipline of children is an example of this. In automatic control, man is essentially removed from the control system, and it controls by itself. The water-level device of the Romans falls into this category.

When people speak of control, they usually have automatic control in mind. From this standpoint, control had its beginnings with the Romans over 2000 years ago. A notable control invention was the flyball governor by James Watt in 1788. This was the crucial contribution made by Watt to the development of the steam engine. By the beginning of the twentieth century, the mathematical foundations for control theory had been laid by such people as Laplace and Fourier. Routh carried out work in analytical dynamics, Kirchhoff in circuit analysis, and Lord Kelvin and Heaviside in physics.

But the development of control theory, per se, and application of it began in the early 1920s. Minorsky's name frequently heads a list of names in the history of control. He was concerned with the automatic steering of ships (1). World War II brought a tremendous impetus for the advancement of control. Engineers and scientists were brought together from many disciplines. Such problems as the automatic bomb sight and control systems for antiaircraft guns required relatively sophisticated theory and equipment. Hence, if the beginnings of control theory mark the beginning of process control, then control dates back only about 40 years.

II. WHAT ARE THE INCENTIVES FOR PROCESS CONTROL?

Although automatic controls will frequently reduce manpower requirements in a plant, this economy factor is usually not the main justification for their installation. The fact is, controls are often a necessity. Standards of quality control for a product may require a degree of control not achievable manually. Sometimes a process may occur so rapidly that it eludes human capabilities. An example of this will be discussed at the end of this chapter concerning the possibility of a midair collision between two airplanes.

The economy factor of controllers has played an important role in some cases. Control has reduced costs by yielding a higher rate of production per dollar of equipment cost when the plant is running, and by reducing shutdown time. Ayres (2) points out that fixed charges continue whether a plant is running or not; hence, downtime can be very expensive. In particular, he cites the case of cracking units in oil refineries. At first, these units could be counted on to run about 87% of the time. In a period of 10 years, average running times rose to 93%. About half of this 6% increase was attributable to improvements in the control systems themselves through new devices and more intelligent use of older ones. The remaining half resulted from changes in operation and design made possible by improvements in automatic control.

If installation of control equipment can result in an improved operating factor, then the payout time for the controls is likely to be quite rapid, possibly in less than a year. This is a faster payout than is normally achieved with processing equipment itself.

Although improvement of the operating factor has been very significant, the biggest dividends from control have been achieved in the area of production capacity per dollar of plant investment. The impact in this area is vividly pointed out by Ayres: "Excluding taxes, the price of motor fuel today is little higher than in 1920, in spite of the shrinking dollar and rising costs, and almost the sole reason is improvement in process equipment made possible by automatic control."

In some cases, a strong argument for automatic control stems from the need for safety. Many simple processes are adequately operated manually, but they might be safer under automatic control.

III. SOME COMMON EXAMPLES OF CONTROL

Many examples of control are found in nature. The leaves of plants tend to turn in the direction of the sun, their motion being controlled by some extremely complex guidance-control system. Another example is the temperature control system of the human body. This system is designed to maintain the temperature at the normal value of 98.6°F. Although there is some variation in temperature with time as well as with position in the body, it is relatively small, especially in some regions, as for example in the brain. An interesting mathematical model of the human thermal system has been developed by Wissler (3). When the body is subjected to a sudden drop in the temperature of its surroundings, the control system senses that it must conserve heat. This is accomplished by reducing the flow rate of blood to the capillaries in the skin. The skin surface then becomes colder, but the internals of the body remain at 98.6°F. This action is reversed when the body is suddenly exposed to warmer surroundings. Blood flow to the capillaries is increased, thus increasing the body surface temperature and increasing the heat removal rate. The ratio of the maximum blood flow rate to the minimum may vary over a hundredfold range in highly vascular regions.

As an introduction to some of the terminology and concepts of control, we shall examine in detail several control systems with which the reader has some experience. One concerns a home-heating system and the other deals with the adjustment of the flow rates of hot and cold water in a shower.

In a home-heating system, the temperature of a room is sensed by the thermostat. When the temperature falls below the desired value, commonly referred to as the "set point" in control terminology, the furnace is turned on. The temperature then rises and when it reaches the set point, the furnace is turned off.

At first glance, it might seem as though perfect control could be achieved. If the temperature were very slightly below the set point, the furnace would be turned on and the temperature of the room would begin to rise. However, as soon as the temperature was very slightly above the set point, the furnace would be turned off and the temperature would begin to fall. This could lead to the conclusion that the temperature oscillates almost imperceptibly about the set point. As it turns out, satisfactory control is achieved, but the temperature may vary a degree or so above and below the set point.

To understand the reason for this, let us suppose that a door is opened, permitting some cold air to enter the room. The temperature may fall a degree or two. The furnace immediately is turned on, but it will take several minutes for the heat to reach the room. In particular, if the system is of the hot-water type, the hot water must be carried to the radiators or heating fins. Then this heat must be transferred to the air. There is a resistance to this heat transfer and the heat capacitance of the mass of air in the room is sizable. Finally, after the room has reached the set-point temperature, the furnace is turned off, but the radiators still contain some warm water, which causes the temperature of the room to continue to rise beyond the set point.

These factors result in some "overshoot" and "undershoot" of the temperature about the set-point temperature. In addition, a "gap" may be incorporated in the thermostat mechanism to insure that the furnace is not being continuously switched on and off. Therefore, the furnace may be on for 10 min, then off for 20, followed by a series of similar cycles.

Although the disturbance to the previously mentioned system was the opening of a door, many others are possible. There could be a sudden drop in the outside temperature, or even a change in the wind velocity. It is important to note that in this temperature control system, the source or cause of the disturbance is really not important. The thermostat cannot distinguish one cause from another, and it would not make any difference if it could.

In the analysis of a control system, it is helpful to draw a diagram indicating the components of the system and their interrelationships. One possible diagram is shown in Figure 1. A box has been used for each component and the boxes are connected by lines that represent signals between

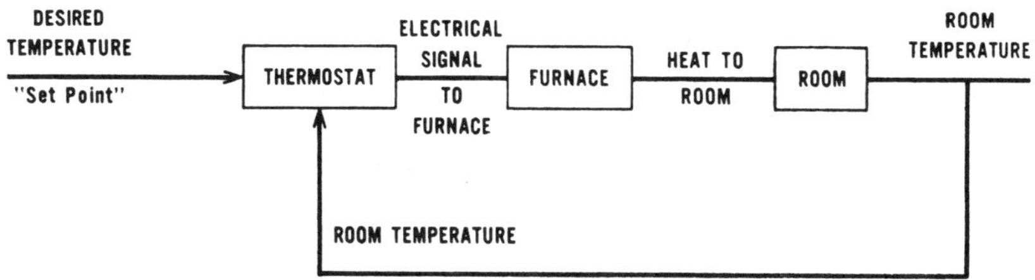

Figure 1.   Block diagram of home-heating system.

them. Other symbols or representations could be used, but the basic structure would be the same as that shown.

As indicated by the figure, the room temperature is sensed and compared with the set point by a "comparator." Here, the thermostat not only serves as the comparator, but also acts as a switch to turn the furnace on and off. An important feature of this system is the "feedback" of the room temperature to the comparator. While feedback is not a universal feature of all control schemes, it is one of the most commonly employed techniques. The reason for this will be explained in Chapter 2.

The control for a home-heating system is completely automatic. An example of a manual control system is the adjustment of the flow rates of hot and cold water in a shower. A flow diagram for the "process" is shown in Figure 2. In most installations, a mixing-T follows closely after the two valves and the combined stream flows up to the nozzle. As indicated, the total distance from the mixing-T to the man's head is about 5 ft. For convenience, we shall assume that the average velocity of the mixed stream is 5 ft/sec. Then, when one of the valves is adjusted, it will take about 1 sec before the new mixture of hot and cold water reaches the man's body.

From the standpoint of controlling the water temperature, the man serves three functions. First, he senses the temperature, perhaps using a finger or toe as the sensor during the initial stages of temperature adjustment, but finally using some appreciable portion of his body during the actual process of taking a shower. Second, he acts as the "comparator," in that he has some desired temperature in mind and he decides if the actual temperature is the desired value. Finally, if the actual temperature is not satisfactory, he must adjust one of the flows; this adjustment constitutes a "control action." We must now consider each of these functions in detail in order to draw a block diagram for the entire system.

The sensing of temperature is not instantaneous, since time is required for the nerves to carry thermal sensations to the brain. Most readers have had the experience of suddenly getting into a cold pool or a hot tub and

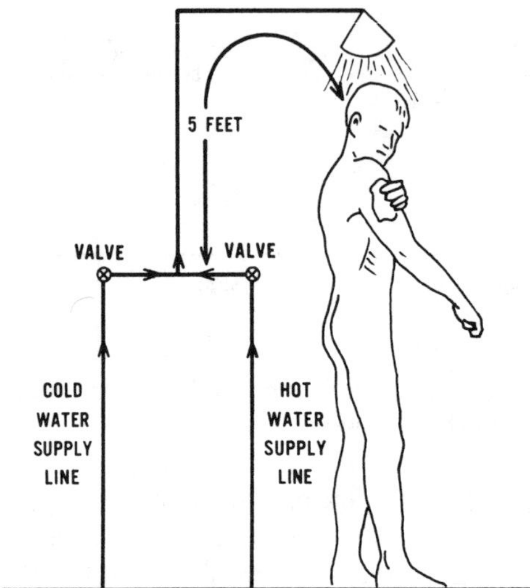

Figure 2. Flow diagram for shower example.

discovering a very short time later that its temperature was shocking. The point is, the sensation is delayed and not immediate.

Once the temperature of the shower water is sensed, the comparison action of the brain is nearly instantaneous. A delay then arises as the brain decides on a course of action. A further delay occurs when the brain sends out its decision of a course of action in the form of a message to a set of muscles. One possible message might be to jump out from under the shower if the water is too hot or too cold. But presuming that the temperature is nearly satisfactory, the command will be for a hand to make some adjustment of either the hot or cold valve. If this adjustment does not result in the desired temperature, the cycle of events is repeated.

A block diagram for the control system is shown in Figure 3. The comparator is indicated by a circle, the usual symbol used in block diagrams for control systems. The difference between the desired temperature and that sensed by the body is the "error." Each of the delays discussed above is indicated by a block. As in the home-heating system, the element of feedback is again present.

Delays sometimes make it difficult to achieve satisfactory control for some processes. The shower example provides a qualitative explanation for this. Suppose the temperature of the water is too low and the hot water valve is opened further. If the valve is opened too much on the first adjustment,

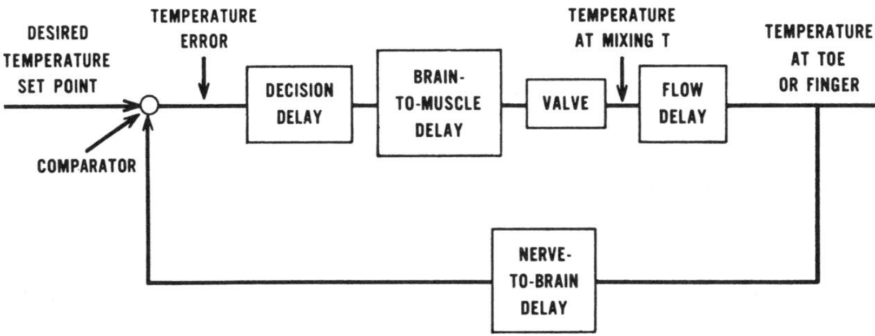

Figure 3.   Block diagram of shower control system.

then a second and perhaps even more adjustments will be necessary. However, because of the delays in the system, a period that is closely related to the sum of the delays will probably separate each individual adjustment of the valve. Hence, if the delays could be reduced or even eliminated, the adjustment period would be reduced as well.

This suggests some changes that might be made in the shower installation to improve the temperature control. From the standpoint of the control system, the man might be removed and replaced by an elaborate automatic temperature control device with a temperature sensor right at the mixing-T. This would eliminate both human and flow delays. However, automatic controls for showers would hardly be justifiable economically. A cheaper way to improve the system would be to locate the valves and mixing-T right next to the nozzle. The flow delay would be eliminated but the location would be inconvenient.

As an extreme example of the significance of "human delays," an interesting illustration is given in Reference 4 of a pilot, who upon bringing his supersonic plane out of a cloud formation, finds that it is on a collision course with another plane. These planes are moving toward each other at a speed of 1800 miles/hr, which means that their closing rate is 1 mile/sec.

From the time that the eyes of the pilot perceive the other plane to the time the message reaches his brain, a delay of about 0.3 sec occurs. It takes him 0.6 sec to realize that there is impending danger, followed by an additional 0.5 sec to decide on a course of action. Finally, another 0.3 sec is consumed in initiating the desired control manipulation. Thus, the total delay of the pilot is 1.7 sec. The plane, in turn, cannot respond instantaneously because of its inertia. If the initial sighting were made when the planes were about 3 miles apart, a collision would occur.

SUMMARY

Several examples of control have been presented in this chapter from rather diverse areas. In the systems that were examined in detail, it was found that there was a comparison made between the set point and the variable being controlled, and the system was said to involve feedback.

Perfect control is usually not possible because processes and elements of a control system do not respond instantaneously. This is true of both automatic and manual control. In cases where very fast control is required, manual control may not be satisfactory because of human delays.

The use of block diagrams was introduced as a way of clarifying the interrelationships between control components. This is the first step in the analysis of a control system.

REFERENCES

1. N. Minorsky, "Directional Stability of Automatically Steered Bodies," *J. Am. Soc. Nav. Eng.*, **34** (1922), 280.
2. E. Ayres, "An Automatic Chemical Plant," in *Automatic Control, A Scientific American Book*, Simon and Schuster, New York, 1955, pp. 41–52.
3. E. H. Wissler, "A Mathematical Model of the Human Thermal System," C. E. P. Symp. Series, **62**, 66 (1966).
4. J. H. Ely, M. Bowen, and J. Orlansky, "Man-machine Dynamics," in C. T. Morgan, J. S. Cook, A. Chapanis, and M. W. Lund (Eds.), *Human Engineering Guide to Equipment Design*, McGraw-Hill Book Company, New York, 1963, pp. 217–245.

PROBLEMS

Problem 1.   There is a rough rule of thumb that, for safety, an automobile driver should be at least one car length behind the car ahead of him for each 10 miles/hr. Is this rule reasonable?

Problem 2.   The New York State Department of Motor Vehicles suggests that motorcyclists be at least 50 ft behind the vehicle in front of them when they are going 20 miles/hr, 100 ft at 30 miles/hr, and more than 300 ft at 40 miles/hr. Are these suggested guidelines reasonable?

Problem 3.   Draw a block diagram for the positioning system of a plant leaf.

Problem 4.   An optional accessory on automobiles is an automatic speed control that enables the driver to set the speed desired. He can then remove his foot from the gas pedal and the car will maintain the desired speed in spite of changes in grade. The driver can override the control should an emergency arise. Draw a block diagram for this automatic control system.

Problem 5. In the shower example of this chapter, the controlled variable was the temperature of the mixed stream. A second variable of equal importance to a person taking a shower is the flow rate of the mixed stream. Draw a block diagram with the total flow rate as the controlled variable. In particular, note that there is an interaction between the temperature control system and the flow control system in that a change in flow rate of either the hot or cold water affects both the temperature and the total flow rate.

Problem 6. The driver of an automobile sometimes "loses control" of it on an icy pavement and the result is a skid and perhaps a spin. Discuss the causes of a spin from the standpoint of control.

CHAPTER   II

Methods of Control

The objective of a control system may be to maintain the temperature of a room constant, or the composition in a reactor constant. There are likely to be several solutions to each of these problems. One engineer may suggest a feedback scheme, another may recommend a feedforward scheme, and a third may advise a combination of the two. The point is that there is no unique answer.

Some of the common methods of control will be discussed in this chapter, but before turning to these, a brief discussion of block diagrams will be useful.

I. BLOCK DIAGRAMS

In Chapter 1, a few block diagrams were drawn, somewhat intuitively, how-ever. Although any given system might be adequately described by a number of different diagrams, a block diagram method has received broad acceptance, especially in process control. Most block diagrams are composed of lines, blocks, comparators, and summers, arranged in a logical order to indicate the components of a control system and their interrelationships.

The lines represent signals and each line represents a single signal. Any quantity that can be measured or calculated can be a signal on a diagram. Some examples are voltage, flow rate, air pressure, temperature, position, and light intensity.

A block in a diagram can be thought of as a "transformer" in the sense that it receives an input signal and transforms or operates on it in some way to produce an output signal. In a simple case, it might amplify the input or attenuate it. In the last chapter, a block was used to signify that an input signal was delayed. This means that the output signal is an exact replica of the input signal, but delayed from the input by a fixed amount of time. More elaborate transformations are common. The output signal might be the integral of the input signal, the derivative of it, or possibly even the sum of the integral and derivative. The main point to remember is that each block performs some transformation of a *single* input into a *single* output. A single signal, of course, may be an input to more than one block.

Comparators and summers are similar. A comparator takes the difference between two signals, whereas a summer adds two signals. Both are represented by circles. The two input signals are indicated by arrows pointing into the circle. If the device is a comparator, one of the two inputs will have a negative sign next to its arrow. If no signs are shown, the device is a summer.

These rules will now be demonstrated as we turn to an examination of the four simplest control methods: (1) open-loop, (2) environmental, (3) feedback, and (4) feedforward.

II. OPEN-LOOP CONTROL

Of the four schemes to be discussed, this one is the simplest. The terminology of "open-loop" becomes reasonable and understandable after a study of the feedback case.

As a simple example of open-loop control, consider the electric timer used for traffic lights at an intersection. Commonly, the light is green for a fixed period of time, then yellow for a short time, and finally red for another fixed period of time; the cycle then repeats itself over and over again. Here the input to the control system is the time duration for each light. The output is the flow of traffic. A traffic engineer calibrates the timing cycle of the lights to the traffic patterns and density at the intersection. If the control is good, then there will be no traffic pileups in any direction. Where traffic loads vary considerably throughout the day and night, it may be difficult to set a cycle that is satisfactory for all times, and some compromise setting will be used. Because of changes in traffic patterns over long periods of time, it may become necessary to adjust the cycle occasionally. However, this adjustment is not part of the normal operation of this type of control system.

As a second example, consider the mixing system shown in Figure 1 for producing a weak salt solution from water and concentrated brine. The water stream flows into the tank at a constant rate of F lb/hr. The brine, of concentration x_b weight percent salt, flows into the tank at a rate of B lb/hr. Since F is constant, the product composition, x_p, depends only upon B. Hence, an indicator, connected to the stem of the valve, can be used to show the product concentration on a calibrated dial. The calibration could be carried out experimentally or calculated on the basis of a mass balance and the flow characteristics of the valve.

A block diagram for the system is shown in Figure 2. As indicated, when the indicator position is changed, the brine rate changes and the result is a change in the product concentration. It is important to note that the product concentration is not compared with the concentration indicated on

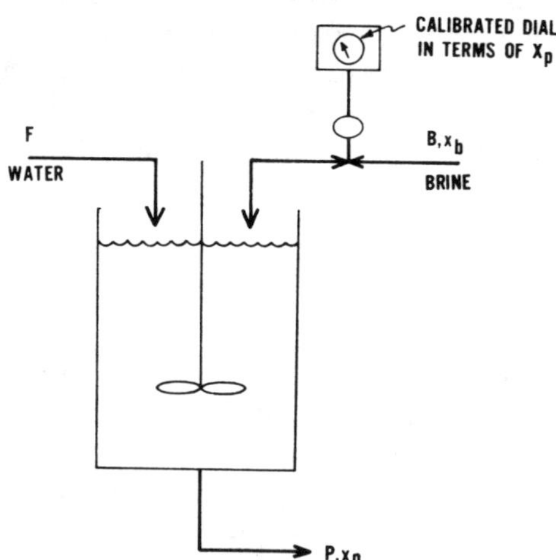

Figure 1. Salt solution mixer under open-loop control.

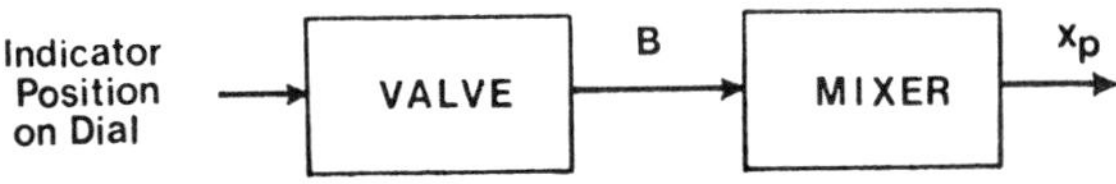

Figure 2.   Block diagram for salt solution mixer under open-loop control.

the dial. This lack of a comparison is the main disadvantage of this control method. When outside disturbances occur, this method is unable to correct for them. For example, if the water rate changed, the dial calibration would no longer be meaningful. Even if the water rate were to remain at the design value, the orifice of the brine valve would gradually wear or erode so that the dial would require recalibration from time to time. The advantage of this method for this process is that it requires only a calibrated valve. The resulting simplicity and economy are advantages common to all open-loop schemes.

III. ENVIRONMENTAL CONTROL

The vulnerability of the open-loop method to disturbances suggests controlling the disturbances themselves to eliminate them. This tactic is referred to as "environmental control." For the brine-water mixer, the most troublesome disturbances to satisfactory open-loop control of x_p would be the water flow rate, F, and the brine concentration, x_b. Environmental control would therefore involve separate control systems for F and x_b. However, as noted previously, the characteristics of the brine valve might change slowly with time and this would affect the calibration. There is little that could be done to avoid this problem except to use a durable valve.

As with open-loop control, the concept of environmental control is basically simple. Since it is based upon the idea of constraining each of the possible disturbance variables to its design value, it could be termed a "brute force" technique. This method can be very effective in achieving good control if all significant disturbance variables are under tight control.

The main disadvantages of environmental control center around its cost and the need for identifying all of the possible disturbances. Since all of these must be controlled, this method may be very expensive if there is a large number of them. Furthermore, it might be ruled out in some cases because one or more of these disturbance variables cannot be controlled, or at least not at reasonable expense. Finally, there is always the possibility that some disturbance variables may be overlooked or unforeseen and this method will be unable to correct for them.

IV. FEEDBACK CONTROL

A. General Concepts

The concept of feedback was introduced in Chapter 1 in connection with the examples of a home-heating system and the adjustment of the water temperature in a shower. To gain a better understanding of this method of control, we shall reexamine the example of the home-heating system. The block diagram for this system is shown in Figure 3. The thermostat not only

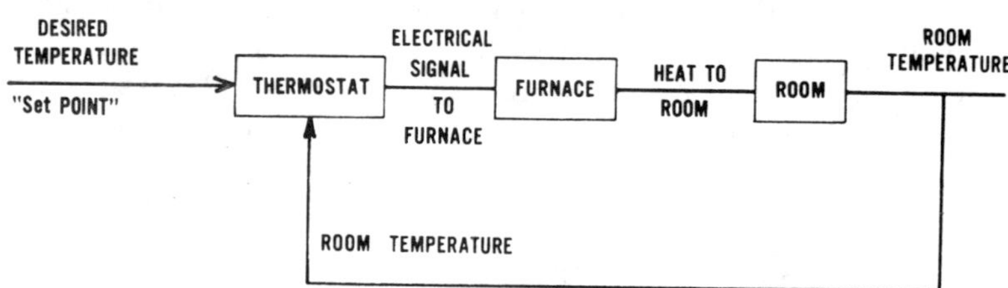

Figure 3.   Block diagram of home-heating system.

serves as the comparator between the set point and the actual room temperature, but also provides the "controller action" in the form of an on-off switch. If the temperature in the room is below the set point, the switch turns the furnace on, and if it is above the set point, it turns it off. For convenience, the furnace and room blocks can be combined into a single "process block" by which the electrical signal to the furnace is transformed into a room temperature. Hence, the block diagram can be reduced to a general form, shown in Figure 4.

There are several universal features of a block diagram for a feedback system. The input or "set point" is the desired value of the output or "controlled variable." The difference between the set point and the controlled

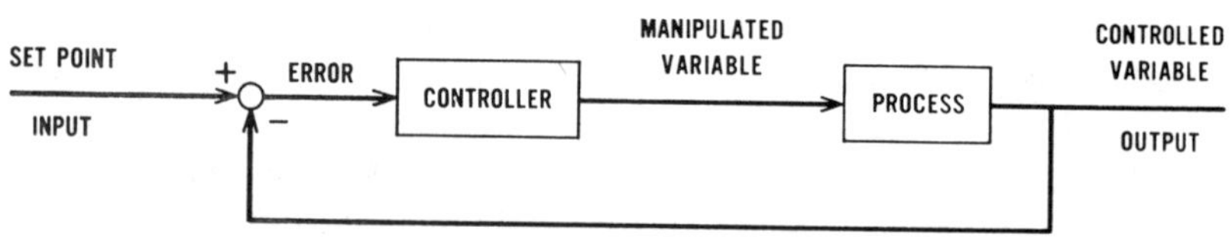

Figure 4.   General feedback block diagram.

variable is the "error." The controller operates on the error and produces an output, which is normally a pneumatic or electrical signal. In most process control systems, this signal actuates a valve, in turn manipulating the flow of some stream. This flow rate is the "manipulated variable" and it is changed in a direction so as to cause the controlled variable to move toward the set point. Besides these variables, there are many others. These are the disturbance variables and are referred to as "load variables." None of these is indicated on Figure 3, but they will appear on most of the block diagrams in future chapters.

The existence of feedback is readily apparent in Figure 4, but the general concept of feedback goes beyond its application to the control of processes. Therefore, a somewhat more formal definition of feedback is in order. Feedback exists when an output variable depends on a variable that is, itself, a function of the output and the input. This occurs in Figure 4. The controlled or output variable depends upon the error, which is a function of the controlled variable and the input or set point.

If the controlled variable were not compared with the set point, the effect would be to "open the loop," and the bottom "feedback line" in Figure 4 would be removed. The resulting configuration would be precisely that in open-loop control and this explains why the block diagram of Figure 2 was referred to as an open-loop situation.

B. Some Examples of Feedback Control

In the home-heating example, there is a continuous comparison of the actual temperature with the set point. Feedback may be present even when the comparison is made only intermittently. Consider, for example, a pilot who sets out on a flight from New York to Chicago. If he merely took off and aimed his plane in the direction of Chicago, he might find himself flying over St. Louis several hours later because of crosswinds. To avoid this possibility, a pilot makes periodic determinations of his location so that he can compensate for winds. Thus, he is using the concept of feedback. Intuitively, it can be seen that the most direct course would be followed if the pilot made essentially continuous checks of his location and corresponding manipulations of the controls.

A motorcycle provides an interesting example of a process that is inherently unstable if uncontrolled but which can be stabilized under feedback control. Consider first the situation when it is being ridden. The rider is the "controller" and the motorcycle itself is the "process." The rider senses deviations or "errors" from the vertical position and he corrects for these by shifting his weight or by moving the handle bars. The motorcycle will not fall over as long as the rider continues to exert control, and therefore the motorcycle is "stable." However, when the rider dismounts, the motorcycle will fall over unless he puts the stand in place. Without the stand, the motorcycle itself is unstable.

The possibility of an inherently unstable chemical reactor has received considerable attention in the chemical engineering literature (1–3). To see how this can come about, recall the rule of thumb that a reaction rate doubles for each 10-degree increase in temperature. If the reaction is exothermic, then as the temperature rises, the rate of heat generation increases. At the same time, the driving force for heat transfer from the reactor to its coolant increases. If the rate of change of heat removal with temperature is less than the rate of change of heat generation with temperature, then the reactor will be inherently unstable. However, with a properly designed controller, the system can be stabilized.

C. Negative versus Positive Feedback

There are two kinds of feedback, negative and positive. The example of the home-heating system can be used to illustrate both types. When correctly installed, if the measured temperature is above the set point, the furnace is switched off, but if below the set point, the furnace is switched on. Thus, the action of the controller is in a direction that tends to move the controlled variable toward the set point, and this is referred to as "negative feedback."

Now suppose that the installer wires the thermostat switch backwards. Further suppose that initially the measured temperature and set point are identical. Then, if the temperature in the room rises very slightly, the furnace will switch on, causing a further rise in the temperature, and the furnace will continue to operate indefinitely. The action of the controller here is in a direction that tends to move the controlled variable further away from the set point, and this is referred to as "positive feedback." Returning again to the initial state where the measured temperature and set point are identical, if the temperature falls slightly, the furnace will remain off and the temperature will continue to fall, again a manifestation of positive feedback. This explains why negative feedback is always used in control applications.

In a process controller, there is a means for "reversing the action" so that negative feedback can always be accomplished with any controller. Hence, when ordering one, it is not necessary to specify whether the controller action be positive or negative.

In electrical engineering, there are many examples of positive and negative feedback. In the early days of radio, regenerative receivers were used in which weak radio signals were greatly amplified through the use of positive feedback. As illustrated in the heating system example, positive feedback can lead to instabilities, and this is exactly what is desired in audio and radio oscillators that are used so extensively in electronic equipment. Negative feedback is used almost universally in high-fidelity amplifiers to improve frequency response. At the same time, it makes the performance of such devices less sensitive to changes in tubes and other components as they age. Automatic volume controls (AVC) and automatic frequency controls (AFC) also incorporate negative feedback principles.

D. Advantages and Disadvantages

Feedback has many advantages over open-loop and environmental control which make it so widely used. The chief advantages arise because the value of the controlled variable is continuously compared with the desired value. This method pays no attention to what the possible disturbances are. In theory, it can compensate for all of them. In all but the most trivial cases, it will involve fewer controllers than environmental control and therefore is likely to be relatively inexpensive.

The main disadvantage of feedback control is a result of the principle upon which it is based—that is, control action to correct for load disturbances does not occur until an error occurs, and errors are just what the control system is attempting to eliminate. Once an error is detected and control action begins, the effect of this action will be retarded by the various delays around the control loop.

Another problem with feedback, somewhat related to the previous one, is that it operates by trial-and-error. If the control action is relatively strong, then there will be a tendency for the controlled variable to "overshoot" the set point, with subsequent oscillations back and forth. This tendency to oscillate or "hunt" is typical of feedback systems. On the other hand, if the control action is relatively weak, the controlled variable will move slowly, and perhaps sluggishly, toward the set point. Some oscillatory behavior may be tolerable and may, in fact, be preferable to nonoscillatory but sluggish behavior.

The disadvantage of not being able to correct for load disturbances until they result in errors has led to the use of feedforward control.

V. FEEDFORWARD CONTROL

A. General Concepts

In feedforward control, the idea is to *detect* disturbances and attempt to correct for them in the system before an error can develop. This is in contrast to environmental control where the object is to *prevent* disturbances from entering the process in the first place. In a completely feedforward system, each major load variable must be measured and compensated for. A different variable might be manipulated for each of these load variables, or perhaps a single one could be used to compensate for several. A common application of feedforward control is "ratio control," discussed in the following example.

B. Example

Consider again the salt-solution mixer shown in Figure 1. For simplicity, suppose that the only load variable is the water flow rate, F, and this variable is to be compensated for through feedforward control. Therefore, F must be measured and deviations from its design value must be compensated for by automatic changes in the valve position.

A physical picture of the feedforward control system is shown in Figure 5. The flow rate is measured here by an orifice meter whose output is fed to a computer or controller. The computer then determines the necessary changes in the position of the valve. As shown in the figure, a calibrated dial has been included in the system to enable changes to be made in the product concentration if desired. Therefore, this system is actually a combination of feedforward and open-loop control.

The corresponding block diagram is presented in Figure 6. Two "process blocks" are needed here since both F and B affect x_p. To understand this better, consider the following equation for x_p as obtained by a salt balance:

$$x_p = \frac{B}{F + B} x_b$$

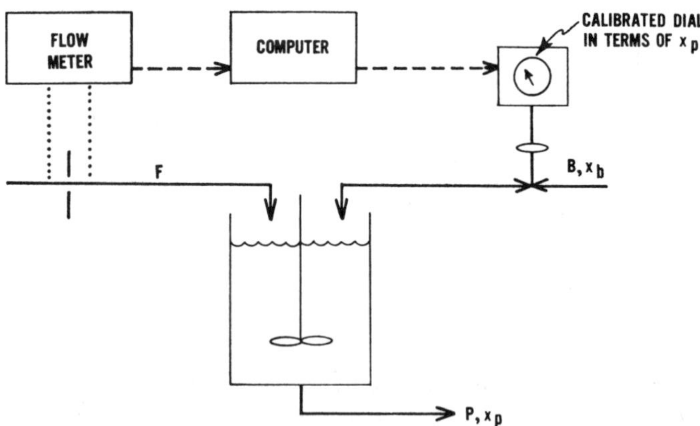

Figure 5.   Salt solution mixer under feedforward control.

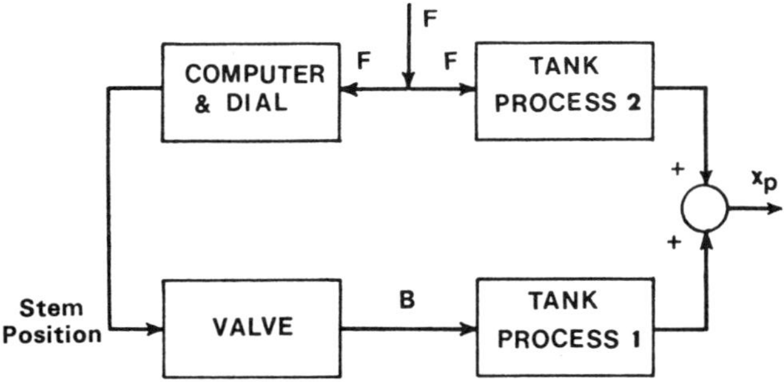

Figure 6.   Block diagram corresponding to Figure 5.

When a change in F occurs, there is a direct effect on x_p, and this is indicated by the block, "Tank Process 2," on the diagram. The computer determines a compensating change in B that affects x_p, through the block denoted "Tank Process 1." The separate blocks are necessitated by the rule that a block can have only one input.

From the physical aspects of the system, it can be seen that the product concentration will remain constant as long as the ratio of the feed streams is maintained constant. This is the task of the feedforward computer here, and explains why this application of feedforward control is sometimes referred to as "ratio control."

The feedforward scheme just discussed would, of course, not compensate for changes in the brine concentration, x_b. If this concentration were measured, then changes in this variable could also be compensated for by the feedforward computer. The manipulated variable could also be the stem position of the brine valve or a control valve could be placed in the waterline to manipulate F.

C. Advantages and Disadvantages

The main advantage of feedforward control is that it detects a load disturbance when it occurs and compensates for it *before* an error develops. This contrasts sharply with feedback control that takes no action until *after* an error occurs. Perfect control is theoretically possible by feedforward control for some systems. Even in those cases where exact compensation is not possible, at least a partial compensation may be obtained, and this can be very advantageous.

There are several disadvantages of feedforward. As in the case of environmental control, feedforward requires the identification of all major disturbances in advance. Then the effects of each of these must be determined, either theoretically or through an experimental calibration. When there are several load variables to be compensated for, a single computer can usually be used for all of them. However, since each load variable must be measured, more equipment will probably be required than for a feedback scheme.

The design of a feedforward system generally requires fairly extensive knowledge of the process dynamics of the process. Even if this information is known, it may turn out that exact compensation is not physically possible. Probably the most serious disadvantage of feedforward control is that the controlled variable is not compared with the desired value. For these reasons, feedforward control is seldom used alone.

VI. COMBINATION SYSTEMS

As previously noted, the feedforward control system was actually a combination of open-loop and feedforward since a calibrated dial was provided to enable an occasional change in the product concentration. Two other combinations will now be illustrated.

A. Feedforward-Feedback

The combination of feedforward with feedback is rather commonly used since these two schemes are basically complementary. That is, the advantages of feedforward can compensate to some extent for the disadvantages of feedback, and vice versa. More specifically, the feedforward portion of the system could be used to attempt to compensate for the most serious or significant load variable, and the feedback portion could take care of the imperfections in the feedforward portion as well as other load variables.

The salt-solution mixer will be used to illustrate this combination. As before, the water rate, F, will be assumed to be a major load variable, and a second load variable will be the brine concentration, x_b. Of the two, F will be assumed to be the most troublesome and therefore will be chosen for feedforward compensation. Therefore, x_b and other load variables will be corrected for by a feedback loop.

A physical picture of the system is shown in Figure 7. The flow meter and controller of the feedforward portion of the system are combined in a single unit. Most valves used in process control are operated pneumatically by air pressure, which operates against a diaphragm with an attached valve stem. The standard range of pressure is between 3 and 15 psig. In an air-to-open valve, the valve is fully opened at 15 psig and fully closed at 3 psig; an air-to-close valve operates in just the reverse manner. When a change in flow

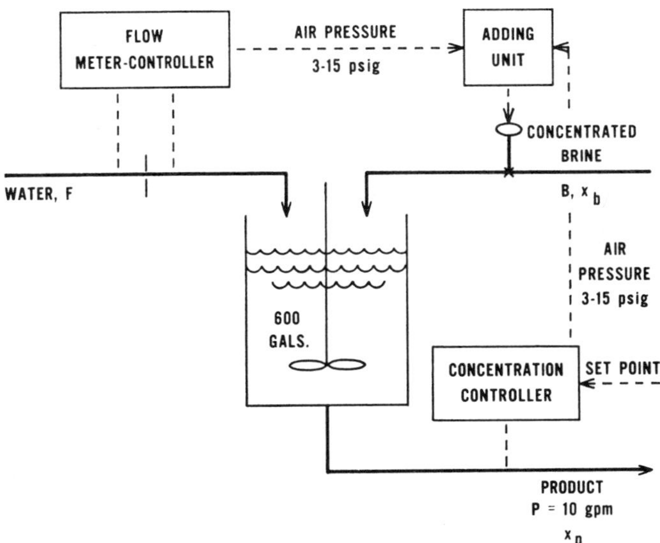

Figure 7.   Salt solution mixer under feedforward-feedback control.

rate, F, is detected by the meter, the flow controller attempts to compensate for the change by an appropriate change in air pressure to the brine valve.

The concentration controller senses changes in product concentration, which is the controlled variable; hence, this is a feedback arrangement. The output of this controller is a pressure that also affects the stem position of the brine valve. Since this valve is positioned by both the feedforward and feedback controllers, an adding unit is used to combine the two effects.

If a change in water flow rate occurs, the feedforward compensation is essentially instantaneous, and if the compensation is exact, there will be no change in the product concentration. If it is not exact, the error will be detected and corrected by the feedback loop in the same way as changes in x_b will be corrected. The corrections by way of the feedback portion of the system are apt to be comparatively slow. Suppose, for example, that the tank has a holdup of 600 gal and the normal flow rate of product is 10 gal/min. Therefore, the holdup time is 1 hr. Since this is in the nature of a delay, the feedback control will tend to be sluggish.

The block diagram for the system is presented in Figure 8. The orientation follows the convention of placing the set point at the far left of the diagram and the actual value of the controlled variable at the far right. The feedback and feedforward portions of the system are evident by the directions of the arrows. It should be noted that there is no direct relationship between the

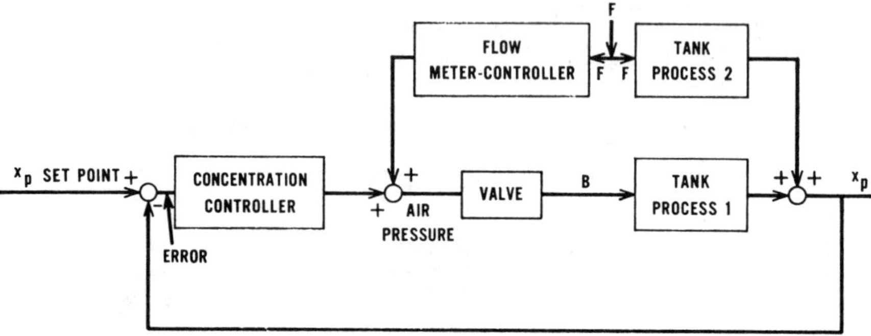

Figure 8.   Block diagram corresponding to Figure 7.

physical drawing of the system and the block diagram. This is usually the case in process control systems.

B. Feedback-Environmental

There are situations in which the control at first glance may appear to be by feedback, without this actually being the case. Examine the system shown in Figure 9. The product composition, x_p, is controlled by controlling both the

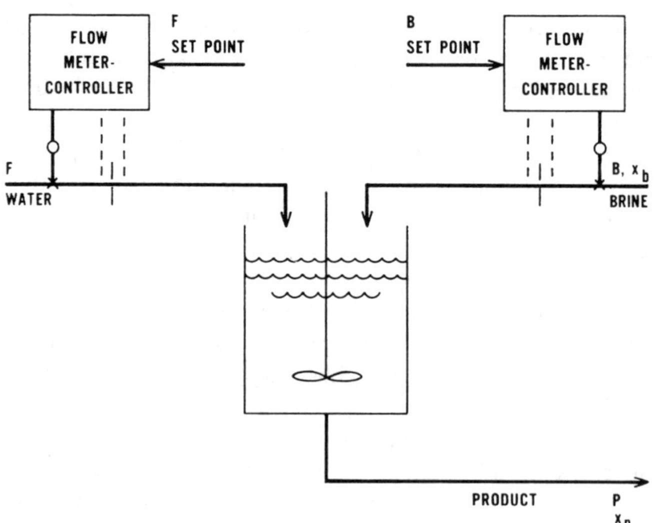

Figure 9.   Environmental control using two feedback loops.

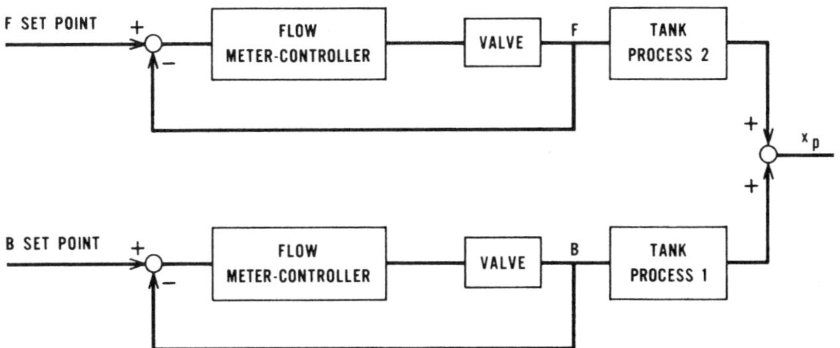

Figure 10.   Block diagram corresponding to Figure 9.

water and brine flow rates with individual controllers. The corresponding block diagram is shown in Figure 10. Two feedback loops are present, yet the overall control method falls into the category of environmental. The key point is that the controlled variable, x_p, is not used as an input to a controller to manipulate its own value. One possible load variable not compensated for in this scheme is the brine concentration, x_b. If this were absolutely constant, then this control system could be adequate though not necessarily. The degree of effectiveness in this case would depend on how good the two feedback loops were for the water and brine flow rates.

SUMMARY

The four main methods of control were presented in this chapter. Each has its advantages and disadvantages, and these must always be weighed against each other in making a choice for a particular process. Questions concerning economics as well as the degree of quality of control desired must be considered. Although probably over 90% of all control systems employ feedback, other methods should always be examined.

In closing, it should be mentioned that not everything is controllable. There are three conditions that must be met for controllability. First, there must be some appropriate variable to manipulate in order to produce a change in the controlled variable. In the case of an automobile, for example, its speed is varied by manipulating the feed rate of gasoline. Second, there must be some means for measuring the controlled variable, or at least for comparing it with some standard. This measurement is accomplished by the speedometer of an automobile. Third, both the response of the system and the measurement of this response must be sufficiently rapid to meet the requirements of the system. In an automobile, the response of the speedometer is usually not a problem, but the response of the automobile can be. For example, the condition of the carburetion and ignition systems can be crucial in obtaining the necessary response when an automobile is attempting to pass another on a two-lane highway.

REFERENCES

1. C. van Heerden, "Autothermic Processes," *Ind. Eng. Chem.*, **45** (1953), 1242.
2. R. Aris and N. R. Amundson, "An Analysis of Chemical Reactor Stability and Control," *Chem. Eng. Sci.*, **7** (1958), 121.
3. T. W. Weber and P. Harriott, "Control of a Continuous-Flow Agitated-Tank Reactor," *Ind. Eng. Chem. Fund.*, **4** (1965), 264.

PROBLEMS

Problem 1.   Turning to Figure 1 and the accompanying discussion of this open-loop control scheme, suppose that the operating parameters are:

$$F = 100 \text{ lb/hr}$$

$$x_b = 0.1 \text{ lb salt/lb brine}$$

Calculate the variation in product composition, x_p, with change in brine rate, B. Is the calibration likely to be linear with dial rotation?

Problem 2.   Frequently in the adjustment of a public address system, the volume control will be turned up too high, causing the system to howl or whistle. Draw a block diagram describing this phenomena and explain what causes the whistle. Is the feedback positive or negative?

Problem 3.   The water storage tank for a toilet incorporates a control system. One end of a lever is attached to a float while the other is connected to the stem of a valve. When the tank is nearly full, the water is automatically shut off. Draw a block diagram of this system. What is the manipulated variable? The controlled variable? List the major disturbance variable or variables.

Problem 4.   Hot water is produced by passing cold water through the tubes of a shell-and-tube heat exchanger with steam in the shell. The temperature of the hot water is controlled by varying the pressure of the steam.

(a)   Draw the block diagram for this system.
(b)   List a number of possible load variables. Which are the most significant?

(c) If a good feedforward control system were to be used, what variables
should be taken into account? Draw a sketch of the physical system
and the corresponding block diagram.

Problem 5.   Hot water is produced by passing cold water through the tubes of
a shell-and-tube heat exchanger with steam in the shell. The temperature of
the hot water is controlled by varying the amount of water bypassed around
the exchanger. A picture of the system is shown in Figure 11.

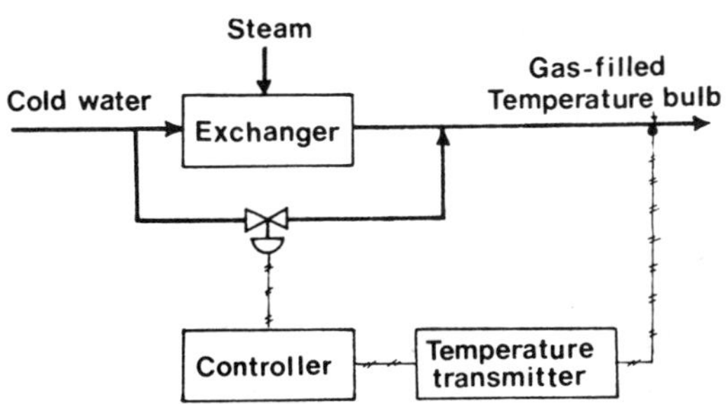

Figure 11.   Physical picture for Problem 5.

(a) List the most probable significant load variables.
(b) Draw a block diagram for this system, carefully identifying the signals
between the blocks.
(c) In an alternate scheme, the temperature of the hot water could be
controlled by manipulating the pressure of the steam. Of the two schemes,
which would give the fastest response?
(d) Suppose that the temperature of the mixed stream is sensed twenty
feet beyond the mixing-T and that the normal velocity of the combined
stream is 4 ft/sec. Indicate how this would change the block diagram
of part *b*.
(e) Draw a physical picture for a control system employing only feedforward control where the measured temperature is that of the entering
cold water and the manipulated variable is the flow rate of the bypass
stream. Draw the corresponding block diagram.

Problem 6.   A compressed air system is to be installed in a new gasoline
station. The air will be used at two tire pump locations and for two hydraulic
lifts.

(a) Identify the *significant* load variables.
(b) Briefly discuss the application of environmental, feedback, and feedforward to this problem. Of the three, which is best for the system?

(c) Draw a physical picture of the best system and the corresponding block diagram.

Problem 7.   Hot water is produced by heat exchange with steam in a shell-and-tube heat exchanger. The steam is on the shell side. The temperature of the water leaving the exchanger is to be closely controlled.

(a) Suggest two possible feedback schemes for achieving the desired result. Draw a carefully labeled block diagram for each of these. List the major load variables for each scheme.
(b) Suppose that the supply pressure of the water is the main load variable. Draw a physical picture and block diagram for a feedforward scheme.
(c) Redraw the diagrams of part *b* but augment them to include feedback.
(d) Suggest an environmental control scheme for this exchanger and draw the corresponding physical picture and block diagram. The only load variables are the water flow rate and the steam pressure.

Problem 8.   Two tanks are connected as shown in Figure 12. Flow rates and levels are indicated. The object is to control the level in the first tank, h_1. In the following table, a measured variable and a manipulated variable are shown for each scheme. Signify which *one* of the four methods of control each scheme falls under. If a particular scheme does not fall into one of the categories, write "none" in the blank space.

Scheme Number	Measured Variable	Manipulated Variable	Method
1	F_1	F_1	
2	F_1	F_2	
3	Nothing	F_1	
4	h_1	F_1	
5	h_2	F_1	
6	h_1	F_2	
7	h_2	F_2	
8	F_2	F_2	
9	F_2	F_1	
10	h_2	F_3	
11	F_4	F_2	
12	h_1	F_3	
13	F_4	F_3	

Problem 9.   A distillation "column" consists of one equilibrium stage as shown in Figure 13.

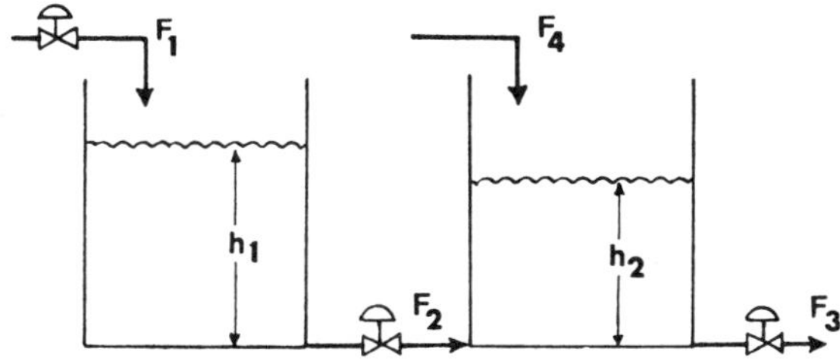

Figure 12. Physical picture of process for Problem 8.

The liquid feed enters at a rate of F_1 lb/hr. Heat is added at a rate of Q Btu/hr. The vapor leaves through a valved line overhead and the liquid leaves through the valved line below. The object is to control the level, h. In the following table, a measured variable and a manipulated variable are shown for each scheme. Signify which *one* of the four methods of control each scheme falls under. If a particular scheme does not fall into one of the categories, write "none" in the blank space.

Scheme Number	Measured Variable	Manipulated Variable	Method
1	None	Q	
2	P	P_{VP}	
3	F_1	F_2	
4	h	P_{VF}	
5	h	P_{VP}	
6	F_2	F_1	
7	h	Q	
8	F_1	F_1	

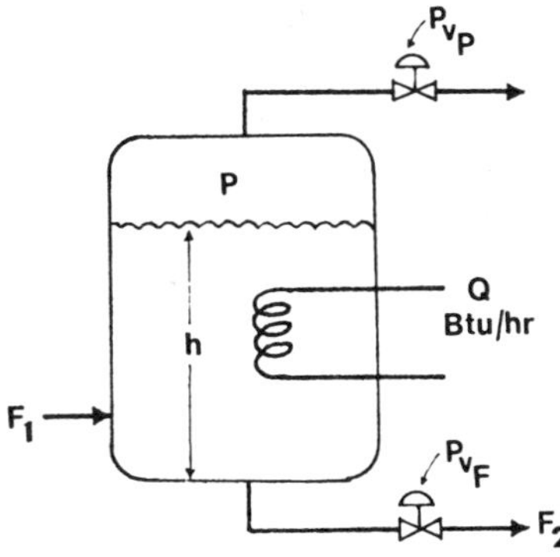

Figure 13. One-stage distillation "column" of Problem 9.

CHAPTER III

Servo Operation and Regulator Operation

A review of the four methods of control described in Chapter 2 reveals that the objectives of the methods are not identical. In the open-loop example concerning the brine mixer, the objective of the controller was to produce a weak salt solution whose composition was set by the calibrated dial. Environmental control was based on the elimination of disturbances, but there was basically no provision in this scheme for making a change in the desired value of the controlled variable. Feedforward control sought to compensate for load disturbances, but also could not enable a desired change in the controlled variable unless combined with open-loop or perhaps feedback control. In the case of feedback we found that the controller action took effect when an error occurred. Since an error could result from either a change in the set point or in a load variable, this method inherently possesses the flexibility to compensate for load disturbances as well as to change the controlled variable when a change in set point is made.

These two possible objectives of a control system—compensation for or elimination of load disturbances on the one hand, and the premeditated adjustment of the value of the controlled variable on the other—are com-

monly referred to as "regulator operation" and "servo operation," respectively. The distinction between the two is of sufficient importance to devote
this brief chapter to it.

I. SERVO OPERATION

As mentioned previously, servo control deals with the response of the system
to changes in the set point. As an illustration, recall the home-heating example
of the previous chapter. Assume that the room has been at the set-point
temperature of 68°F for some time. If the set point is quickly moved to 70°F,
the movement will require not more than 1 sec. On a time scale of several
hours, the motion of the set point appears as a "step change" as shown in
Figure 1a. The time origin has been arbitrarily located at the instant at which
the set point is moved.

If the control system is perfect, the room temperature will jump immediately to 70°F, as shown in Figure 1b. This response is impossible because
of the limited heat producing capacity of the furnace. The room temperature

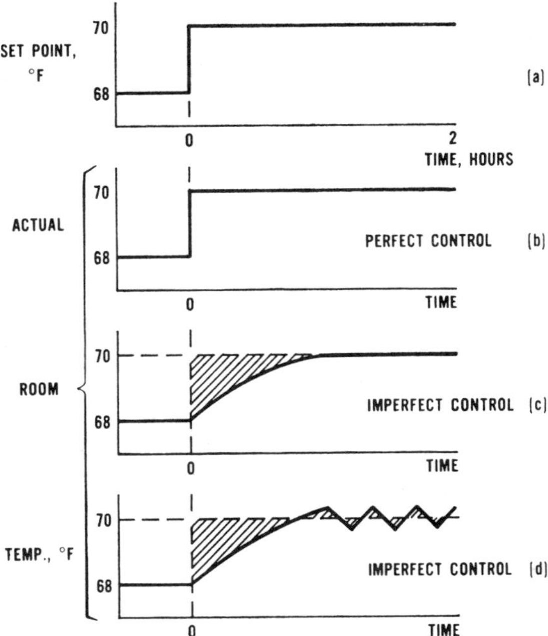

Figure 1. Set point and actual room temperature versus time.

might reach 70°F in a few minutes, but it could require a considerably longer time than this, depending on the type of heating system. The actual temperature could follow the path shown in Figure 1c, but that of Figure 1d is more likely for reasons that will be apparent after Chapter 6 has been studied.

The error is the difference between the actual temperature and that commanded by the set point. This error has been shaded into Figures 1c and 1d. This suggests that one index of the "quality" of control might be the time integral of the error. However, for the behavior in Figure 1d, if the oscillations were to continue, negative and positive areas would cancel. If this oscillatory behavior were considered undesirable, then alternate indices of the quality might be the integral of the absolute value of the error or the integral of the square of the error. All three of these criteria have been used by various investigators.

II. REGULATOR OPERATION

Regulator control concerns the behavior of the control system in warding off the effects of changes in load variables. The interest is in the response of the controlled or output variable to changes in the load variables. The goal is to have the controlled variable remain at the set-point value in spite of all possible changes in load variables.

In the home-heating example, two possible load variables are the outside temperature and wind velocity. Suppose the outside temperature drops rather abruptly from 30 to 20°F. It might take a few minutes for this to occur, but again on a scale of hours, it appears as a step. This is shown in Figure 2a.

If the control system is perfect, the room temperature will not change at all, but will remain at the set point, as illustrated in Figure 2b. Since perfect control is not possible, the actual behavior may be similar to one of the forms given in Figures 2c and 2d.

As shown in Figure 3, in broadest terms both the set point and the load variables are inputs that tend to cause changes in the controlled variable. We have seen that servo operation calls for the system to *respond* to one particular input and that regulator operation calls for it to *ignore* all of the rest of the inputs. Intuition suggests that this may not be physically possible and that compromises in design of the control system may be necessary.

III. SOME EXAMPLES

Chapter 1 noted that a strong impetus for the development of control theory was the need for automatic bomb sights and gun controls in World War II. Consider an antiaircraft gun following a plane. The object is to aim the gun

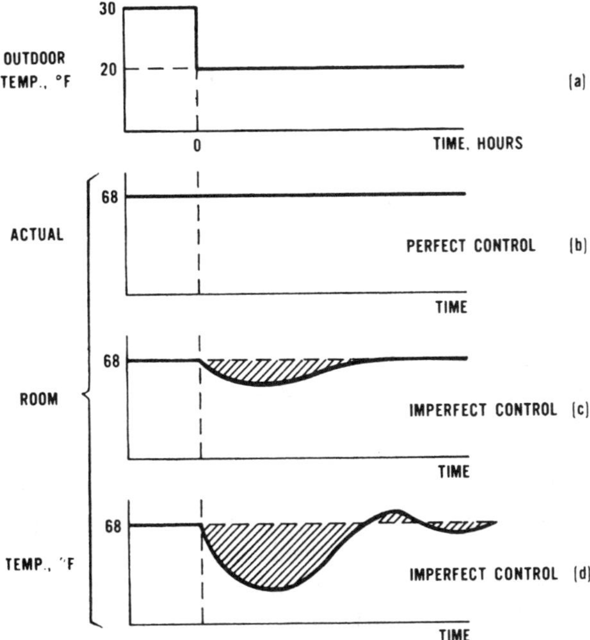

Figure 2. Outdoor and indoor temperatures versus time.

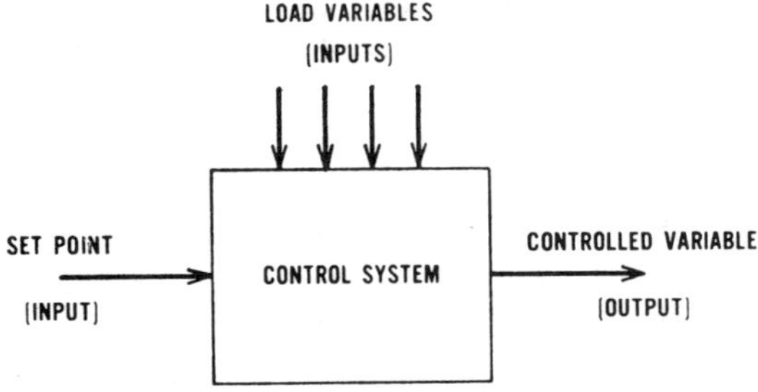

Figure 3. Set point and load variables both cause changes in the controlled variable.

in a direction so that its missile will hit the plane. Clearly, this falls into the category of servo control.

Another example of servo control concerns orbiting a space ship around a planet. A computer determines if the direction of the ship is such that the goal will be achieved. If not, it calculates a midcourse correction that is transmitted to the ship as a command.

The rotor used for a television antenna is another example of servo control. The dial on the control is moved to the desired direction and when the antenna reaches this position, the rotor motor stops. "Servos" of this type are quite common.

In a somewhat broader sense, a high-fidelity amplifier is an example of a servo controller. The purpose of this unit is to transform a weak input voltage into an output having relatively high power. The output should be an exact replica of the input with only a change in scale. The input voltage actually amounts to a constantly varying set point. For example, if a pure sine wave input voltage is used, this means that the "set point" is changing sinusoidally. If perfect servo operation is achieved, the output voltage and power will vary sinusoidally at the same frequency. Possible load variables might be changes in components as they age with time, or fluctuations in the supply voltage.

In the area of process control, servo control is a main objective in a batch reactor. The reactor is charged with reactants and then some heating "recipe" is followed. Ideally, the reactor temperature should exactly follow this recipe, which amounts to a prescribed variation of the set point with time.

By far, however, the main interest in process control is in regulator action. To most people, the term, "process control" really means regulator control, and the vast majority of chemical control systems are installed with this objective in mind. This is certainly the case with a home-heating system. The control system is supposed to guard against the effects of cold fronts, strong winds, opening and closing of outside doors, quality of furnace fuel, and countless other possible load variables. Of course, sometimes the set point is changed but this is a rarity compared with the large number of load changes.

This same idea prevails in most process control systems. In the start-up of a processing unit, it is important to reach design conditions as quickly as possible. The behavior during this period is related to that in servo operation. However, once design conditions are attained, interest focuses upon the regulator operation of the system.

Since most control systems do have some means for carrying out both servo and regulator operation, the question naturally arises as to whether a system that is very good in servo operation will also be very good in regulator operation. Unfortunately, the answer is "no," since the very properties of a process that are beneficial to its behavior in servo operation are detrimental to its behavior in regulator operation. A simple example may help to explain this. Consider a space ship traveling to a planet. The controlled variable is the direction of the ship and the load variables are the glancing blows that it occasionally shares with small meteors. The regulator operation of the guidance system is concerned with the effects of these collisions upon the direction of the ship. If the ship is large, the effects of these collisions will be small because of the large mass and inertia of the ship. The servo operation of the system will be important when a midcourse correction is made in the direction of the ship. A relatively large thrust will be required to accomplish this because of the large inertia. On the other hand, a small ship would suffer severe changes in direction as a result of the collisions while a midcourse correction would require a relatively small thrust.

In electrical engineering, most of the interest has been in servo control, whereas in process control, attention has centered upon regulator control. In spite of this difference in emphasis, the theory for both types of operation is closely related.

SUMMARY

The distinction between and objectives of servo and regulator control have been examined in this chapter. As pointed out, process control is usually concerned with regulator control. However, as we consider the theoretical and quantitative aspects of control in the next and succeeding chapters, we shall find that servo and regulator control are so closely related that neither will be ignored at the expense of the other.

PROBLEMS

Problem 1.   In a home-heating system, what factors are most important in its performance under servo control? In its performance under regulator control?

Problem 2.   Complex control problems arise in driving an automobile. Sometimes these problems are not properly solved and the results are disastrous. Probably the two main problems are steering and speed. Consider each of these. Are there any feedback loops involved? If so, draw them, indicating the nature of the set point and possible load variables. Do the control systems for steering and speed have any elements in common—that is, are there any interactions between the two systems?

CHAPTER   IV

Analysis of Control Systems by Steady-State Calculations

The quantitative evaluation of a control system is determined mainly from its unsteady-state behavior. There are a number of specialized techniques for obtaining information about various aspects of this. Some of these will be presented in later chapters. However, considerable insight about control methods can be gained just by making some steady-state calculations.

Figure 1 illustrates what can be learned from a steady-state analysis. Two cases are presented, one for a change in set point and the other for a change in load variable. In both cases, the controlled variable is initially constant. Referring to Figure 1a, the time origin is placed at the instant at which a change in set point is made. The system responds and the controlled variable moves from its original steady-state value to a final steady state. Although the figure indicates that the controlled variable begins to change immediately upon making the change in set point, this will not always be the case. A steady-state analysis cannot predict the path of the controlled variable between the two steady states. To indicate this fact, a dashed line has been drawn in for one possible transient path. An infinite variety of transient curves could be drawn. In an ideal system, the difference between the two steady states would be the difference between the two set-point values, and the transient portion of the curve would vanish.

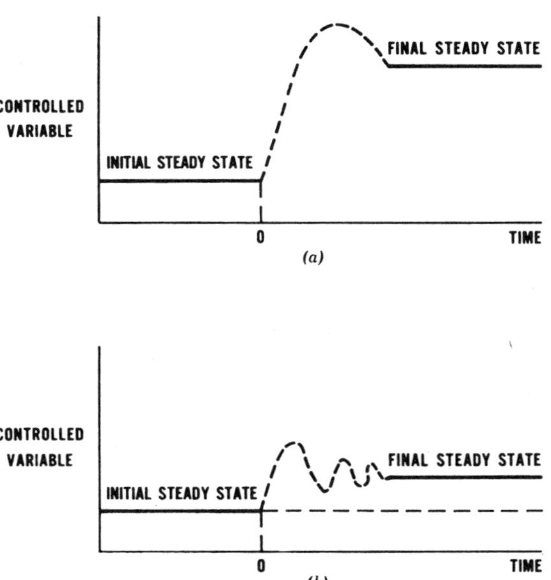

Figure 1. Steady-state and transient behavior of controlled variable (*a*) when a set point change is made; (*b*) when a load change occurs.

The situation in the case of a change in load variable is different, as stressed in the last chapter. When a change of load variable occurs, ideally the initial and final steady-state values will be the same and there will be no transient in between them. As the graph is drawn in Figure 1*b*, the initial and final steady-state values are nearly the same, although a slight displacement is shown. As in Figure 1*a*, the indicated transient curve is only one of an infinite number of possibilities. The difference in the two steady states is an important measure of the imperfection of the control.

In this chapter, steady-state analysis will be applied first to feedback, and then to feedforward. Open-loop control is actually a degenerate case of feedback in which there is no return loop, so it will not be specifically discussed. Environmental control requires no mathematical development.

I. FEEDBACK

Some important properties of feedback can be learned from the following very simple illustration.

A. An Elementary Illustration

Consider the simple block diagram shown in Figure 2. The "process" merely consists of an element that multiplies the error by a "gain" constant, K_o. We shall first examine the behavior of this system to a change in set point, and then to a change in a load variable.

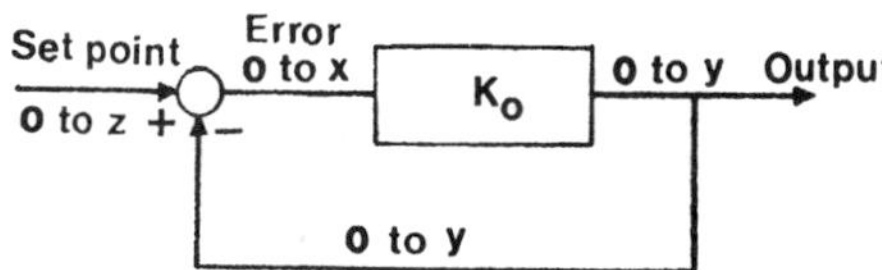

Figure 2.   Block diagram for a change in set point.

1. Change in Set Point

As indicated in Figure 2, initially all values of the variables around the loop, as well as the set point, are zero. Then the set point is moved to z. The error takes on a value, x, and the output takes on a value, y. The equations describing this system are:

$$x = z - y \tag{1}$$

$$y = K_o x \tag{2}$$

Our interest is in the value of the output. Eliminating x, we find

$$y = \frac{K_o}{1 + K_o} z \tag{3}$$

This value of y is attained instantaneously upon the change in set point, so for this particular system, there is no transient between the initial and final values of the output. If the system were ideal, the change in output would be the same as the change in set point. From Equation 3 it is apparent that this ideal is approached as K_o approaches infinity.

2. Change in Load Variable

Turning now to the case for a change in load variable, refer to the block diagram shown in Figure 3. As in the previous example, all variables are

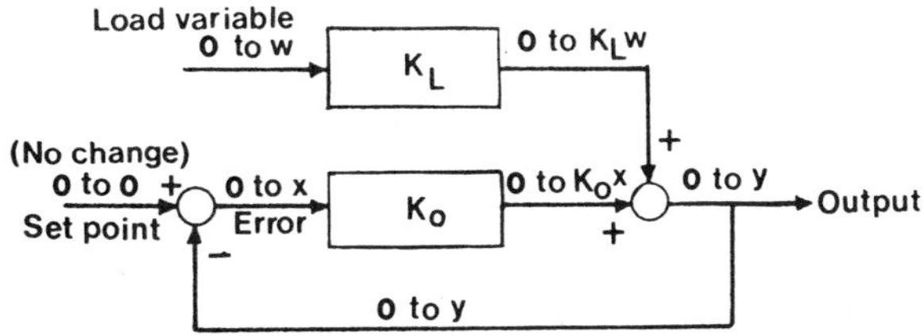

Figure 3.   Block diagram for a change in load variable.

initially zero. Then a change in load variable occurs from zero to w, the set point remaining fixed at zero. Again, the process has the property of multiplying the error by K_o; similarly, it multiplies the load variable change by the "load gain," K_L. From the diagram, the following equations can be written:

$$y = K_o x + K_L w \tag{4}$$

$$x = -y \tag{5}$$

Hence

$$y = \frac{K_L}{1 + K_o} w \tag{6}$$

Since this is the case of regulator action, the ideal result would be no change in output in spite of the change in load variable. Equation 6 indicates that either a small value of K_L or a large value of K_o will be in the right direction.

The preceding illustration was very abstract. The following example shows that a practical situation can be made to fit the abstract form. Furthermore, the origin of the gain factors, K_o and K_L, will be developed.

B. A Practical Example—Stirred-Tank Heater

1. Physical Description of the System

This example concerns the production of hot water from cold water in a continuous-flow, stirred-tank exchanger. Heat is supplied by condensing steam in a jacket surrounding the tank. In normal operation, the entering water has a temperature of 60°F and leaves the tank at 160°F. The purpose of the control system is to maintain the leaving stream at the temperature indicated by the set point. A diagram of the system is presented in Figure 4.

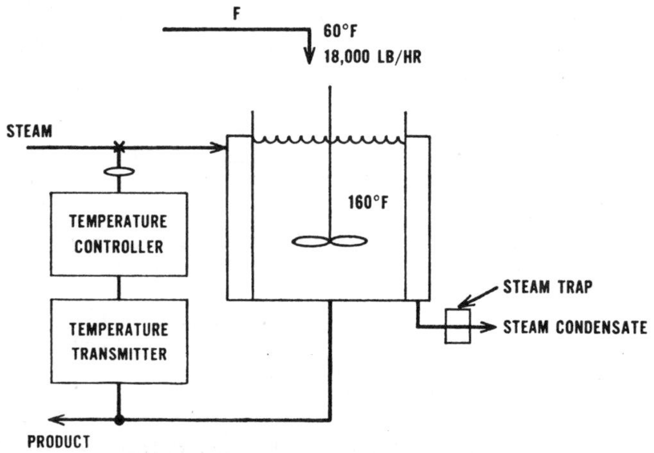

Figure 4. Physical picture of stirred-tank heater system.

a. PHYSICAL DIMENSIONS AND OPERATING CONDITIONS

Before going into the actual control calculations, the physical aspects of the system will be discussed. The tank dimensions and holdup are:

$$D = \text{tank diameter} = 6 \text{ ft}$$
$$H = \text{liquid depth and steam jacket height} = 8 \text{ ft}$$
$$M = \text{tank holdup} = 14,100 \text{ lb}$$

The normal or design operating conditions are:

$$F = \text{feed rate} = 18,000 \text{ lb/hr of water}$$
$$W_S = \text{steam flow rate} = 1865 \text{ lb/hr}$$
$$\lambda = \text{heat of condensation of steam} = 965 \text{ Btu/lb}$$
$$T_F = \text{feed temperature} = 60°F$$
$$T_S = \text{set-point temperature} = 160°F$$
$$T = \text{tank temperature} = 160°F$$

In this illustration, the load variable will be the feed temperature; other possible load variables such as the steam supply pressure and the flow rate of water will be assumed invariant.

b. FLOW RATE OF STEAM AND THE RATE OF HEAT TRANSFER

The steam is throttled through a pneumatically actuated control valve. In general, the flow rate of steam depends on both the upstream and downstream pressures as well as on the area of the valve opening. As previously mentioned, in this illustration it is assumed that the upstream pressure is constant so this eliminates one of the variables. In thermodynamics, it is shown that for a gas when the ratio of the downstream pressure to the upstream pressure is less than the critical pressure ratio, the velocity in the valve opening reaches sonic velocity and is independent of the downstream pressure. Therefore, by making the assumption of critical flow here, the problem is considerably simplified since the flow rate of the steam is then only a function of the valve opening or area, and this in turn is set by the pressure in the valve motor. Furthermore, since all of the steam in the jacket is condensed, this means that the rate of heat transfer is also only a function of the pressure in the valve motor.

c. PNEUMATIC CONTROLLER

A standard pneumatic controller will be used for this system. Both electronic and pneumatic controllers are available, but the vast majority are pneumatic. The motive power for the controller is a 20-psig air supply, analogous to the 110-V alternating current supply used for many home appliances and lighting. The input signal to the controller is an air pressure of 3 to 15 psig, which usually originates from a "transmitter." The output of the controller is a pressure of 3 to 15 psig.

As was the case for the thermostat of the home-heating system, the controller also incorporates the comparator, which takes the difference between the set point and the input signal. This difference or "error" is then acted upon by the controller "action" to produce the controller output. Several kinds of standard actions are available. In a "proportional" controller, the type to be used in this example, the controller action merely multiplies changes in the error by a constant "gain factor" to produce corresponding changes in the output signal to the valve. For example, if the gain is two, then for a fixed set point, a 1-psi change in input signal will result in a 2-psi change in the output signal.

d. TEMPERATURE TRANSMITTER

Since the input to the standard controller is an air signal, a device is required to convert the tank temperature measurement into an air signal covering the range between 3 and 15 psig. A temperature transmitter is used for this. This device is simply a transformer, converting °F to psig. A temperature transmitter may have a gas-filled metal bulb as the sensor. The pressure of the gas in the bulb increases with temperature, and this pressure is in turn converted by the transmitter to the required air signal.

Two specifications must be made in purchasing a transmitter. The first is the "span," which is the difference between the maximum and minimum temperatures of the unit. The maximum temperature would correspond to a transmitter output of 15 psig and the minimum would correspond to an output of 3 psig. The second specification is the "range," which gives the limits within which the span must lie. For example, the following are some spans and range limits offered by one manufacturer.

Span = 50°F	Span = 100°F	Span = 200°F	Span = 400°F
Range Limits, °F	Range Limits, °F	Range Limits, °F	Range Limits °F
−375 to −225°	−250 to +50°	−300 to +100°	0 to 1000°
−250 to −100°	0 to 300°	0 to 400°	
−125 to +25°	250 to 550°	300 to 700°	
0 to 150°	500 to 800°	600 to 1000°	
125 to 275°	700 to 1000°		
250 to 350°			

Suppose a unit were chosen with a span of 50°F and with range limits of 0 to 150°F. This means that the transmitter could be set to operate from 0 to 50°F, 10 to 60°F, or 100 to 150°F, just to name a few of the infinite number of possibilities.

In this water-heating process, it would be desirable to have an indication of the tank temperature if it fell as low as the feed temperature of 60°F, or if it rose to the boiling point of 212°F. Therefore, the minimum span is 152°F. Looking at the preceding table, we see that a suitable transmitter would be one with a span of 200°F and range limits of 0 to 400°F. It would be reasonable to adjust the span to cover the range from 50 to 250°F. Although there is flexibility in this regard, the controller will be assumed to be of the recording type. Charts can be purchased with temperatures already printed on them, and one with a lower limit of 50°F and an upper limit of 250°F would be a standard item. Therefore, although the exact location of the span is not important from a control standpoint, practically speaking, the span should be set in accordance with standard charts which are available.

e. VALVE

Chapter 2 mentioned that the standard pressure range of a valve motor is between 3 and 15 psig and that there are two types, air-to-open and air-to-close valves. The choice is usually dictated by considerations of safety. If the air supply for the plant should fail, then the system should "fail safe." In the present example, an air-to-open valve will be used to avoid the possibility of boiling should the air supply fail.

Other specifications for the valve include its diameter and materials of construction, but for control, the main concern is with the change in flow rate produced by a given change in input pressure to the valve motor. In general terms, this effect is described by the "valve trim." The choice of trim depends on the type of process, no one type being suitable for all processes.

Among the several trims available, two of the more common ones are "linear" and "equal percentage." With a linear valve, the change in valve opening or area is proportional to the change in the pressure to the valve. On the other hand, the equal percentage type has the property that a 1% change in actuating pressure to the valve causes a constant percent change in valve opening. For example, for a "5% valve," a 1% change in pressure results in a 5% change in valve opening over the range of operation of the valve. Mathematically, this characteristic is given by the equation

$$\frac{dA_v/A_v}{dP_V} = k$$

where A_v = valve opening
$\quad\quad P_V$ = pressure to the valve motor
$\quad\quad k$ = constant

It must be emphasized that if the pressure to the valve exceeds 15 psig or falls below 3 psig, the valve trim becomes irrelevant since the valve is either fully opened or fully closed, depending on the type of valve.

In this particular example, a 5%, equal percentage valve will be used. As was pointed out earlier, the steam flow rate is proportional to the area of the opening because critical flow exists.

2. Block Diagram for the Control System

The block diagram for the system can now be drawn and is presented in Figure 5. The signal between each block is indicated. At the normal feed temperature of 60°F, the tank temperature, T, is 160°F and the steam flow rate, W_S, is 1865 lb/hr. The output pressure of the transmitter, P_T, can be calculated as follows. The span is set with the lower limit at 50°F and the upper limit at 250°F, as discussed previously. The output pressure is linear

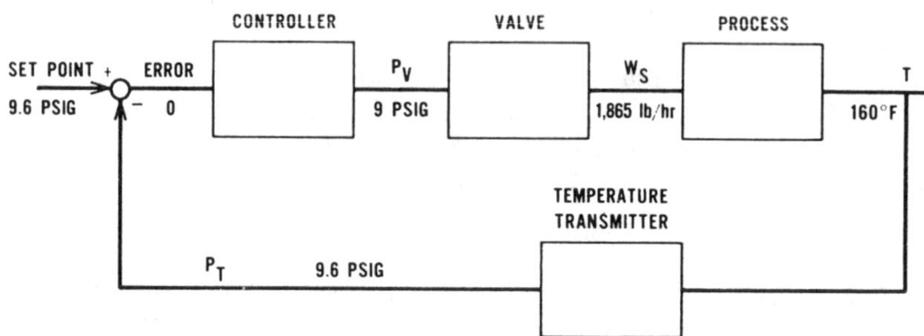

Figure 5.   Simple block diagram for tank-heater system.

with temperature between 3 and 15 psig. Therefore, when T is 160°F, P_T may be found by interpolation:

$$P_T = 3 + \frac{160 - 50}{200} (15 - 3) = 9.6 \text{ psig}$$

The set point on the controller is usually represented by a pointer on the chart, which in this case is marked off in degrees from 50 to 250°F. In terms of temperature then, the set point in this example is at the design value of 160°F. However, the controller input is a pressure, and this pressure is compared with that represented by the position of the set point. Since the input pressure is 9.6 psig at a temperature of 160°F, it follows that the "set-point pressure" is also 9.6 psig.

The normal pressure to the valve will depend upon the size of the valve selected. The normal opening in turn is determined by the steam supply pressure and the design rate of heat transfer. One valve might have the required area at a pressure of 12 psig, and another might have the same opening at 4 psig. Since the valve position will be changing to compensate for various load disturbances and changes in set point, it is reasonable to select one that has the desired normal opening when the pressure is 9 psig. This results in equal flexibility in both directions, so 9 psig will be used in this example.

3. An Upset in Feed Temperature

Having described the various components in the system, we are now in a position to study its steady-state behavior. The effect of a change in feed temperature will be considered first.

a. MATHEMATICAL CHARACTERIZATION OF THE COMPONENTS

To find out how effective the control system is, an arbitrary change in feed temperature will be assumed—say a 10°F rise from 60 to 70°F. As a basis for comparison, it should be noted that if there were no controller, the steam rate would not change. The normal steam rate supplies 965×1865 or 1,800,000 Btu/hr, which is sufficient to raise the temperature of 18,000 lb/hr of water, 100°F. Hence, with no control at all, each change in feed temperature would be reflected eventually by exactly the same change in product temperature. The change would be gradual rather than instantaneous because of the large mass of water in the tank. On the other hand, perfect control would result in no change in tank temperature.

The block diagram of Figure 5 has four blocks plus the comparator. Each of these must be described by an equation or transformation.

(1) *Process*

The input to the process block is the steam rate and the output is the tank temperature. These quantities are related through the following energy balance:

$$Fc_p(T - T_F) = W_S\lambda \tag{7}$$

where c_p = heat capacity of water = 1.0 Btu/lb, °F.
The flow rate, F, is the normal value of 18,000 lb/hr and the heat of condensation is 965 Btu/lb. The feed temperature is the newly assumed value of 70°F. Substituting these values into Equation 7, we obtain

$$18,000(T - 70) = W_S(965) \tag{8}$$

(2) *Valve*

Recalling the previous discussion of the valve, a 5% change in flow is produced by a 1% change in pressure to the valve. Mathematically, this means

$$\frac{dW_S/W_S}{dP_V/(15 - 3)} = 5 = K'_V \tag{9}$$

Note particularly that the percentage change in flow is based on the flow at the point at which the change occurs; in contrast, the percentage change in valve pressure is based on the pressure range of the valve, which is 12 psi.

If the change in pressure to the valve is small, then the differential changes above can be approximated by small delta changes, ΔW_S and ΔP_V; furthermore, at the normal conditions, W_S is 1865 lb/hr and P_V is 9 psig. Therefore

$$\frac{\dfrac{\Delta W_S}{1865}}{\dfrac{\Delta P_V}{12}} = 5 = K'_V \tag{10}$$

where $\Delta P_V = P_V - 9$ and $\Delta W_S = W_S - 1865$. $\tag{11}$

Hence, substituting Equation 11 into Equation 10 and rearranging, we obtain

$$W_S = 1865 + K'_V(1865)\frac{\Delta P_V}{12} \tag{12}$$

The first term on the right represents the normal steam rate and the second, the deviation due to the change in steam pressure to the valve. It is important to note that because of the approximation involved in going from Equation 9 to Equation 10, Equation 12 is a linearized result. This will make later calculations approximate to some extent.

(3) *Comparator and Controller*

As noted earlier, the comparator takes the difference between the set point and the input pressure from the transmitter. In this example for a change in feed temperature, the set point remains constant at the set-point pressure of 9.6 psig. Therefore, the error, e, is given by the following equation:

$$e = 9.6 - P_T \tag{13}$$

The proportional action of the controller then magnifies this error by the gain factor, K_C. Therefore

$$\Delta P_V = (P_V - 9) = K_C e \tag{14}$$

where K_C = controller gain, psi/psi.

(4) *Transmitter*

As before, the transmitter output is obtained by interpolation:

$$P_T = 3 + \frac{T - 50}{200}(12) = 0.06T \tag{15}$$

(5) *Simultaneous Solution of Equations*

The five equations describing the system are Equations 8 and 12 to 15. These involve the five unknowns, T, W_S, P_T, e, and P_V, and the controller gain, K_C, which is arbitrary. These equations are easily solved simultaneously for the tank temperature, T, to yield

$$T = \frac{3171 + 7460K_C}{18.65 + 46.63K_C} \tag{16}$$

If K_C is 1, for example, then T is 162.86°F. If K_C is increased to 5, T decreases to 160.74°F. As K_C approaches infinity, the expression above approaches 7460/46.63 or 160.00°F, which corresponds to perfect control.

b. OFFSET

The undesirable feature of this control system is that the ultimate new value
of tank temperature is not the same as the one specified by the set point
unless K_C is infinite. The difference between the final and original values of
temperature is termed the "offset." Since the offset here is reduced by
increasing the controller gain, this suggests using the highest possible controller gain. However, there is a limitation on the allowable gain because of
the transient behavior of the system. If the gain is too high, the system will be
unstable as evidenced by oscillations of increasing magnitude. This fact is
not revealed by this steady-state analysis.

Offset is a major drawback of proportional control. Unless very high
gains can be used, the magnitude of the offset may be unacceptable. It may
have occurred to the reader that the tank temperature could be forced back
to 160°F by "resetting" the set point. To illustrate this, suppose K_C is 1,
which resulted in a tank temperature of 162.86°F. We wish to find out what
set-point pressure and corresponding set-point temperature will result in a
tank temperature of 160°F. The calculation begins with Equation 8, but
with T equal to 160°F. The new value of ΔP_V can then be found from Equation 12. In Equation 13, the set point is no longer 9.6 psig but rather is the
unknown to be solved for; P_T is 9.6 psig since the tank temperature is 160°F.
The result is that the set point must be 9.36 psig, which corresponds to a set-
point temperature of 156.0°F.

Several interesting points come out of this calculation. The first concerns
the magnitude of the "new offset," which is required to negate the effect
of the old offset. Since the original offset was 2.86°F, it might be suspected
that by lowering the set point to $160 - 2.86$°F or 157.14°F, the temperature
would fall to 160°F. But the actual correction needed is 4.0°F rather than
2.86°F. So if offset were observed in an actual system, the "correction" to
remove the offset would be found by trial and error.

A second point that deserves emphasis is that some offset will normally
be present in a system under proportional control. In this example, if the
only possible load variable is the feed temperature, then there is only one feed
temperature that produces no offset and that is 60.0°F; all other feed temperatures will result in offset. Although the offset can be removed by manually
resetting the set point, this would require the constant attention of the
operator, and then this is likely to be unsatisfactory because of the erratic
character of the load changes, not to mention the transient behavior of the
system. Clearly, the solution to this problem lies in some kind of "automatic
reset" action. As we shall see later, this action is commonly incorporated in
controllers.

Lest the reader be misled by the preceding paragraph, it is possible, although unlikely, that the effect of one load variable can cancel that of another. Thus, if feed temperature and feed rate are both load variables, an increase in feed temperature could be negated by an increase in feed rate. For example, if the feed temperature increased from 60 to 70°F, the tank temperature would still remain at 160°F if there were a corresponding rise in feed rate from 18,000 to 20,000 lb/hr. This fact is easily obtained from an energy balance on the tank. Thus, this illustrates a situation in which the process is not operating at design conditions and yet there is no offset. This is such an exceptional situation that it would seem to have little bearing on the previous assertion—namely, that some offset will normally be present in a system under proportional control.

Having demonstrated mathematically that offset is a property of proportional control, it is worthwhile formulating an explanation for this in words. When the process is operating with all flow rates, temperatures, and other variables exactly at their design values, there is no error since the controlled variable is right at the set point. Now, if there is a change in one of the load variables, then there must be some change in the manipulated variable to compensate for this change. In our example, the feed temperature changed so that the steam rate had to change in a direction to attempt to compensate for this. But there can be no change in the manipulated variable unless there is an error, and this error is the offset.

c. DEVIATION VARIABLES SIMPLIFY CALCULATIONS

The simultaneous solution of Equations 8 and 12 to 15 led to Equation 16, an expression for the tank temperature in terms of the controller gain. While this procedure is usable, it is to some extent cumbersome. It would become less and less satisfactory as the complexity of the system increased. The purpose of this section is to simplify the procedure in order to take advantage of the illustrations that culminated in Equations 3 and 6.

(1) *Process*

The desired simplification is achieved through the use of deviation variables. The idea behind this is that any variable can be expressed as the sum of its normal, original steady-state value plus the deviation from this state. We shall begin by illustrating this for the process. In general terms the energy balance for the tank is

$$Fc_p(T - T_F) = W_S \lambda \tag{17}$$

Any of these factors could vary. The feed flow rate, F, is a possible load variable; the tank temperature, T, is the controlled variable; the feed temperature, T_F, is the load variable we are considering; the steam rate, W_S, is the manipulated variable; and finally, the heat of condensation, λ, could vary if the quality of the steam was variable. However, in this illustration, only T, T_F, and W_S vary and are written as follows:

$$T = \bar{T} + \Delta T \tag{18}$$

$$T_F = \bar{T}_F + \Delta T_F \tag{19}$$

$$W_S = \bar{W}_S + \Delta W_S \tag{20}$$

The symbols with bars denote that these are the original steady-state values and the delta quantities represent the final steady-state deviations from the initial steady-state values. In this example, $\bar{T} = 160°F$, $\bar{T}_F = 60°F$, and $\bar{W}_S = 1865$ lb/hr. Two energy balances can be written, one for the initial steady state and the other for the final steady state:

$$\text{Final} \qquad Fc_p(\bar{T} + \Delta T - \bar{T}_F - \Delta T_F) = (\bar{W}_S + \Delta W_S)\lambda \tag{21}$$

$$\text{Initial} \qquad Fc_p(\bar{T} - \bar{T}_F) = \bar{W}_S\lambda \tag{22}$$

Subtracting the initial equation from the final, we obtain

$$Fc_p(\Delta T - \Delta T_F) = \Delta W_S\lambda \tag{23}$$

(2) *Valve*

From Equation 10, the deviation equation for the valve is

$$\Delta W_S = \frac{K_V' \, \Delta P_V \bar{W}_S}{12} \tag{24}$$

(3) *Comparator and Controller*

By definition

$$e = P_S - P_T \tag{25}$$

Each of these variables can be expressed in terms of deviation variables as follows:

$$e = \bar{e} + \Delta e \tag{26}$$

$$P_S = \bar{P}_S + \Delta P_S \tag{27}$$

$$P_T = \bar{P}_T + \Delta P_T \tag{28}$$

Therefore, Equation 25 can be written

$$\bar{e} + \Delta e = \bar{P}_S + \Delta P_S - \bar{P}_T - \Delta P_T \tag{29}$$

Hence
$$\bar{e} = \bar{P}_S - \bar{P}_T \tag{29a}$$

and
$$\Delta e = \Delta P_S - \Delta P_T \tag{29b}$$

Initially there is no error, which means that $\bar{e}$ is zero. Furthermore, in this part of the problem there is no change in set point, so ΔP_S is zero. Therefore

$$\Delta e = -\Delta P_T \tag{30}$$

From Equation 14 and the fact that $\bar{e}$ is zero, it can be seen that the equation for the action of the controller is

$$\Delta P_V = K_C \, \Delta e \tag{31}$$

(4) *Transmitter*

Finally, it is evident from Equation 15 that the deviation of the transmitter pressure, ΔP_T is

$$\Delta P_T = 0.06 \, \Delta T \tag{32}$$

(5) *Simultaneous Solution of Equations*

The five equations describing the system are Equations 23, 24, and 30 to 32. These involve the five unknown deviations, ΔT, ΔW_S, ΔP_T, Δe, and ΔP_V. Combining these equations results in the following expression for ΔT:

$$\Delta T = \frac{-K_C K_V' \overline{W}_S \lambda (0.06)}{(12) F c_p} \Delta T + \Delta T_F \tag{33}$$

As before, $K_V' = 5$, $\overline{W}_S = 1865$ lb/hr, $\lambda = 965$ Btu/lb, $Fc_p = 18{,}000$ Btu/hr, °F, and $\Delta T_F = 10°$F. When $K_C = 1$, ΔT is calculated to be $2.86°$F. Using Equation 18 and the fact that $\overline{T} = 160°$F, we find that $T = 162.86°$F, which is consistent with the value obtained from Equation 16.

d. RELATIONSHIP OF DEVIATION VARIABLES TO BLOCK DIAGRAMS

Having demonstrated that the method of deviations yields the same answer as the first method, we will now see how this concept can be related to block diagrams. The analysis begins with the process equation, Equation 23, but written in a slightly rearranged form:

$$\Delta T = 1 \, \Delta T_F + \frac{\lambda}{F c_p} \Delta W_S \tag{34}$$

This equation reveals nothing previously unknown, but the form of the equation is significant. First, it indicates that if there is no change in steam flow, a 1°F change in feed temperature will result in an identical change in tank temperature. To emphasize this point, the factor of unity has been placed in front of ΔT_F. Second, it shows that if there is no change in feed temperature, the change in tank temperature is equal to the factor $\lambda/F c_p$ times the change in steam flow rate.

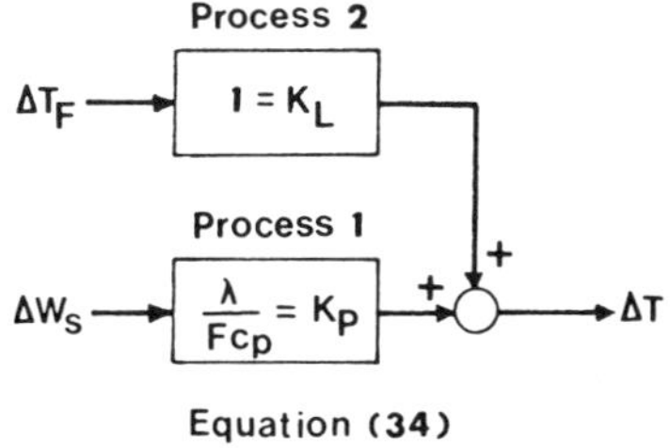

Figure 6. Block diagram representation of Equation 34.

Recalling that one rule of block diagrams was that a block can have only one input and one output, this equation can be represented by two blocks and a summer as indicated in Figure 6. Each block symbolically represents the transformation of the change in input into the change in output. In the present case, the use of deviation variables has made the "transformer" for a change in T_F simply a multiplying factor of unity; for changes in steam flow rate, the transformer is the constant multiplier, λ/Fc_p. It is customary to refer to these multipliers as "gain factors," a term that has been used previously in this chapter. The gain factor for the load variable will be given the symbol, K_L, and that for the manipulated variable, K_P. As defined here, the units of a gain factor must be those of the output divided by those of the input.

Similarly, the remaining elements and the comparator can be represented as shown in Figure 7. When there is no change in set point, as in the example we are considering, ΔP_S is zero, but it has been included in Figure 7 for completeness.

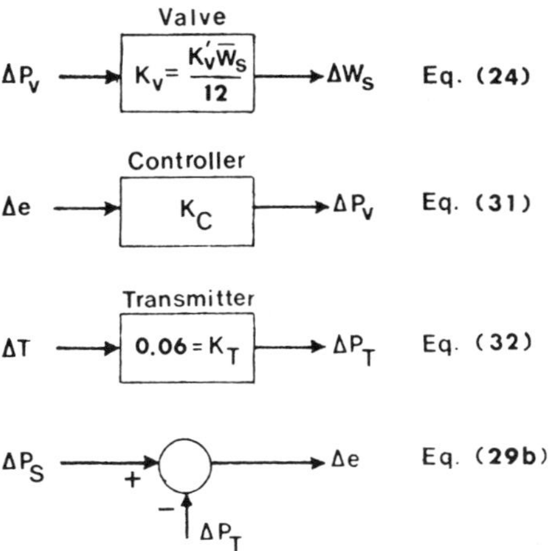

Figure 7. Block diagram representations of Equations 24 and 30–32.

When Equations 23, 24, and 30 to 32 were solved simultaneously to find ΔT, the result was Equation 33, which can be regrouped as follows:

$$\Delta T = -(K_C)\left(\frac{K_V'\overline{W}_S}{12}\right)\left(\frac{\lambda}{Fc_p}\right)(0.06)\,\Delta T + 1\,\Delta T_F \tag{33}$$

Referring to Figures 6 and 7, we see that this equation can be rewritten in terms of the gain factors for each block:

$$\Delta T = -K_C K_V K_P K_T\,\Delta T + K_L\,\Delta T_F \tag{35}$$

Solving for ΔT, we obtain

$$\Delta T = \frac{K_L}{1 + K_C K_V K_P K_T}\,\Delta T_F \tag{36}$$

The block diagram for Equation 34 can be combined with that for the valve since the output of the valve block is ΔW_S and the input for Process 1 is also ΔW_S. The resulting combination is presented in Figure 8. This idea can be extended to include the other blocks with the result shown in Figure 9. The gain factor inside each block multiplies the signal fed into it. When ΔP_S is zero, it can be seen that this block diagram is really a representation of Equation 35.

Equation 36 gives the offset for a change in feed temperature. It should be noted that the product of all of the gain factors in the loop appears in the denominator. This product is often referred to as the "open-loop gain" or "overall gain" and is designated K_o. The origin of this terminology is apparent from Figure 9. Suppose that the line connecting the output from the transmitter to the input of the controller were broken, in effect "opening the loop" at the comparator. Then the change in the error, Δe, would be amplified by the product of each of the individual gains in the loop as it passed through to the output of the transmitter.

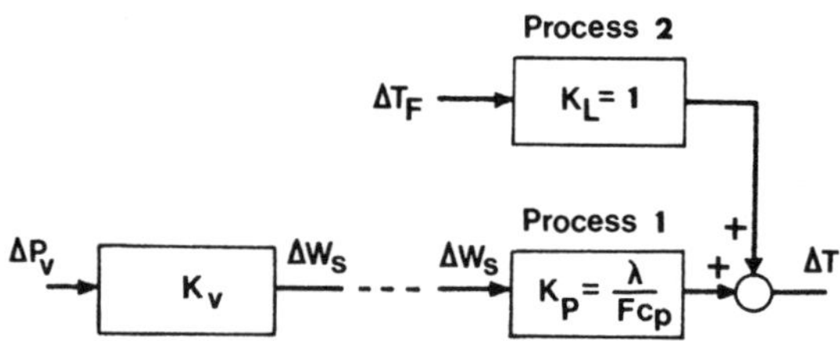

Figure 8.   Combination of block diagrams of Equations 24 and 34.

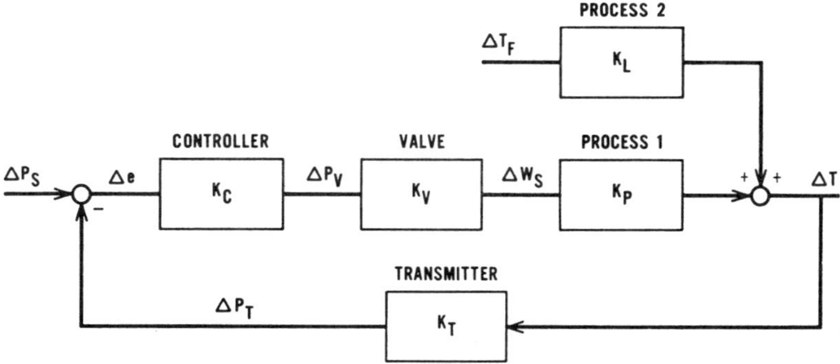

Figure 9.   Final block diagram combining those of Equations 24, 30, 31, 32, and 34.

In terms of the overall gain, Equation 36 can be written

$$\Delta T = \frac{K_L}{1 + K_o}\, \Delta T_F \tag{37}$$

where K_o = overall gain = $K_C K_T K_V K_P$.

Equation 37 can, of course, be generalized for cases in which more or less elements are involved than those in this example.

4. A Change in Set Point

Now we shall consider the response of the system to a change in set point when there is no change in a load variable. The initial condition of the system is that at which all of the variables are at their design values and the set point is at 160°F. The expression for the change in tank temperature, ΔT, for a given change in set point temperature, ΔT_S, is obtained from the previous equations involving deviation variables. The change in set-point temperature is related to the corresponding change in set-point pressure through Equation 32. Either Equation 23 or 34 can be used for the process, but with ΔT_F equal to zero since there is no change in feed temperature. Equation 31 for the controller, Equation 24 for the valve, and Equation 32 for the transmitter remain unchanged. However, Equation 29b must be used for the comparator since ΔP_S is not zero. Also, the gain definitions for K_V, K_T, and K_P are used for simplification. The result is

$$\Delta T = \frac{K_C K_V K_P K_T}{1 + K_C K_V K_P K_T}\, \Delta T_S = \frac{K_o}{1 + K_o}\, \Delta T_S \tag{38}$$

The numerical values of the gains in this example are:

$$K_P = \frac{\lambda}{Fc_p} = \frac{965}{18{,}000 \times 1.0} = 0.0536 \; \frac{°F}{\text{lb steam/hr}}$$

$$K_V = \frac{K_V' \overline{W}_S}{12} = \frac{5 \times 1865}{12} = 777 \; \frac{\text{lb steam/hr}}{\text{psi}}$$

$$K_T = 0.06 \; \frac{\text{psi}}{°F}$$

If the controller gain, K_C, is arbitrarily given a value of 1 psi/psi, then the overall gain is

$$K_o = K_C K_V K_P K_T = 2.5 \text{ unitless}$$

Now suppose that the set point is moved from 160°F to 170°F. Thus, ΔT_S is 10°F and Equation 38 yields a change in tank temperature, ΔT, of 7.14°F, *not* the 10°F change made in the set point. In this case there is an offset, too, but this time it is 10.00°F − 7.14°F, or 2.86°F. This number will be recognized as the magnitude of the offset found in the example for a change in feed temperature. However, these two offsets happen to be the same only because the gain factor for changes in feed temperature is unity. This deserves further clarification.

It was shown that for a change in load variable, the offset is given by

$$\Delta T = \frac{K_L}{1 + K_o} \Delta T_F = \text{offset for load variable change} \qquad (37)$$

The desired offset is zero since ideally there should be no difference between the initial and final steady states. A different situation arises when a change in set point is made. If the control is perfect, the change in the controlled variable will be exactly equal to the change in the set point. Any deviation from this is termed the offset, as illustrated in Figure 10. The offset is equal

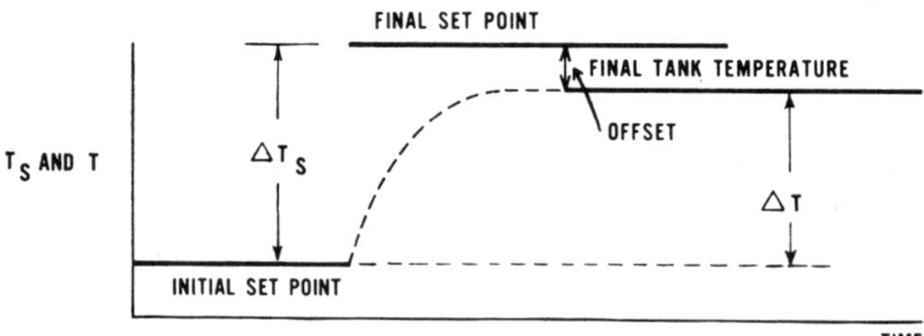

Figure 10.   Offset for a change in set point.

to $T_S - T$, which is the same as $\Delta T_S - \Delta T$. Using Equation 38, the offset is given by the expression

$$\Delta T_S - \Delta T = \frac{1}{1 + K_o} \Delta T_S = \text{offset for set point change} \qquad (39)$$

Comparing Equations 37 and 39, we see that when K_L is unity, as it was in this example, the offset for a change in the load variable will be the same as for the same change in set point. This explains why the offset of 2.86°F was found in the example for a 10°F change in feed temperature as well as for a 10°F change in set point.

From Equations 37 and 39, it can be seen that the offset is reduced for changes in load variables and set point as the overall gain is increased. An increase in the overall gain is usually accomplished by increasing the controller gain.

5. Combining the Transmitter and Controller on the Block Diagram

As discussed earlier, although the input to the controller is a pressure in the range between 3 and 15 psig, the chart on the controller is chosen to match the span of the transmitter and is graduated in the units of the controlled variable. This is convenient since the values of the set point and the controlled variable can then be read directly off the chart.

From the standpoint of the block diagram, this has the effect of moving the transmitter block from its position in the "feedback portion" of the loop to a new location in the "forward portion" of the loop. Since the gains of the transmitter and controller each multiply changes in their respective inputs, the blocks for the transmitter and controller can be combined into a single block that has a gain, $K_T K_C$. Hence, the diagram of Figure 9 is changed to that of Figure 11. The change in set point is now ΔT_S and the change in error

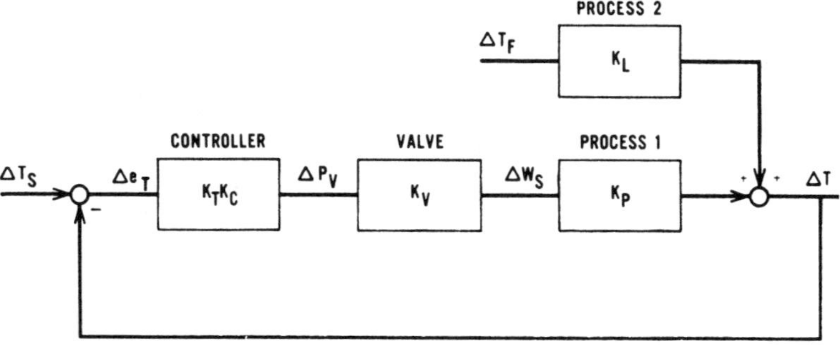

Figure 11.   Block diagram combining transmitter and controller.

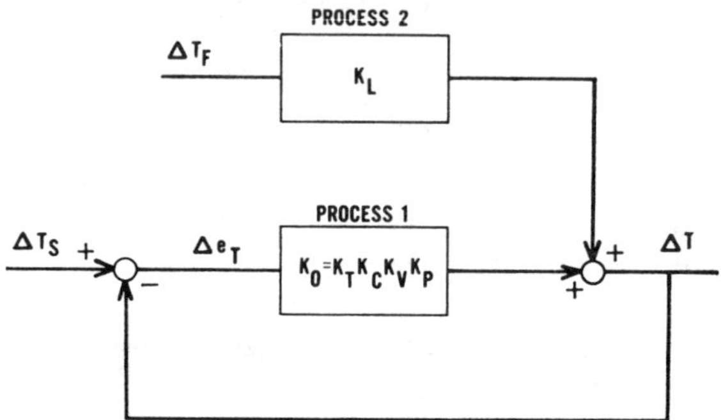

Figure 12.   Block diagram of Figure 11 reduced to simplest form.

is Δe_T, and these changes have the units of temperature. This rearrangement has no effect upon any of the previous results because the gains multiply. For example, with the configuration of Figure 11, it is easily shown that the change in tank temperature for a change in feed temperature is still given by Equation 37, and Equation 38 still holds for changes in set point.

When there is no element or block in the return path as in Figure 11, the system is said to have "unity feedback." This configuration is helpful in the physical interpretation of the diagram since it is no longer necessary to think of the set point and controlled variable in terms of pressures. In a steady-state analysis, it is always permissible to combine the transmitter and controller so that a unity feedback diagram is obtained. In a transient analysis, this rearrangement can be made without serious error if the response of the measuring element and transmitter is fast compared with those of the other elements in the loop. This assumption is usually justifiable. A more comprehensive discussion of this can be found in Section I.C of Chapter 13.

Because the various gains multiply their respective inputs, Figure 11 can be reduced to the form shown in Figure 12. This diagram is basically the same as the very simple diagrams of Figures 2 and 3. Furthermore, Equation 37 for the effect of a change in feed temperature on the tank temperature is in exactly the same form as Equation 6, and Equation 38 for the effect of a change in set point on the tank temperature is in exactly the same form as Equation 3. Hence, the goal of adapting the tank problem to the simple illustration has been completed. Basically, the key to the reduction was the introduction of deviation variables.

6. Dimensions of Gain Factors

It was shown below Equation 38 that the overall gain, K_o, was unitless. A glance at a feedback system will quickly show that this must always be true;

otherwise, the comparator would be comparing two things that were not comparable because they had different units.

The units of the individual blocks are simply the ratio of the units of the output to those of the input. In setting up a block diagram and calculating gains, a test of dimensional consistency can be made. If the product of all gains in the loop is not dimensionless, then an error has been made.

II. FEEDFORWARD CONTROL

The example of the stirred-tank heater can be extended to illustrate a steady-state analysis for a feedforward control system. In the feedback illustration, it was assumed that the only load variable was the feed temperature, T_F, and it was further assumed that the feed rate was constant at 18,000 lb/hr. Now, if the feed rate is a variable, the effects of this on the tank temperature can be compensated for through feedforward control.

Physically, the system amounts to the measurement of the feed rate, followed by appropriate manipulation of the steam rate to maintain the tank temperature constant. A physical picture of the system is shown in Figure 13, and the corresponding block diagram is given in Figure 14. In this analysis, it will be assumed that the tank level remains constant in spite of changes in feed rate so that the product rate is the same as the feed rate.

In Figure 14, the signals between the blocks are the deviations in the variables. These changes are multiplied by the gain factors in the blocks. Having demonstrated the validity of this procedure in the case of feedback,

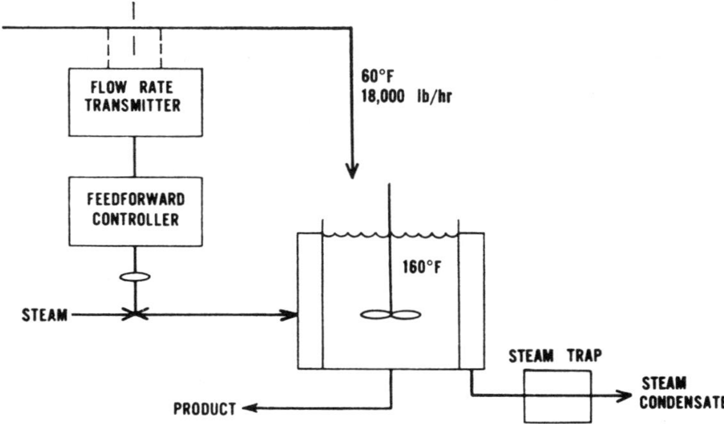

Figure 13.   Feedforward control for stirred-tank heater.

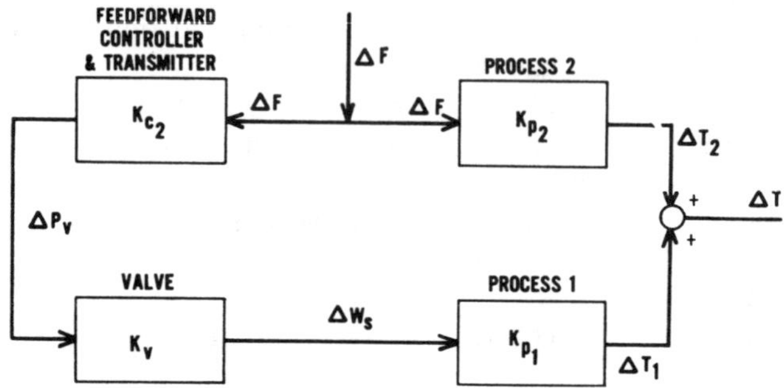

Figure 14.   Feedforward block diagram for Figure 13.

it should be apparent that the technique carries over to feedforward. The transmitter and controller are combined in a single block.

As indicated on the diagram, a change in feed rate has a direct effect on the tank temperature through the process itself. The entering feed is normally at 60°F and leaves the tank at 160°F. Hence, a higher feed rate will cause the tank temperature to fall if the steam rate remains constant. Therefore, if the feedforward controller is properly adjusted, this increase in feed rate will be accompanied by a compensating increase in steam rate. These effects have been labeled ΔT_1 and ΔT_2 on the diagram; their sum, ΔT, is the net effect on the tank temperature. The object is, of course, for ΔT to be zero. Mathematically, this means that the following equation must hold:

$$K_{C_2}K_V K_{P_1} = -K_{P_2} \tag{40}$$

The process gain, K_{P_2}, relates a change in feed rate to a change in tank temperature with the steam rate being held constant. This gain can be calculated from an energy balance on the tank. The initial steady-state energy balance is:

$$\bar{F}c_p(\bar{T} - \bar{T}_F) = \lambda \bar{W}_S \tag{41}$$

The final steady-state energy balance is

$$(\bar{F} + \Delta F)c_p(\bar{T} + \Delta T - \bar{T}_F) = \lambda \bar{W}_S \tag{42}$$

When the first equation is subtracted from the second, the result is

$$\bar{F}\,\Delta T + \bar{T}\,\Delta F + \Delta F\,\Delta T - \bar{T}_F\,\Delta F = 0 \tag{43}$$

For small changes, the second-order term can be neglected. Rearrangement yields

$$\Delta T = \left(\frac{\bar{T}_F - \bar{T}}{\bar{F}}\right)\Delta F \tag{44}$$

The gain factor, K_{P_2}, is the ratio of $\Delta T/\Delta F$:

$$K_{P_2} = \frac{\bar{T}_F - \bar{T}}{\bar{F}} = \frac{60 - 160}{18,000} = -0.00555°F, \text{ hr/lb feed} \tag{45}$$

The process gain, K_{P_1}, is identically the same gain as K_P in the feedback illustration since it transforms a change in steam rate into a change in the tank temperature. For simplicity, the same valve will be used here as was used in the feedback case. Therefore

$$K_V = 777 \text{ lb steam/hr, psi}$$

$$K_{P_1} = 0.0536°F, \text{ hr/lb steam}$$

Using Equation 40, we find that K_{C_2} is 1.33×10^{-4} psi/lb feed/hr.

SUMMARY

The steady-state analysis of a feedback system has shown that if there is only one load variable, there will always be some offset unless the system is operating at the normal design condition. While it is possible for the effects of several load variables to cancel one another, such an occurrence is rather unlikely.

The introduction of deviation variables simplified the calculations for the steady-state behavior of systems. Furthermore, they led to a quantitative interpretation for block diagrams. It was pointed out that gain factors are sometimes approximations, and therefore analyses should usually be restricted to small changes in the variables.

PROBLEMS

Problem 1. For the feedback example of this chapter, develop an expression for the change in set-point temperature, ΔT_S, required to compensate for a change in feed temperature, ΔT_F, so that there is no change in the tank temperature. Express the result in terms of the various K-type gain factors that were used in the illustration. It was pointed out in the example of Section I.B.3.b that a change in set point of $-4°F$ exactly compensated for a change in the feed temperature of $+10°F$. Verify this using the expression obtained above.

Problem 2. The level in a continuous-flow mixing tank is controlled by a proportional controller. The design depth is 8 ft. The overall gain of the feedback loop is 9.0 and the load variable gain for changes in feed rate, K_L, is 5 ft/gpm.

(a) If the set point is moved from 8.0 to 8.5 ft, what will the eventual depth be?

(b) With the set point at 8.5 ft, if the feed rate increases from its design rate of 100 gpm to a new value of 105 gpm, what will the eventual depth be now?

(c) With the final steady state of part *b* established, if the set point is moved to 8.3 ft, what will the eventual depth be?

(d) With the final steady state of part *c* established, if the feed rate drops to 102 gpm, what will the eventual depth be?

Answers: (a) 8.45 ft
(b) 10.95 ft
(c) 10.77 ft
(d) 9.27 ft

Problem 3. A schematic of a blending process for two aqueous streams is shown in Figure 15. Stream 1 is blended with Stream 2, forming a well-blended stream whose normal temperature is 66.7°F. The normal temperatures and flow rates are also indicated in the figure.

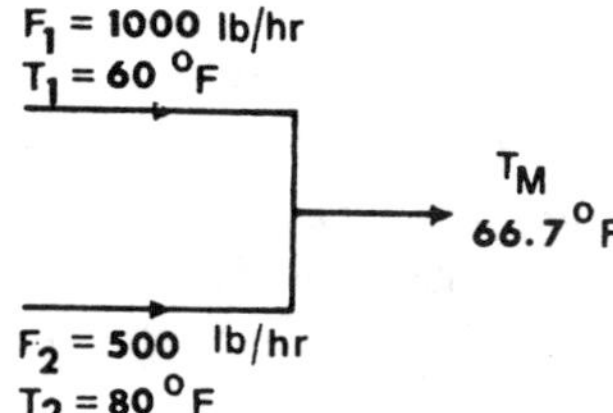

Figure 15.   Blending process for Problem 3.

(a) It is desired to control T_M by a feedback scheme using F_2 as the manipulated variable. Draw a physical picture of the control system and also a block diagram, labeling each of the blocks and signals.

(b) Using a heat capacity of 1 Btu/lb, °F for each stream and an energy balance, the following table was prepared to indicate the effects of F_1 and F_2 on T_M:

F_1, lb/hr		F_2, lb/hr	
	400	500	600
1100	65.33°F	66.25°F	67.06°F
1000	65.71	66.67	67.50
900	66.15	67.14	68.00

The temperature transmitter has a span of 20°F and is set between 60 and 80°F. The valve gain is 120 lb/hr/psi. If the controller gain is 15 psi/psi, find the ultimate change in T_M when F_1 changes from its normal value of 1000 lb/hr to a new value of 1050 lb/hr.

(c) Suppose that the process is operating at the normal values. If the set point is then moved from 66.7°F to 73°F, what will be the ultimate value of T_M?

Answers: (b) −0.0208°F
(c) 72.41°F

Problem 4. A tank-flow heater with steam in the jacket is under feedback temperature control using the steam flow rate as the manipulated variable. The temperature transmitter has a span of 100°F and is set between 100 and 200°F. The proportional controller has a gain of 5 psi/psi. The normal pressure to the valve is 8 psig and the valve is an air-to-open type. The normal temperature of the tank is 170°F and the normal feed temperature is 65°F.

(a) The set point is moved from 170°F to 175°F and the tank eventually comes to a new steady-state temperature of 174.14°F, which was measured with a highly accurate thermometer.

(1) What is the offset?
(2) What is the overall gain of the system?
(3) What is the gain of the valve-tank combination?
(4) What is the pressure to the valve in the final steady state?

(b) The set point is returned to 170°F and the tank eventually returns to 170°F. The feed temperature rises to 85°F and the tank eventually reaches a new steady-state temperature of 173.45°F.

(1) What is the offset?
(2) What is the pressure to the valve in the final steady state?

Problem 5. A hot solution is continuously produced in a stirred tank containing a heating coil. The temperature in the tank is sensed by a gas-filled bulb attached to a transmitter. The pneumatic output of the controller is sent to an air-to-open valve that controls the steam flow rate to the coil.

Some tests have shown that a 1-psi change in pressure to the valve causes a change in heat input of 1000 Btu/hr. Furthermore, a change of heat input of 1000 Btu/hr causes a 2.0°F change in tank temperature. These gains can be assumed constant. The set-point temperature is 160.0°F.

(a) The control system is operating well with both the tank temperature and the set point at 160°F. The pressure to the valve is 8.0 psig. Your boss tells you to put the set point at 165°F, the temperature at which he wants the process to operate. It would be just your luck to have something go wrong in front of your boss—the transmitter jams so that the output is frozen at a pressure corresponding to a temperature of 160°F! You, of course, are unaware of this, at least for quite some time. What does the actual temperature in the tank do if the gain of the transmitter-controller combination is 1 psi/°F?

(b) Your boss becomes most impatient, not to mention yourself, after watching nothing happen on the recorder chart after an hour or so. You finally put a thermometer in the tank. Noting the temperature, you conclude that the transmitter must be jammed. After cleaning it, the temperature indicator begins to move. Does it pass by 165°F? If so, where does it eventually stop? What is the eventual valve pressure?

(c) Returning to part *a*, suppose that the transmitter-controller gain were 5 psi/°F. At the instant the set point is moved from 160°F to 165°F, calculate the change in pressure to the valve. What problem arises here which did not occur in part *a*?

(d) In spite of the complication of part *c*, can the problem of part *b* be answered if the transmitter-controller gain is 5 psi/°F? If so, find the eventual temperature and valve pressure.

$$\textit{Answers: (a) It rises to } 170°F$$
$$\text{(b) It stops at } 163.33°F$$
$$P_V = 9.67 \text{ psig}$$
$$\text{(d) } 164.55°F; \ 10.25 \text{ psig}$$

Problem 6.   For the feedback example of this chapter in Section I.B, suppose the feed rate, F, increases from 18,000 to 18,100 lb/hr. Find the resulting value of the tank temperature when $K_C = 1.0$ psi/psi.

$$\textit{Answer: } 159.841°F$$

Problem 7.   A cylindrical tank is 20 ft high and 8 ft in diameter. Water is fed continuously into the top and leaves through a valved line. The valve is a pneumatically operated one, of the air-to-open type. A picture of the process is shown in Figure 16. Some steady-state tests have been made on this process

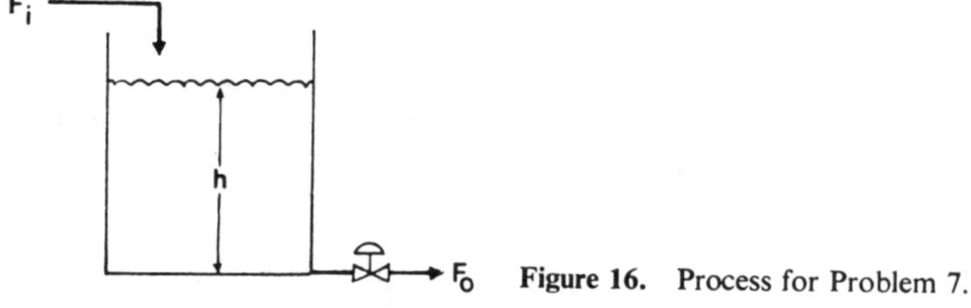

Figure 16.   Process for Problem 7.

correlating the flow rate through the valve with the liquid depth in the tank and with the pressure to the valve. The results of these tests are presented in the following table:

Liquid Depth, h, ft	Valve Pressure, P_V, psig					
	5	7	8	9	10	15
	Flow Rate Through Valve, F_o, gpm					
0	0.0	0.0	0.0	0.0	0.0	0.0
1	10.0	20.0	25.0	30.0	35.0	60.0
2	14.1	28.3	35.4	42.4	49.5	84.8
3	17.3	34.6	43.3	52.0	60.6	103.9
4	20.0	40.0	50.0	60.0	70.0	120.0
5	22.4	44.7	55.9	67.1	78.3	134.2
6	24.5	49.0	61.2	73.5	85.7	147.0
7	26.5	52.9	66.1	79.4	92.6	158.7
8	28.3	56.7	70.7	84.9	99.0	169.7
9	30.0	60.0	75.0	90.0	105.0	180.0
10	31.6	63.2	79.1	94.9	110.7	189.7

The level in the tank is to be controlled by feedback using a proportional controller as shown in Figure 17. The level will be measured by a level transmitter. This device consists of a float connected to a mechanism by which the position of the float is translated into a pressure in the range between 3 and 15 psig. The span of the transmitter is 20 ft, which is set between 0 and 20 ft. The output of the controller will actuate the valve whose characteristics are described by the data given above.

(a) Draw a block diagram for this control system. Label the signal between each block. Show the major load variable, F_i, on the diagram.

(b) The normal flow rate into the tank is 75 gpm and the normal valve pressure is 8 psig. If there is to be no offset at this condition, where must the set point be?

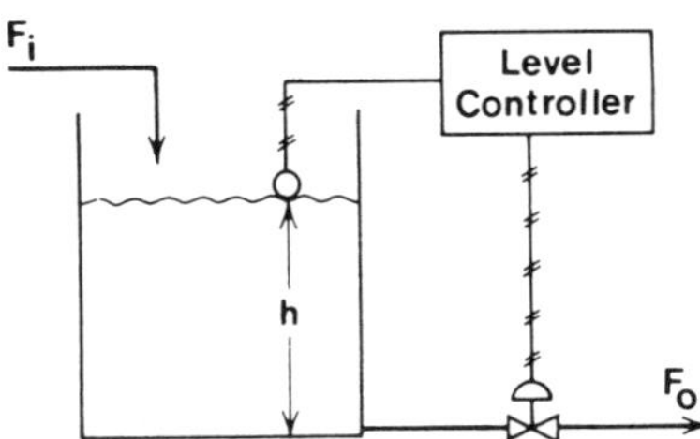

Figure 17. Level control system for Problem 7.

(c)   Calculate the load gain factor, K_L, for changes in F_i.

(d)   Calculate the valve gain, K_V, which relates the change in leaving flow rate to a change in valve pressure.

(e)   Calculate the process gain, K_P, which relates the change in level to a change in leaving flow rate.

(f)   Calculate the transmitter gain.

(g)   If the controller gain is 5 psi/psi, find the eventual level if F_i increases from 75 to 80 gpm.

(h)   For the conditions of part g, find the eventual valve pressure, P_V.

(i)   For the conditions of part g, what must the eventual change in the leaving flow rate be? What fraction of this change is the result of the change in valve position and what fraction is the result of the change in level?

Answers: (c) 0.238 ft/gpm (g) 9.102 ft

(d) 15.0 gpm/psi (h) 8.306 psig

(e) −0.238 ft/gpm (i) 5 gpm; 91.8%; 8.2%

(f) 0.6 psi/ft

Problem 8.   A schematic of a blending system for two aqueous streams is shown in Figure 18. Stream 1 is blended with Stream 2 to form a well-blended stream whose normal temperature is 66.7°F. Controller 1 is actuated by a thermometer bulb that senses the temperature of the combined stream and controls the flow rate of Stream 1. Controller 2 is actuated by a temperature bulb in Stream 2. When the temperature of this stream varies, Controller 2 attempts to compensate.

(a)   Draw a block diagram of this system and label all signals.

(b)   Is there any feedforward in this system? If so, where?

(c)   Calculate each of the following gains assuming that the specific heat of each stream is 1.0 Btu/lb, °F.

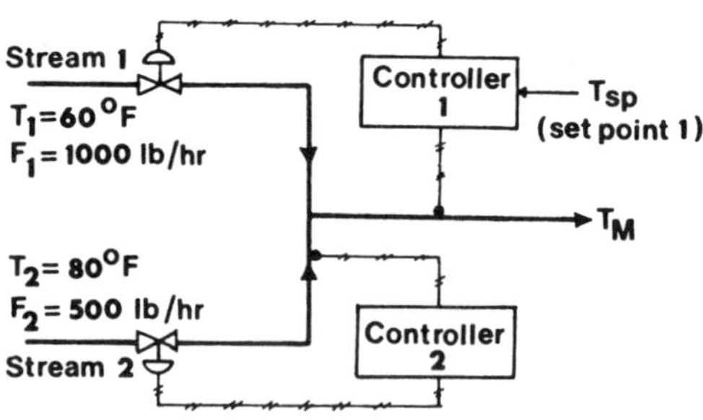

Figure 18.   System for Problem 8.

(1) K_{P_1}, which relates changes in T_M to changes in F_1.
(2) K_{P_2}, which relates changes in T_M to changes in F_2.
(3) K_{L_1}, which relates changes in T_M to changes in T_1.
(4) K_{L_2}, which relates changes in T_M to changes in T_2.

(d) Suppose that $K_C K_T = 2$ psi/°F and $K_{V_1} = 20$ lb/hr/psi. What would the change be in T_M if T_1 increased 5°F?

(e) Suppose that K_{V_2} is 20 lb/hr/psi and T_2 increased 1°F. What should the value of K_{C_2} be for perfect control?

$$
\begin{aligned}
\textit{Answers:} \quad &\text{(c)} \ (1) \ -0.00444 \ °F/lb/hr \\
&\qquad (2) \ 0.00888 \ °F/lb/hr \\
&\qquad (3) \ 2/3 \ °F/°F \\
&\qquad (4) \ 1/3 \ °F/°F \\
&\text{(d)} \ 2.83°F \\
&\text{(e)} \ -1.875 \ psi/°F
\end{aligned}
$$

CHAPTER   V

An Introduction To
The Unsteady State

Chapter 4 was concerned with the calculation of steady states following changes in set point and load variables. The required calculations were purely algebraic. Assuming that the steady-state behavior of a control system is satisfactory, most of our interest will be concerned with the transient behavior between steady states. Information concerning this is derived from differential equations.

It is useful to think of the control system as consisting of two parts—the process itself, and the control components. The latter include the controller and its auxiliary equipment, namely the transmitter to measure the controlled variable and the control valve or other actuator to change the manipulated variable. When a change in set point or a load disturbance occurs, it is the purpose of the control components to force the controlled variable to follow an acceptable path between the initial and final steady states. Once the process and control components are installed, the controller parameters must be matched or "tuned" to exact the desired behavior from the process.

The unsteady-state analysis of a control system must consider the dynamic behavior of the process and control components in terms of their differential equations, and the overall behavior of a system that is described by the set of these equations and the relationships between them. The behaviors of the components themselves fall in the general area of "process dynamics" and the system behavior is classed under the heading of "control theory." The middle portion of this text is concerned with process dynamics. Some aspects of control theory have already been presented in the first four chapters and much more will be given throughout the remaining chapters.

I. TWO IMPORTANT PROBLEMS OF CONTROL THEORY

Most simple control theory deals with two problems. The first might be termed the "stability problem" and the second, the "transient problem." These will be discussed here briefly before examining the unsteady-state behavior of a simple process and control system.

A. Stability Problem

It was mentioned earlier that after the installation of a process and its control components, the controller must be tuned to obtain satisfactory dynamic behavior of the controlled variable. If the controller action is relatively weak, the response of the controlled variable to changes in load variables or in set point will be slow. The response can be quickened by increasing the strength of the controller action. However, there is a limit as to how strong this action can be, for if it is too strong, the system will become unstable. The stability problem is usually concerned with establishing the stability limits for the controller parameters.

The definition of stability depends on whether the system is "linear" or "nonlinear." For linear systems, to which this text is devoted, a system is considered stable if its response is bounded for all bounded inputs. Thus, if some disturbance causes the controlled variable to oscillate with growing amplitude, it is unstable. Obviously, the operability limits of the system will eventually be reached with the result that the control valve will continue to cycle back and forth between fully closed and fully opened, or perhaps in an extreme case, an explosion might occur.

There are a number of methods for the determination of the stability of linear systems and several will be discussed in later chapters.

B. Transient Problem

The transient problem deals with the path followed by the controlled variable as it moves from one steady state to another. One might be interested in the behavior for either a change in set point or a change in a load variable. The path will not only depend upon the initial and final states of a disturbance but also upon the path followed by the disturbance itself.

Three types of disturbances are particularly interesting and useful in studying the response of systems. These are shown in Figure 1. The step

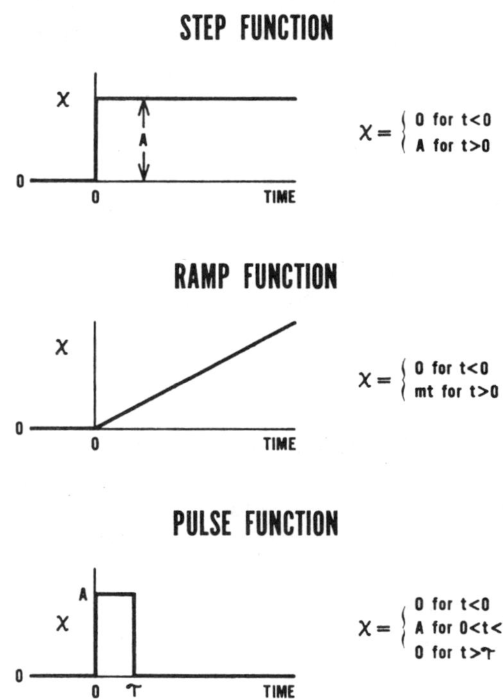

Figure 1.   Three common input functions used for studying the response of systems.

function was already mentioned in Chapter 3. Quite frequently a "unit step" change, which has a magnitude of unity, is used. The response to a step disturbance is a relatively good test of the control system in its ability to follow an abrupt change in set point or to correct for an abrupt change in a load variable. A ramp input would indicate the ability of the system to follow gradual changes in an input. For example, batch operations are often programmed so that the set point changes gradually in accordance with a prescribed processing recipe.

The pulse function can be considered to be a positive step function followed by a negative step function of the same magnitude at a later time. A unit pulse is one whose area, $A \times \tau$, is unity. If this area is held constant while the time duration, τ, approaches zero, the magnitude, A, approaches infinity and the resulting function is called a unit impulse. The response to a unit impulse is of great theoretical significance and some industrial testing of processing equipment has employed the pulse function.

Because of its mathematical simplicity as well as its ability to reveal offset, the step function has been used extensively in studying the transient response of systems. It will be used in most of the analyses of this book.

II. RESPONSE OF A STIRRED-TANK HEATER

In Chapter 4, the steady-state behavior of a continuous-flow, stirred-tank heater was studied under proportional control. We turn now to an examination of the transient response of the system to step changes in set point and in the load variables of feed temperature and feed rate. Before an analysis of the system can be made, the process itself must be examined. A picture of the tank is shown in Figure 2.

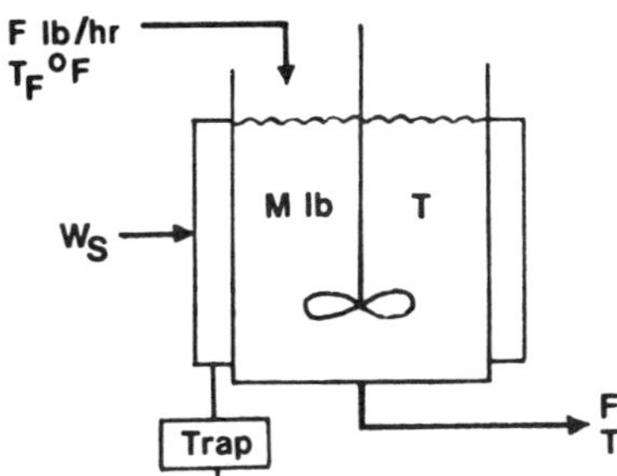

Figure 2.   Stirred-tank heater.

A. Behavior of Heater Alone to a Change in Steam Flow Rate

Since this is an unsteady-state problem, the unsteady-state macroscopic energy balance is the starting point (1, 2). The fluid within the tank is chosen as the system. For this problem, basically the energy balance says that the rate of change of internal energy of the liquid in the tank is equal to the rate of heat transfer from the jacket plus the difference between the enthalpy of the entering and leaving streams. The holdup of liquid in the tank is assumed constant. For a liquid, a change in its internal energy is essentially equal to the corresponding change in its enthalpy. Furthermore, the choice of an enthalpy datum plane is unimportant since it drops out of the energy balance. Therefore, the energy balance can be simplified to the following form:

$$\text{heat input} \quad - \text{ heat output} = \text{accumulation}$$

$$Fc_p T_F + W_S \lambda - \quad Fc_p T \quad = \quad Mc_p \frac{dT}{dt} \tag{1}$$

where F = feed rate, lb/hr
  c_p = heat capacity, Btu/lb, °F
  T_F = feed temperature, °F
  W_S = steam rate, lb/hr
  λ = heat of vaporization of steam, Btu/lb
  T = tank temperature, °F
  M = mass of fluid in the tank, lb
  t = time, hr

In the analysis of the system for a change in set point, the only variables are the tank temperature and the steam rate. The steam rate is a variable since it will be used as the manipulated variable in the control system. For convenience in the analysis, it will be assumed that the system is initially in a steady-state condition with all parameters at their design or "normal" values. Then, at some time designated as "time zero," there is a change in the input variable, W_S.

In the last chapter, the concept of deviation variables was introduced and found to be advantageous. The same advantages accrue in an unsteady-state analysis. The deviations in an unsteady-state problem are time variant and represent the deviations from the original steady states. Thus, we set

$$W_S = \overline{W}_S + \Delta W_S$$
$$T = \overline{T} + \Delta T$$

As before, the symbols with bars are the normal or original steady values, which are constants. Introducing these expressions into Equation 1, we obtain

$$Fc_p \overline{T}_F + (\overline{W}_S + \Delta W_S)\lambda - Fc_p(\overline{T} + \Delta T) = Mc_p \frac{d(\overline{T} + \Delta T)}{dt} \tag{2}$$

The symbol, $\overline{T}_F$, has been substituted for T_F merely to emphasize that it is invariant and remains at its original value. Also, on the right-hand side, $d\overline{T}/dt$ is zero since $\overline{T}$ is a constant.

Prior to time zero when the system is at its original steady state, the energy balance equation is

$$Fc_p \overline{T}_F + \overline{W}_S \lambda - Fc_p \overline{T} = 0 \tag{3}$$

When Equation 3 is subtracted from Equation 2, the result is

$$\lambda \, \Delta W_S - Fc_p \, \Delta T = Mc_p \, \frac{d \, \Delta T}{dt} \tag{4}$$

The significant aspect of this equation is that all of the variables appear only as deviations; hence, this equation can be referred to as the "deviation equation." For convenience, we shall drop the Δ's in front of the symbols with the understanding that W_S and T represent deviations:

$$\lambda W_S - Fc_p T = Mc_p \, \frac{dT}{dt} \tag{5}$$

Interestingly, if this equation is compared with Equation 1, we find that the only difference is that the first term of Equation 1 is missing. The implication is that we could have written Equation 1 in the first place and said that *it* represented the deviation equation, but dropping the first term, since T_F exhibits no deviation. This apparent coincidence occurs here because Equation 1 was linear in the factors chosen as variables. In general, it is advisable to go through the steps previously demonstrated since this procedure avoids mistakes in those cases where nonlinear factors are present.

During the remaining part of this example, only the deviation equation will be used, and the Δ's will be dropped for convenience. Equation 5 is an ordinary, first-order, linear differential equation. As will be shown in Chapters 8 and 9, many pieces of equipment can be "modeled" approximately by equations of this type. The equation can be rearranged to the following "standard" form:

$$\frac{M}{F} \frac{dT}{dt} + T = \frac{\lambda}{Fc_p} W_S \tag{6}$$

The temperature, T, is the controlled or dependent variable, and W_S is the manipulated or input variable. The two factors, M/F and λ/Fc_p, are constants so the equation has the following general form:

$$\tau \frac{dy}{dt} + y = Kx \tag{7}$$

The factor, τ, which has the units of time, is commonly referred to as the "time constant." As will be shown below, the factor, K, is the gain factor for changes in the input, x. The solution of this equation requires the specification of the initial condition on y, which we shall denote as y_i. The input function,

x, in the most general case could be any function of time. However, most of the studies in this text will assume step inputs, so the input here will be a step input of magnitude x. Therefore, the solution of Equation 7 is

$$y = y_i e^{-t/\tau} + Kx(1 - e^{-t/\tau}) \tag{8}$$

The symbol, e, in this solution is the base of natural logarithms. The same symbol is used for an error but this should cause no ambiguity since the natural base is only used to indicate exponential functions. The ultimate or final steady-state value of y, denoted as y_f, is found by taking the limit as time approaches infinity. The result is that y_f is equal to Kx.

Now suppose that y is defined to be a deviation variable and at time zero this deviation is zero. Therefore, y_i is zero and the ultimate change in y is y_f. Furthermore, x is the magnitude of the step input and is therefore the ultimate change in x. It will be recalled from Section I.B.3.d of Chapter 4 that a gain factor is defined as the ultimate change in an output variable divided by the ultimate change in an input variable. It follows that K must be the gain factor.

Equation 8 can be written in a somewhat more useful form that is relatively easy to remember. Since $y_f = Kx$, it can be expressed as follows:

$$\frac{y - y_i}{y_f - y_i} = 1 - e^{-t/\tau} \tag{9}$$

The fraction on the left is the "fractional response" of the output and has the form illustrated in Figure 3. In particular, at a time equal to τ, the fractional response is 0.632. Therefore, if the response of a process is known to be first order or suspected to be so, the time constant can be found by determining the time at which 63.2% of the ultimate change occurs.

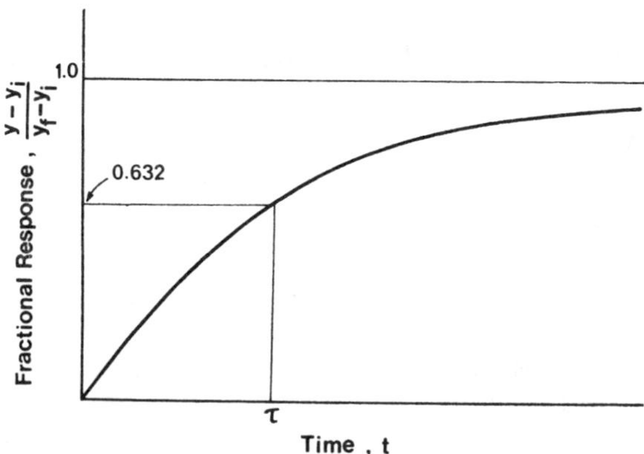

Figure 3.   Fractional response of a linear, first-order process to a step input.

In terms of the time constant and gain factor, Equation 6 can be written

$$\tau \frac{dT}{dt} + T = K_P W_S \tag{10}$$

where

$$\tau = \frac{M}{F}, \text{ hr} \tag{11}$$

and

$$K_P = \frac{\lambda}{F c_p} \text{ °F/(lb/hr)} \tag{12}$$

The gain factor has been given a subscript, P, to indicate that this is the gain factor for changes in the manipulated variable. It is the same gain, of course, as was found for this process in the steady-state analysis of Chapter 4.

Equation 10 is a deviation equation and, as mentioned previously, it is assumed that at the initial steady state, the deviation in T is zero. At time zero, suppose that a step change in steam rate occurs. Equation 8 then indicates that the deviation in temperature will respond in accordance with the following equation:

$$T = K_P(1 - e^{-t/\tau})W_S \tag{13}$$

where W_S is the magnitude of the step change in steam rate.

B. Response of Heater Under Proportional Control to a Change in Set Point

Having examined the response of the tank temperature for the process alone, we now turn to a consideration of its behavior when the temperature is placed under proportional control. A physical picture of the system is shown in Figure 4. The system is the same as that treated in Chapter 4.

A partial block diagram for the system is presented in Figure 5. The diagram is incomplete in the sense that no load variables are indicated. These will be considered later. All symbols between the blocks represent deviations of the variables from their original steady-state values. The transmitter and controller have been combined in a single block to yield a unity feedback configuration. It has been assumed that the response of the steam flow rate to a change in the valve pressure is instantaneous so the valve is characterized by a pure gain, K_V. The differential equation for the process is indicated in the process block.

We shall now determine the time functionality of the deviation in tank temperature, T, to a change in set point, T_S. The deviation in the error is

$$e_T = T_S - T \tag{14}$$

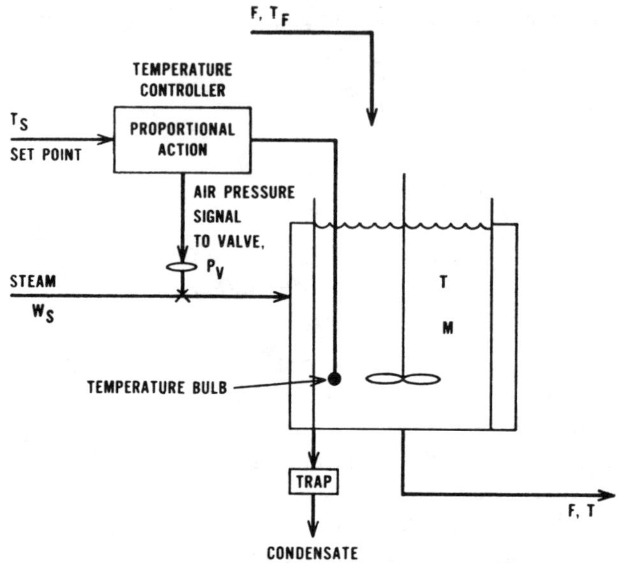

Figure 4.   Stirred-tank heater with temperature control system.

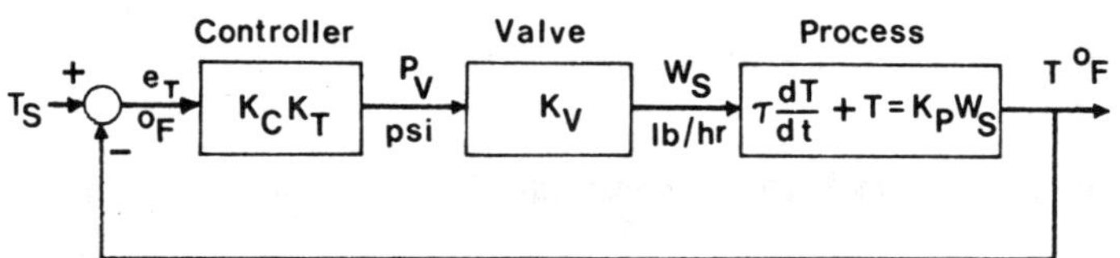

Figure 5.   Partial block diagram of temperature control system.

The deviation in the steam rate is

$$W_S = K_C K_T K_V e_T \tag{15}$$

When these equations are combined with Equation 10 for the process, the following differential equation results:

$$\tau \frac{dT}{dt} + T = K_C K_T K_V K_P (T_S - T) \tag{16}$$

The product, $K_C K_T K_V K_P$, is the open-loop gain or overall gain, K_o. Since this is a linear ordinary differential equation, it can be rearranged into the standard form of Equation 7:

$$\frac{\tau}{1 + K_o} \frac{dT}{dt} + T = \frac{K_o}{1 + K_o} T_S \tag{17}$$

where

$$K_o = K_C K_T K_V K_P \tag{18}$$

Prior to the change in set point, the deviation in T is zero. Assuming a step change in set point, we see from Equation 8 that the response in the deviation in tank temperature is given by the following equation:

$$T = \frac{K_o T_S}{1 + K_o} \left\{ 1 - \exp\left[-\frac{t}{\tau/(1 + K_o)} \right] \right\} \tag{19}$$

where T_S is the magnitude of the step change in set point. The deviation, T, then has the form shown in Figure 3. However, it must be emphasized that the time constant for the "closed-loop" *system* is $\tau/(1 + K_o)$, whereas the time constant of the *process* is τ. The gain factor for the closed loop is $K_o/(1 + K_o)$, which agrees with that given in Equation 38 of Chapter 4.

As pointed out in Chapter 4, a high value of K_o is desirable since it reduces the offset resulting from a change in set point. Equation 19 indicates that a high value of K_o is also desirable from the standpoint of the speed of response, since this makes the system time constant small. The only adjustable factor is the controller gain, and presumably the controller has some maximum gain setting. In theory, if the gain could be set at infinity, the tank temperature would instantaneously and exactly follow the change made in the set point. Even if a controller with a gain of infinity were possible, there are physical limitations that make such ideal behavior impossible. Nonetheless, the idea is directionally correct in that the higher the gain, the faster the response. This result is in line with physical intuition, since a high gain means that strong corrective action is applied by the controller and this tends to reduce the error.

No instability is predicted for this control system under any conditions. That is, even if an infinite overall gain did exist, the controlled variable is always bounded and nonoscillatory. In later chapters, it will be shown that real systems always have an upper limit in overall gain beyond which oscillations would occur. This existence of a maximum is not predicted by the model used here because of its simplicity. The differential equation written for the tank is not an exact representation of the true behavior of the tank. The valve has been represented as a constant-gain element, implying that it reacts instantaneously to changes in valve pressure. These are just two of the numerous approximations that have been introduced. However, these approximations and assumptions only become significant when the model is applied for extreme conditions, as for example when it is used for a system having a very high overall gain.

C. Response of Heater Under Proportional Control to Changes in Load Variables

We shall now examine the response of the system to a change in feed temperature, T_F, and to a change in feed rate, F. This requires first finding the differential equation describing the response of the process to each of these changes. The starting point in the analysis is again the energy balance for the tank that was given in Equation 1:

$$Fc_pT_F + W_S\lambda - Fc_pT = Mc_p\frac{dT}{dt} \tag{1}$$

The four variables are F, T_F, W_S, and T. Therefore, the differential equation contains two nonlinear factors, FT_F and FT. Each of the variables is expressed as the sum of its original steady-state value plus an unsteady-state deviation, with the result that Equation 1 can be written

$$(\bar{F} + \Delta F)c_p(\bar{T}_F + \Delta T_F) + \lambda(\bar{W}_S + \Delta W_S)$$
$$- (\bar{F} + \Delta F)c_p(\bar{T} + \Delta T) = Mc_p\frac{d\,\Delta T}{dt} \tag{20}$$

If this is multiplied out, two terms appear that contain the product of two deviations, namely $c_p\,\Delta F\,\Delta T_F$ and $c_p\,\Delta F\,\Delta T$. This analysis is restricted to small deviations and therefore these second-order terms can be dropped. The effect of dropping these terms is to make the equation linear in the deviations.

The next step is to write the original steady-state equation:

$$\bar{F}c_p\bar{T}_F + \lambda\bar{W}_S - \bar{F}c_p\bar{T} = 0 \tag{21}$$

When Equation 21 is subtracted from Equation 20, the following deviation equation results:

$$\bar{F}c_p\,\Delta T_F + c_p\bar{T}_F\,\Delta F + \lambda\,\Delta W_S - \bar{F}c_p\,\Delta T - c_p\bar{T}\,\Delta F = Mc_p\frac{d\,\Delta T}{dt} \tag{22}$$

Dropping the Δ's for convenience with the understanding that the variables now represent deviations from the original steady state, we obtain

$$\bar{F}c_pT_F + c_p\bar{T}_FF + \lambda W_S - \bar{F}c_pT - c_p\bar{T}F = Mc_p\frac{dT}{dt} \tag{23}$$

Comparing this deviation equation with Equation 1, we see that the two are not the same; this dissimilarity occurs because Equation 1 is nonlinear in the variables chosen. Equation 23 can be arranged to the standard form:

$$\frac{M}{\bar{F}}\frac{dT}{dt} + T = \frac{\lambda}{\bar{F}c_p}W_S + T_F + \frac{\bar{T}_F - \bar{T}}{\bar{F}}F \tag{24}$$

The process time constant, τ, is the same as that given in Equation 11 and it is evaluated at the normal flow rate, $\bar{F}$. Similarly, the process gain factor,

K_P, is the same as that given in Equation 12 and is evaluated at the normal flow rate. The load gain factor for changes in feed temperature is

$$K_{T_F} = 1 \tag{25}$$

The gain factor for changes in flow rate is

$$K_F = \frac{\overline{T}_F - \overline{T}}{\overline{F}} \tag{26}$$

These load gain factors are consistent with those found in Chapter 4 for this process. The subscript of each gain factor indicates the load variable to which it applies.

Notice that the time constant is the same for changes in both load variables as well as for changes in the manipulated variable. Hence, if the heat of vaporization were a variable, the time constant for this variable would also be τ.

Equation 24 indicates that the response of T is linear in the three variables, W_S, T_F, and F. Therefore, the individual effects of each of these three inputs are additive. The differential equation for a change in W_S alone is found by setting T_F and F equal to zero; this results in Equation 10, which was previously given. An analogous treatment results in the following two equations for the load variables:

$$\tau \frac{dT}{dt} + T = K_{T_F}T_F \tag{27}$$

$$\tau \frac{dT}{dt} + T = K_F F \tag{28}$$

The response of the *process alone* to a step input in each of the three variables is first order with a time constant, τ, and each input has its own gain factor.

It is now possible to augment the block diagram of Figure 5 to include the load variables. The complete diagram is shown in Figure 6. Each process block has been given a number to indicate that it corresponds to a specific process input.

We shall now consider the response of the *system* to a change in feed temperature. Therefore, the deviations in set point and feed rate are taken as zero. Equation 24 then reduces to

$$\tau \frac{dT}{dt} + T = K_{T_F}T_F + K_P W_S \tag{29}$$

The deviation equation for the comparator is

$$e_T = -T \tag{30}$$

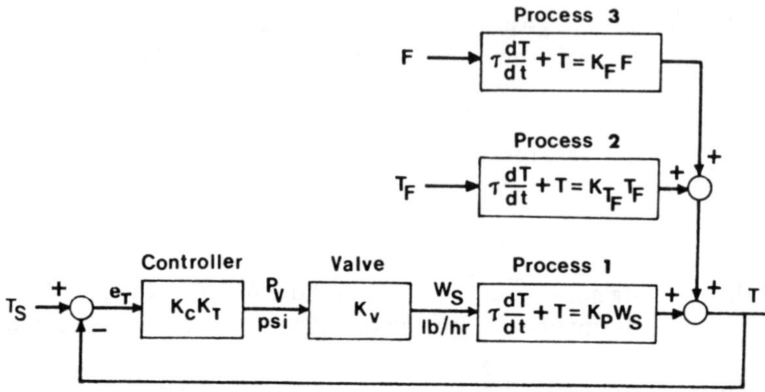

Figure 6.   Complete block diagram of temperature control system.

Only T appears since there is no change in the set point, T_S. Equation 15 applies for the controller, transmitter, and valve:

$$W_S = K_C K_T K_v e_T \tag{15}$$

Thus, combining Equations 15, 29, 30, and 18, we obtain

$$\tau \frac{dT}{dt} + T = K_{T_F} T_F - K_o T \tag{31}$$

Rearrangement to the standard form yields

$$\frac{\tau}{1 + K_o} \frac{dT}{dt} + T = \frac{K_{T_F}}{1 + K_o} T_F \tag{32}$$

The solution for a step change in T_F is

$$T = \frac{K_{T_F} T_F}{1 + K_o} \left\{ 1 - \exp\left[-\frac{t}{\tau/(1 + K_o)} \right] \right\} \tag{33}$$

where T_F is the magnitude of the step change in feed temperature. The system time constant is again $\tau/(1 + K_o)$ and the gain factor is $K_{T_F}/(1 + K_o)$, which is consistent with Equation 37 of Chapter 4.

By analogy, it can be seen that the response for a step input in F would be

$$T = \frac{K_F F}{1 + K_o} \left\{ 1 - \exp\left[-\frac{t}{\tau/(1 + K_o)} \right] \right\} \tag{34}$$

where F is the magnitude of the step change in feed rate.

As with a change in set point, the transient response improves and the offset decreases as the overall gain increases.

SUMMARY

The unsteady-state behavior of a process was examined, first for the process alone, and then when the process was part of a control system under proportional control. The process was described by a simple "first-order model" and was assumed to be the only dynamic element in the system. The process was characterized by a time constant, τ, and a gain factor for each of its input variables. When the process was part of a control system, the time constant was reduced by a factor of $1/(1 + K_o)$ and the gain factors for changes in set point and load variables were reduced by the same factor. Thus, the transient response improved as the controller gain was increased and furthermore, the offset was reduced. Although the model used for this process was simple and simple relationships were employed for the controller and valve, the results of this example have qualitative significance for systems considerably more complex, especially if the controller is of the proportional type with a relatively low gain.

It was pointed out that in a real control system, there is an upper limit on the proportional gain, usually dictated by stability problems. Therefore, some offset will usually result from proportional action. To eliminate offset and to improve the transient response of systems, other controller actions are often used. These are the subject of Chapter 6.

REFERENCES

1. R. B. Bird, W. E. Stewart, and E. N. Lightfoot, *Transport Phenomena*, John Wiley and Sons, Inc., New York, 1960, pp. 458–460, 473–475.
2. J. M. Smith and H. C. Van Ness, *Introduction to Chemical Engineering Thermodynamics*, Second Ed., McGraw-Hill Book Co., Inc., New York, 1959, pp. 36–37.

PROBLEMS

Problem 1. As will be discussed in Chapter 9, the dynamic behavior of a mercury-filled thermometer can be approximately described by a linear, first-order differential equation. The dependent variable is the height of the mercury column and the height is calibrated in terms of degrees. The input variable is the actual temperature of the fluid whose temperature is being measured. Suppose the time constant of a given thermometer has been found to be 0.1 min when it is placed in a constant temperature bath. Assume that the thermometer has been in one bath at 90°F for a considerable length of time so that it indicates the temperature of the bath to be 90°F. If it is then quickly moved to a second bath at 100°F, how long will it be before the thermometer indicates a temperature of 98°F?

Answer: 0.161 min

Problem 2. The level in a tank is controlled by a proportional controller as shown in Figure 7. The corresponding block diagram is presented in Figure 8. The normal set point and level are 8.0 ft. If the set point is moved to 9 ft, how long will it take for the level to reach 8.3 ft?

Answer: 1.56 min

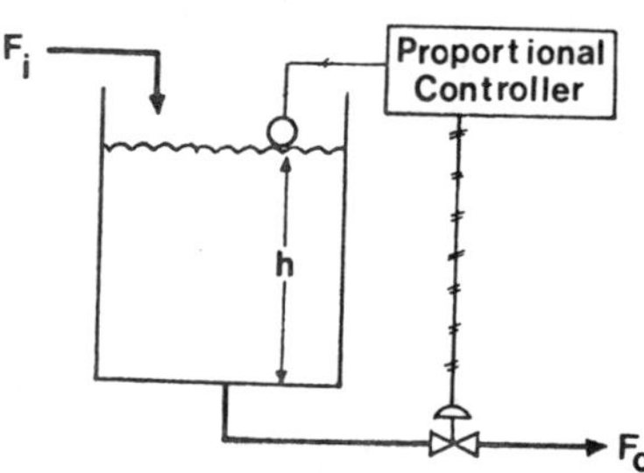

Figure 7. Physical picture of level control systems for Problems 2 and 3.

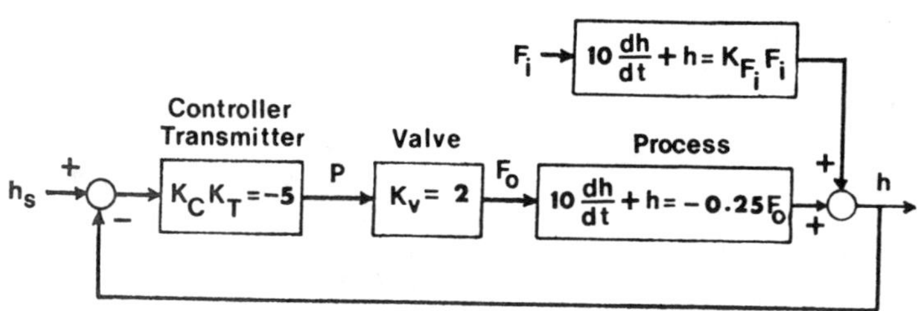

Figure 8. Block diagram for Problem 2.

Problem 3. The level in a tank is controlled by a proportional controller as shown in Figure 7. The controller gain is 3 psi/psi, the valve gain is 10 gpm/psi, and the process gain is -0.5 ft/gpm. The span of the transmitter is 10 ft and is set between 0 and 10 ft. The time constant of the tank alone is 20 min. At the normal flow rate of 100 gpm, the level is 8.0 ft and there is no offset.

(a) The gain factor for changes in the load variable, F_i, has not been given. How is it related to the process gain just given? Physical intuition should provide the answer.

(b) If F_i suddenly increases from 100 to 110 gpm, what will the ultimate level be and the corresponding offset?

(c) With the flow rate at the normal value of 100 gpm, the set point is suddenly moved to 8.5 ft. What will the ultimate level be and the corresponding offset?

(d) The flow rate is at 100 gpm and the set point is at 8.0 ft. Suppose that simultaneously F_i jumps to 110 gpm and the set point is moved to 8.5 ft. What will the ultimate level be?

(e) The level and set point are at 8.0 ft and F_i is 100 gpm. If F_i suddenly jumps to 110 gpm, what will the level be after 0.5 min?

(f) The level and set point are at 8.0 ft and F_i is 100 gpm. If the set point is suddenly moved to 8.5 ft, what will the level be after 0.5 min?

(g) Following along from part f, the set point is now moved to 7.5 ft. What will be the level 0.5 min later? In other words, the set point moves in accordance with the following schedule:

Time, t (min)	Set point, h_s (ft)
0	Moved from 8.0 to 8.5
0.5	Moved from 8.5 to 7.5
1.0	Still at 7.5. What is h at this time?

Answers: (b) $h = 8.263$ ft; offset $= 0.263$ ft
(c) $h = 8.474$ ft; offset $= 0.026$ ft
(d) 8.737 ft
(e) 8.0994 ft
(f) 8.179 ft
(g) 7.932 ft

Problem 4. A feedback control system for a special application is represented by the block diagram of Figure 9. Note particularly that the output pressure of the controller is proportional to the square of the error, e. Furthermore, the range of the output is unusual in that it runs between 0 and 50 psig and the valve has a corresponding range.

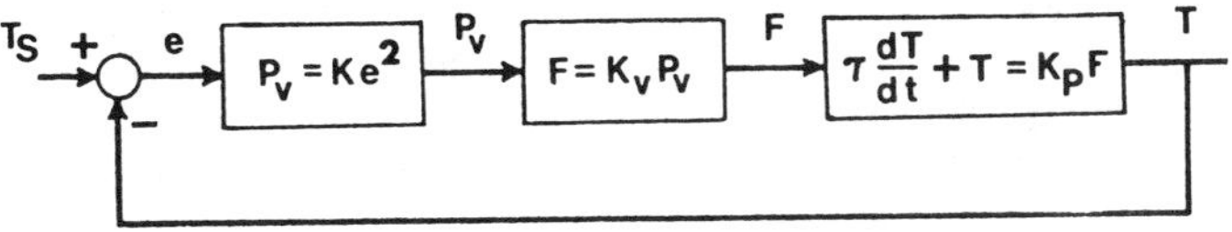

Figure 9. Block diagram for Problem 4.

(a) Linearize the behavior of the controller about the normal value of e, which is 4. Then develop an expression for the time functionality of T for a unit step change in T_S. The numerical values of the gain constants are $K = 2$; $K_V = 2$; and $K_P = 0.1$.

(b) What is the ultimate change in T predicted by the linearization?

(c) What is the actual ultimate change in T? Calculate the percentage error incurred by the linearization.

Answers: (b) 0.7619

(c) 0.7671; 0.678%

Problem 5.   The theoretical development in this chapter was based on the idea of a continuous measurement of the controlled variable and corresponding continuous comparison with the set point and corrective action by the controller. In direct digital control as well as in manual control, both the measurement and corrective action are performed periodically. This intermittent type of control is often referred to as "sampled-data control." A block diagram for such a system is shown in Figure 10. The property of feedback is present, but it is not continuous feedback in the sense of the illustration of this chapter.

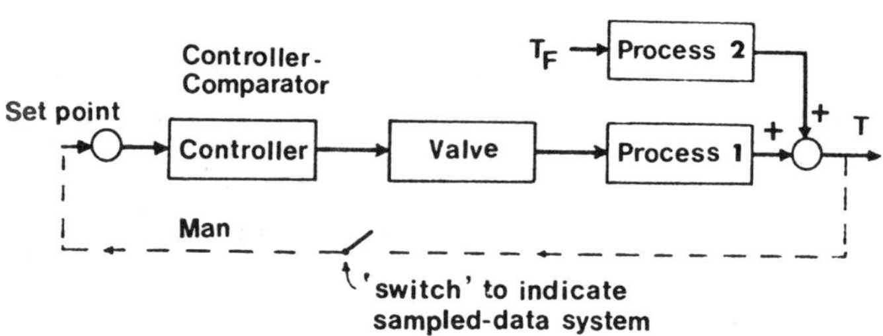

Figure 10.   Block diagram for Problem 5.

Suppose that the process is a stirred-tank heater. Ordinarily the temperature of the liquid is controlled by a conventional continuous feedback system, but the temperature transmitter has been removed for repairs. The operator has been instructed to measure the temperature in the tank every 15 min with a mercury thermometer and then to attempt a manual correction of the valve pressure. The controller has a "manual" switch that enables the operator to adjust the valve pressure by the set-point indicator. When the set point is at the bottom of the scale, the valve pressure is 3 psig and the valve is fully closed; when the set point is at the top of the scale, the pressure is 15 psig and the valve is fully opened.

The normal feed temperature is 60°F and the normal tank temperature is 160°F. The gain factor for changes in feed temperature is 1.0°F/°F. The rate of steam flow and corresponding rate of heat transfer are proportional to the valve pressure. When the valve is fully opened, the steam rate is just sufficient to bring the water to its boiling point of 212°F, but with no boiling, provided, of course, that the feed rate and feed temperature of the water are at their design values.

(a) Suppose that at 10:00 a.m., the process is operating in a steady-state condition with the feed at 60°F and the tank temperature at 160°F. The operator takes a reading of the temperature at this time and since the temperature is right at the desired value, he makes no change in the valve pressure. Then at 10:05 a.m., the feed temperature suddenly jumps to 65°F. The operator is, of course, unaware of this. What temperature does he measure in the tank at his next reading at 10:15 a.m.? The time constant for the tank alone is 15 min.

(b) Not having much feeling for the sensitivity of the process, the operator reduces the valve pressure by 1.0 psi at 10:15 a.m. What temperature does he measure at his 10:30 a.m. reading?

(c) What pressure adjustment should the operator have made at 10:15 a.m. to ultimately bring the temperature back to 160°F, assuming that the feed temperature remains at 65°F? If he had made this adjustment, what temperature would he have measured at 10:30 a.m.?

Answers: (a) 162.43°F
(b) 156.05°F
(c) −0.395 psi; 160.89°F

Problem 6. Referring to the example discussed in this chapter, one of the possible load variables was the feed rate, F, to the tank. Suppose it were desired to compensate for this variable in a feedforward arrangement.

(a) Draw the block diagram for the feedforward system, fill in the blocks, and label the signals.

(b) Determine the characteristics of the feedforward computer needed for exact, unsteady state compensation.

Problem 7. This problem concerns the stirred-tank heater discussed in this chapter. Suppose the set point is at 100°F and the process is in a steady-state condition with no offset. At time zero, the set point is abruptly moved to 110°F. The temperature in the tank is then measured with a very accurate thermometer at 5-min intervals over a period of 45 min. The results of these measurements are:

Time from Start, min	Tank Temperature, °F
0	100.0
5	100.9
10	101.6
15	102.2
20	102.7
25	103.1
30	103.4
35	103.7
40	103.9
45	104.0

(a)   Based on the theory developed in this chapter, find the time constant of the system, the time constant of the process, and the overall loop gain.

(b)   Suppose that the preceding change had been made, but that at the end of 15 min, the set point had been returned to 100.0°F. What would the temperature in the tank be 10 min after this second change in set point?

Answer: (b) 101.45°F

CHAPTER VI

Controller Actions

Some aspects of the control of a home-heating system were examined in Chapters 1 to 3. It was noted that if the temperature of the room was less than the set point, the thermostat turned the furnace on, and if greater than the set point, it turned the furnace off. This type of control action is commonly referred to as "on-off."

In Chapters 4 and 5 we considered the quantitative behavior of a simple control system in which the controller action was proportional to the error

"

detected by the comparator. We found that proportional control could lead to offset that may not be tolerable.

Both on-off and proportional control have advantages of simplicity, but their limitations and disadvantages have led to the use of other actions such as integral, derivative, and various combinations of these. In this chapter, all of these actions will be described and the effects of some of them will be illustrated for some simple systems.

I. PROPORTIONAL ACTION

A. Description of Action

As mentioned in Chapter 4, the basic property of proportional action is that the change in output is proportional to the change in the error which is detected by the comparator of the controller. If the set point is fixed, then the change in output is proportional to the change in input from the transmitter. For electronic controllers, the input and output are either currents or voltages. However, since most controllers in chemical plants are pneumatic, this discussion will be in the context of pneumatic units. Accordingly, the proportional action is defined by the following deviation equation:

$$P = K_c e \tag{1}$$

where P = change in controller output pressure, psi

K_c = controller gain, psi/psi

e = change in error, psi

In some texts, the controller and transmitter are considered as a single unit to obtain a unity feedback configuration. When this is done, the controller gain is the product, $K_c K_T$, as defined in Chapters 4 and 5, and has the units of pressure divided by those of the controlled variable. Correspondingly, the units of the error are those of the controlled variable. In these introductory chapters, the gains of the controller and transmitter will be separately designated, but in later chapters, the symbol, K_c, may be used to indicate the product of the controller and transmitter gains. There should be no ambiguity, however, since it is possible to tell which definition is being used from the units given for K_c or from the context of the discussion.

B. Recording Charts and Gain Dial Calibrations

In many standard pneumatic controllers, provision is made for the recording of the controlled variable as measured by the transmitter. Two types of charts are in common use. One is circular and revolves with time. The main advantage of this type is that the complete record is visible at all times. Although the time lines on the chart are curved, the graduations for the controlled variable are uniformly spaced. These graduations cover a total width of 4 in. in the radial direction. The indicator pen is at its lowest position when the input pressure from the transmitter is at 3 psig, and is at its maximum position at 15 psig, but the chart graduations are calibrated in terms of the measured variable, as was explained in Chapter 4. The second type of chart is a strip chart, which comes in a roll. The chart width for this type is also 4 in. The time lines are straight and the graduations for the controlled variable are uniformly spaced.

There are three common ways of calibrating the gain dial for proportional action. One way is to calibrate the dial in terms of K_C as used in Equation 1, so that the dial graduations have the units of psi/psi. Suppose, for example, that K_C is **-1 psi/psi,** that the set point is at the midpoint of the chart, and that the output pressure has been adjusted so that it has a value of 9 psig when the indicated error is zero. Then, as the input pressure is increased from 3 to 15 psig, the output pressure will correspondingly increase from 3 to 15 psig if the controller is positive acting, or decrease from 15 to 3 psig if the controller is negative acting. The indicator pen moves a total of 4 in. while the output changes 12 psi, so the "sensitivity" is the 12-psi change in the output pressure divided by the 4 in. of pen travel, or 3 psi/in. Since 100% of the chart width must be traversed by the indicator pen to cause the output to move from 3 to 15 psig, the "bandwidth" or "proportional band" is said to be 100%.

To further clarify this, suppose K_C is increased to **-4 psi/psi.** Then a 1-psi change in input will result in a 4-psi change in output. A 1-psi change in input causes the indicator pen to move $\frac{1}{12} \times 4$ in. or $\frac{1}{3}$ in. The sensitivity is then the 4-psi change in output divided by $\frac{1}{3}$ in. or 12 psi/in. Since the indicator must move only 1 in. to change the output pressure from 3 to 15 psig, the proportional band is 25%.

Chapters 2 and 4 noted that the standard input pressure range of control valves is 3 to 15 psig. Therefore, if the controller output exceeds these limits, "saturation" is said to have occurred. Although the maximum controller output may be as high as 20 psig and the minimum output as low as 0 psig, the *useful* control range is still 3 to 15 psig because of the operability limits of the valve. In the illustration that follows, it will be assumed that the *actual* range of the controller is also 3 to 15 psig.

The practical significance of saturation and its relationship to the controller gain is best shown by an example. Consider a controller connected to a temperature transmitter having a span of 200°F, which has been adjusted to cover the range between 50 and 250°F. Assume that the set point and the measured temperature are both at 150°F. This means that the output pressure corresponds to the normal or design value. Standard controllers are provided with a means for adjusting this pressure to any desired value in the range between 3 and 15 psig. For equal flexibility in both directions, assume that the design pressure has been set at 9 psig.

If the gain of the controller is -1 psi/psi, then the valve is completely opened at a temperature of 50°F and completely closed at 250°F; thus, the proportional band is 100%. Changing K_c first to -2 psi/psi and then to -6 psi/psi progressively narrows the proportional band, as shown in Table 1. When the temperature falls outside the band, the output is constant.

TABLE 1

Measured Temperature, °F	Transmitter Output Pressure, psig		Controller Gain, K_C, in psi/psi		
			-1	-2	-6
250.0	15		15	15	15
233.3	14		14	15	15
216.7	13		13	15	15
200.0	12		12	15	15
183.4	11		11	13	15
166.7	10	Prop.	10	Prop. 11	Prop. 15
150.0	9	Band =	9	Band = 9	Band = 9
133.3	8	100%	8	50% 7	16.7% 3
116.7	7		7	5	3
100.0	6		6	3	3
83.4	5		5	3	3
66.7	4		4	3	3
50.0	3		3	3	3

Temperature Transmitter Relationship covers the first two columns; *Controller Output Pressure, psig, Set Point = 150°F* covers the Controller Gain columns.

Alternate expressions for the controller gain are also apparent from the table. When the controller gain is -1 psi/psi, the gain can be expressed as 12 psi/200°F or 0.06 psi/°F; when -2 psi/psi, as 12/100 or 0.12 psi/°F; and when -6 psi/psi, as 12/33.3 or 0.36 psi/°F. Beyond the proportional band, the gain in all cases is zero since no change in output can occur when the input changes, unless the input changes sufficiently to move the controller back into its band.

The gains used in the preceding illustration are quite moderate. However, one commercial unit has a gain range from 0 to 200 psi/psi. The upper value corresponds to a proportional band of 0.5%. Referring back to the illustration, this means that a change of 1°F would cause the valve to go from a fully opened to a fully closed position. Therefore, with a proportional band of less than 1%, moderately small changes in load variables could easily result in saturation. The resulting behavior of the proportional controller is then almost that of an on-off controller, which is described next.

II. ON-OFF ACTION

The simplest and most inexpensive type of controller is an on-off controller. Although the name, "on-off," implies that the manipulated variable is either on or off, it applies in a more general sense to any control system in which the manipulated variable can take on only two values. However, in the vast majority of cases, the two values of the manipulated variable are zero and the maximum value. Home-heating systems operate in this way, and so do constant temperature baths used in laboratories.

A. Stirred-Tank Heater Example

The continuous-flow, stirred-tank heater will be used to illustrate the behavior of a simple process under on-off control. For reasons that will be apparent shortly, little is learned if the analysis at the start is confined to the region of normal operation. Therefore, we shall begin by considering the "start-up" behavior of the process.

The physical picture of the system is exactly the same as that in Figure 4 of Chapter 5, except that the controller is an on-off unit rather than the proportional one shown. We shall assume that the set point is at 150°F. Therefore, for a standard, air-to-open valve, the output of the on-off controller is 15 psig when the temperature is less than 150°F, and 3 psig when it is greater than 150°F. The block diagram is shown in Figure 1. The symbol

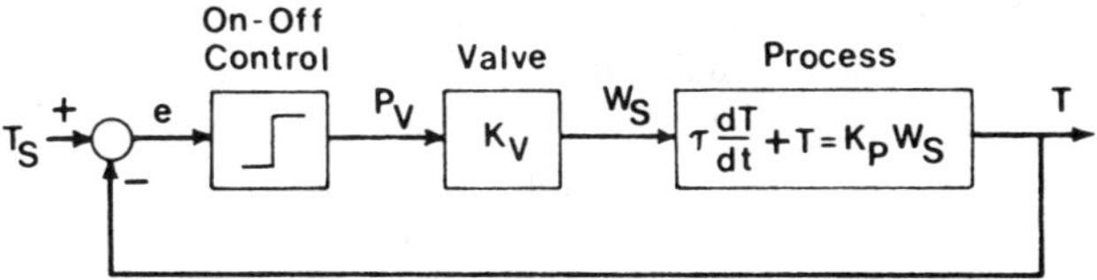

Figure 1. Block diagram for on-off control of tank temperature.

inside the controller block indicates that the controller action is on-off. As in Chapter 5, the differential equation describing the behavior of the tank temperature to changes in steam flow rate is placed in the process block.

The start-up procedure for this system is as follows. Before turning on the steam, the water flow rate is adjusted to the design value and a valve in the outlet line is set to maintain the liquid level in the tank at the desired point. At time zero, the main steam valve is opened. The control valve will be completely opened since the temperature of the water in the tank is below the set point.

In this development, our interest is not in the deviation of the tank temperature from its normal value, but rather in the actual tank temperature. Hence, total differential equations will be used rather than deviation equations. For the tank itself, two equations apply, one when the steam is on and the other when it is off. During the on-period, Equation 1 of Chapter 5 applies:

$$Fc_pT_F + W_S\lambda - Fc_pT = Mc_p\frac{dT}{dt} \tag{2}$$

During the off-period, there is essentially no heat exchange between the tank and the jacket, so Equation 2 becomes

$$Fc_pT_F - Fc_pT = Mc_p\frac{dT}{dt} \tag{3}$$

For simplicity, we shall assume that at the instant the steam valve opens, the steam flow rate jumps to a constant value. This assumption is justified if critical flow occurs in the valve, which was the case in Chapters 4 and 5. We will further assume that at the instant the steam is shut off, the jacket temperature falls instantaneously to the temperature of the water in the tank. This assumption is reasonable if the metal wall between the jacket and the tank fluid is thin and has a reasonably high thermal conductivity.

Rearrangement of Equation 2 into the standard form yields

$$\frac{M}{F}\frac{dT}{dt} + T = T_F + \frac{\lambda}{Fc_p}W_S \tag{4}$$

The right-hand side of this equation involves only constants. Referring back to the discussion of Equations 7 and 8 of Chapter 5, it follows that the right-hand side is the value to which the tank temperature would ultimately rise if the steam remained on. We shall suppose that the steam valve has been sized so that for the normal feed rate and feed temperature, this value is 212°F and no boiling occurs. Hence, Equation 4 reduces to

$$\frac{M}{F}\frac{dT}{dt} + T = 212 \tag{5}$$

When Equation 3 is rearranged to the standard form, the result is

$$\frac{M}{F}\frac{dT}{dt} + T = T_F \tag{6}$$

The right-hand side is the feed temperature, which is assumed to be constant at 60°F. This is the value to which the temperature would ultimately fall if the steam remained off. Therefore, Equation 6 becomes

$$\frac{M}{F}\frac{dT}{dt} + T = 60 \tag{7}$$

When Equations 5 and 7 are compared, we see that the time constants for heating and cooling are the same. This occurred because of the simplicity of the mathematical models chosen for the heating and cooling periods. Usually in on-off control, the time constant for the on-period will be different from that of the off-period.

Returning to the analysis of the start-up problem, the tank is initially at 60°F. Therefore, using Equation 8 of Chapter 5, the solution of Equation 5 is

$$T = 212 - 152e^{-t/\tau} \tag{8}$$

where $\tau = M/F$, the time constant

This equation applies until the temperature reaches 150°F, when the controller switches into the off-position. If the time constant for heating were 30 min, then from Equation 8, it can be shown that the set-point temperature will be reached in 26.9 min. As soon as the steam is switched off, the tank temperature begins to drop in accordance with Equation 7. For an initial condition of $T = 150$°F at $t = 26.9$ min, the solution of Equation 7 is

$$T = 60 + 90e^{-(t-26.9)/\tau} \qquad t > 26.9 \tag{9}$$

However, the cooling will no more than begin when the comparator will sense that the temperature is below the set point and the steam will be switched back on. Clearly, in the limit, the temperature will remain infinitesimally close to the set point, as shown in Figure 2. Mathematically, it amounts to an oscillation of the temperature about the set point at an infinitely high frequency and at an infinitely small amplitude. In effect, the offset is zero. If the controller had not been present, after reaching 150°F, the temperature would have proceeded along the dashed line, which becomes asymptotic to 212°F.

It is interesting to recall the effect of proportional control on this system. It was shown in Chapter 5 that if the gain is infinite, the offset is zero. Hence, for this example, on-off control achieves the same result as proportional control at infinite gain.

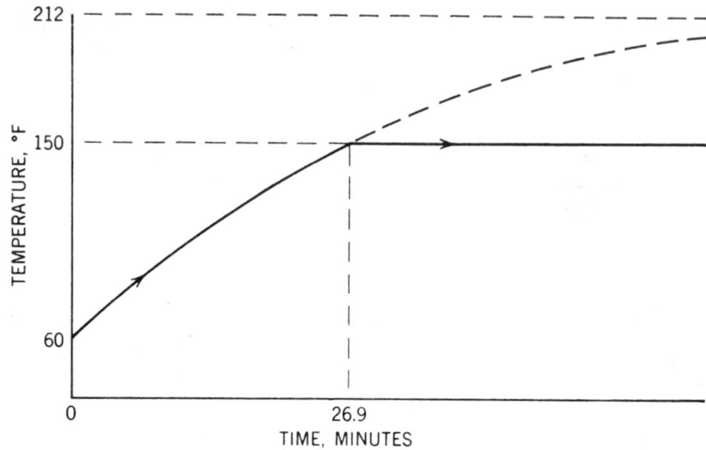

Figure 2.   Response of tank temperature during start-up under on-off control.

B. Implications for Proportional Control

This illustration brings out a very important point about proportional control. Suppose the process were under proportional control rather than on-off. We shall assume that all components of the system are the same as those of the on-off system, with the exception of the controller. What would be the behavior of the proportional system under the same start-up situation?

The start-up is initiated by moving the set point from 50 to 150°F. The upper limit of this move is based on this being the desired value of the temperature and on the assumption that this is the design condition; according to the theory of Chapters 4 and 5, at all other positions of the set point, there would be some offset. The lower limit of 50°F results from the assumption that the chart span of the controller is 200°F, set between 50 and 250°F. Therefore, with the set point at the lower limit of the range, the steam valve will be closed if the bandwidth is 100% or less. For this illustration, the bandwidth will be assumed to be 30%.

Upon moving the set point from 50 to 150°F, the theory of Chapter 5 would seem to predict that the temperature would follow the curve shown in Figure 3a. The corresponding time constant would be $\tau/(1 + K_o)$. The temptation is to suggest that as the controller gain is increased, the time constant will decrease, ultimately approaching zero as the controller gain approaches infinity. The fallacy in this argument lies in the fact that Chapter 5 applies only as long as the system is linear. In particular, the controller is linear only when the tank temperature is inside the proportional band, which has been assumed to be 30%. Thirty percent of the span is 60°F and it will be

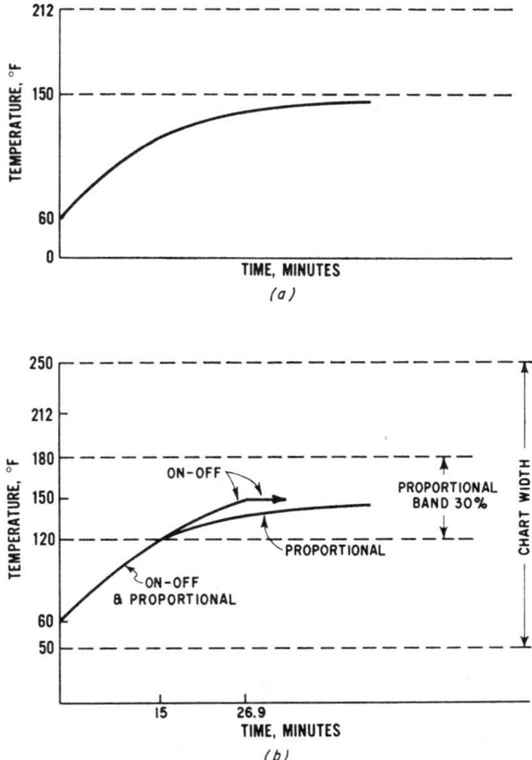

Figure 3. Response of tank temperature during start-up under proportional control. (*a*) Theoretical path. (*b*) Actual path.

assumed here that this band is centered about the set point of 150°F; therefore, the lower limit of the proportional band is 120°F and the upper limit is 180°F. Thus, the controller is only "in control" in this band. Below 120°F, the valve will be fully opened and above 180°F, it will be fully closed.

Combining these ideas, we see that the start-up for the proportional system occurs in two stages. In the first, the temperature rises from 60 to 120°F with the steam valve wide open. The tank temperature follows the same path here as the on-off system did, and the time constant is τ. When the temperature reaches 120°F, the proportional action takes over. From Equation 8, this occurs at a time of 15 min. The temperature path then changes to another first-order curve but with a time constant of $\tau/(1 + K_o)$, and the temperature ultimately approaches 150°F.

The equations of the paths for the on-off and proportional systems in the range of 120 to 150°F are easily obtained using Equation 8 of Chapter 5.

Both have the same initial condition of 120°F, and allowance is made for the fact that time zero is located at the point at which the start-up commenced. The ultimate temperature for the on-off system is 212°F, so the equation is

$$T = 212 - 92e^{-(t-15)/\tau} \qquad \text{on-off system} \qquad (10)$$

The ultimate temperature for the proportional system is the set point temperature of 150°F, so the equation is

$$T = 150 - 30e^{-(t-15)/[\tau/(1+K_o)]} \qquad \text{proportional system} \qquad (11)$$

If the derivatives of these equations are evaluated at the point where the second stage commences, the slope for the on-off system is found to be $92/\tau$ and that for the proportional system is $30(1 + K_o)/\tau$. It can be shown that these two slopes must be identical. Furthermore, since the valve in the proportional system reduces the steam rate within the proportional band while the valve is wide open for the on-off system, the proportional curve must fall below the on-off curve, as indicated in figure 3*b*.

C. Practical Aspects of On-Off Control

The preceding theoretical treatment indicates that on-off control is ideal for a process described by a first-order differential equation. However, the controller is required to continuously switch between the on- and off-positions at an infinite rate and correspondingly, the valve is required to move back and forth between fully opened and fully closed. This is physically impossible because of the air requirements of the controller and valve, and also because of the inertia of all moving parts. Apart from these physical limitations, the mathematical model used for the process was rather simplified. It did not account for the thermal capacitance of the tank wall and the various thermal resistances. As a result of these, when the valve closes, the tank temperature will continue to rise somewhat beyond the set point. Conversely, when the temperature falls below the set point, the valve will open, but the temperature will continue to fall for some time. The result is that the temperature will oscillate with finite amplitude and period as indicated in Figure 4. The magnitude of the first "overshoot" beyond the set point will probably exceed those of succeeding undershoots and overshoots, and a few oscillations will occur until one cycle looks the same as the next.

The slow oscillations of an actual system under on-off control may be tolerable. If there are sufficient "inertial effects" of the type just mentioned, the valve will switch positions at a sufficiently low frequency that the wear and tear on the valve will not be excessive. When these effects are lacking, an on-off controller incorporating a "differential gap" may be used. The gap straddles the set point and produces the effect of hysteresis in the on-off

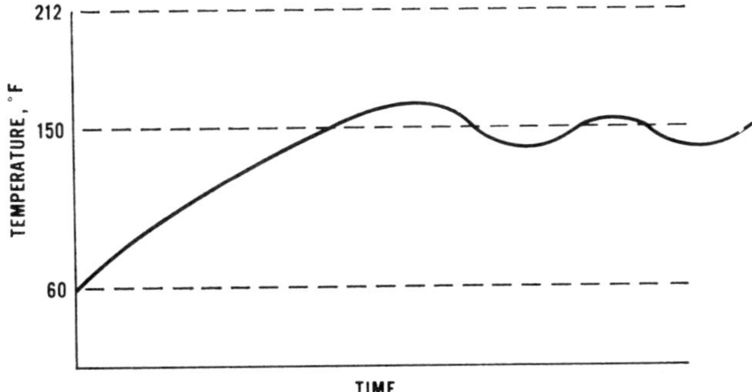

Figure 4. Actual response of tank temperature during start-up under on-off control.

controller. For example, if the gap is 2°F, then when the temperature reaches 151°F, the valve is closed, and when the temperature falls to 149°F, the valve is opened again. Clearly, this has the effect of lengthening the cycle and increasing the amplitude.

III. INTEGRAL ACTION

The major drawback of proportional action is the offset that usually occurs when the system is not operating at design conditions. It was shown in Chapter 4 that this offset can be removed in either of two ways. First, a gain of infinity results in no offset but as we shall see later, unsteady-state considerations usually prevent using very high controller gains. Second, if the set point is properly "reset" after a load change, the newly established offset will force the measured variable to the desired value. However, the magnitude of this purposely introduced offset is not easily predictable, and would normally have to be found by trial and error.

With proportional control, corrective action continues as long as there is an error, but the action ultimately reaches a constant value that is proportional to the offset. To remove offset, a different kind of corrective action is required, one that continues to grow or increase with time as long as the error does not decrease to zero or change its sign. This property will eventually force the controlled variable back to the set point. Although there are many types of controller actions that can remove offset in this manner, "integral action" is the most widely used in process control. It is sometimes referred to as "reset action" since it accomplishes the same result as the purposely introduced offset referred to previously in connection with proportional action alone.

A. Description of Action

With integral action, the controller output is proportional to the integral of the error. Mathematically it is described by the following equation:

$$P = \frac{1}{T_i} \int e\, dt \qquad (12)$$

where P = change in controller output pressure, psi
  T_i = integral time, time units
  e = change in error, psi

If the derivative of both sides of Equation 12 is taken with respect to time, it can be seen that an alternate description of reset action is that the rate of change of the output is proportional to the error, the proportionality constant being $1/T_i$.

B. Flow Process Example

Integral action is a dynamic element since it integrates the error with respect to time. To focus in on the effect of this behavior on the control of a process, we shall choose a process whose dynamic behavior is negligible compared with that of the controller. Flow processes typically respond rapidly to changes in the position of a control valve, so this type of process is a suitable choice.

Specifically, refer to the physical picture of a flow control system shown in Figure 5 and to the corresponding block diagram in Figure 6. The flow rate, F, is controlled by an integral controller. An orifice meter is used for the measurement of the flow rate and a differential pressure transmitter converts the pressure drop across the meter into a pressure in the range

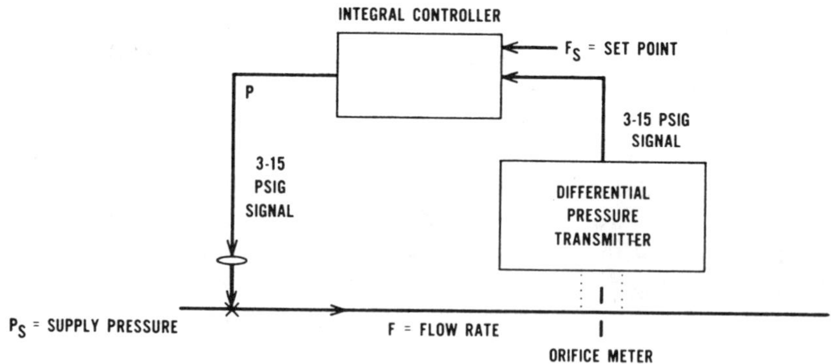

Figure 5.   Physical picture of flow control system.

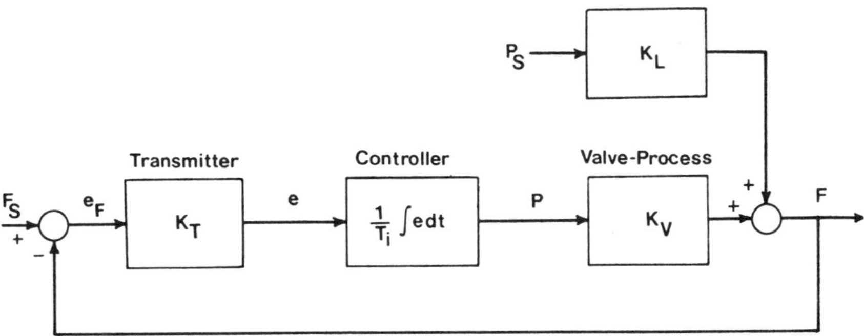

Figure 6.   Block diagram of flow control system.

between 3 and 15 psig. The integral controller then manipulates the position of the control valve.

Although the orifice meter and transmitter are physically in the feedback portion of the loop, they have been shown in the forward portion so that the set point and recording chart are in terms of the units of the controlled variable. For convenience in the analysis that follows, the transmitter and controller blocks have not been combined here, although they will be in future examples. It is assumed that the orifice meter and transmitter are characterized by a transmitter gain, K_T, and that the flow rate is related to the valve pressure by a gain, K_V. Finally, the load variable is assumed to be the supply pressure, P_S. Since the flow rate responds nearly instantaneously to changes in this pressure, changes in flow rate are related to changes in supply pressure by the load gain factor, K_L.

The analysis here will be confined to the determination of the response of the system to a step change in the supply pressure, and the response to a change in set point is the subject of Problem 9 at the end of this chapter. The equations describing the system are conveniently obtained by referring to the block diagram of Figure 6. Considering F_S, e_F, e, P, F, and P_S to be deviations from their normal or design values, the following equations can be written:

$$F = K_V P + K_L P_S \qquad (13)$$

where P_S is the magnitude of the step change in supply pressure

$$e_F = F_S - F = \text{error in terms of flow rate} \qquad (14)$$

$$e = K_T e_F = \text{error in terms of pressure} \qquad (15)$$

$$P = \frac{1}{T_i} \int_0^t e \, dt' \qquad (16)$$

where $t' = $ dummy variable

Since there is no change in set point, $F_S = 0$. Therefore, when Equations 13 to 16 are combined, the result is

$$F = - \frac{K_V K_T}{T_i} \int_0^t F \, dt' + K_L P_S \tag{17}$$

This equation can be solved by differentiating it first with respect to time. The term, $K_L P_S$, is a constant, since P_S is the magnitude of the step input. Hence

$$\frac{dF}{dt} = - \frac{K_V K_T}{T_i} F \tag{18}$$

This linear, first-order differential equation has the following general solution:

$$F = ce^{-(K_V K_T / T_i)t} \tag{19}$$

where c is a constant determined by the initial condition on F. The value of this constant can be determined in the following way. Prior to the step change in P_S, F is zero. However, it was originally assumed that F responds instantaneously to changes in P_S, so when there is a step change of magnitude, P_S, F must instantaneously increase by an amount, $K_L P_S$. This behavior is satisfied by Equation 17, which shows that F approaches $K_L P_S$ as time approaches zero through positive values since the integral ultimately vanishes. Equation 19 must also satisfy this condition and it will if c is equal to $K_L P_S$. Hence, the desired solution is

$$F = K_L P_S e^{-(K_V K_T / T_i)t} \tag{20}$$

A graph of F versus time is presented in Figure 7. The error, e_F, which is the mirror image of the flow deviation, is also shown.

Equation 20 indicates that the deviation in F approaches zero as time approaches infinity, so there is no offset. The error, e_F, is $-F$, so the integral of the error from time zero to infinity can be obtained by integrating the negative of Equation 20. The resulting value of the integral is $-K_L T_i P_S / (K_V K_T)$. As would be expected from physical intuition, the integral is proportional to the load gain factor and the magnitude of the step input, and also is proportional to the integral time, T_i.

One of the important properties of integral action is revealed by the time functionality of the controller output pressure, P. Using Equations 14 to 16 and 20, together with the fact that $F_S = 0$, the controller output pressure is given by the equation

$$P = \frac{K_L P_S}{K_V} \left[e^{-(K_V K_T / T_i)t} - 1 \right] \tag{21}$$

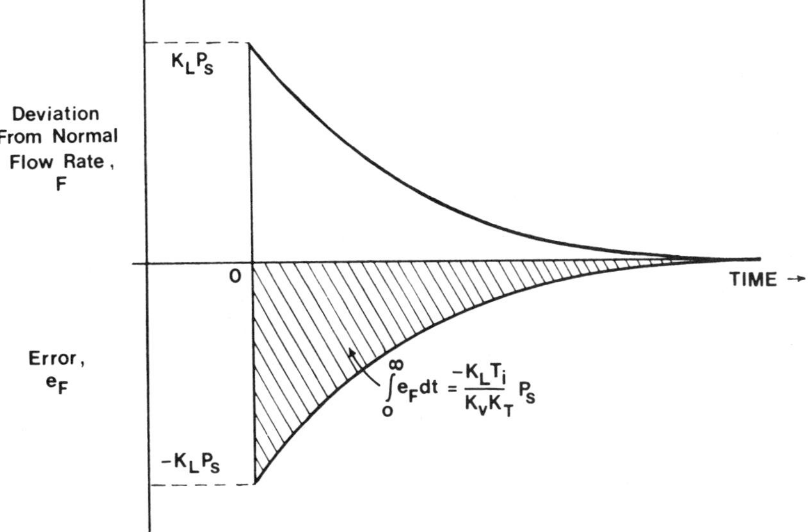

Figure 7. Deviation and error versus time.

This function is plotted in Figure 8. The response is in the typical form for the step response of a first-order system. The graph indicates that 63.2% of the ultimate controller effect is reached in the time, $T_i/K_v K_T$, reflecting the desirability of a low value of T_i. The key point is that the controller output pressure changes slowly since time is required for the controller action to integrate the error. It should be noted that the

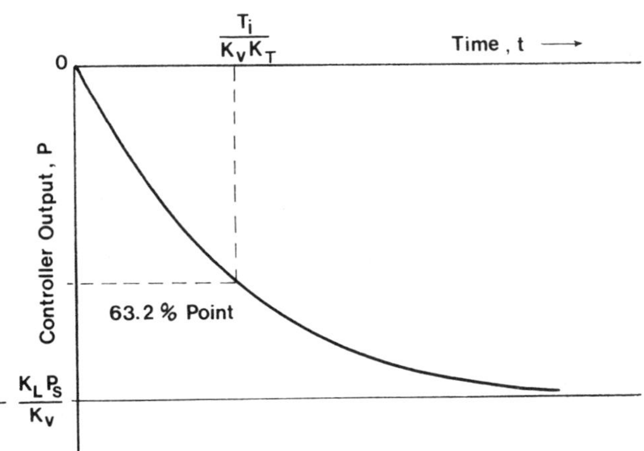

Figure 8. Controller output versus time.

controller output pressure ultimately changes by $-K_L P_S/K_V$ psi. The initial and final errors are zero, so we see that there is no direct relationship between the error and the controller output. For this reason, integral action is sometimes referred to as "floating control."

For this particular example, the optimum integral time is zero since this results in an error integral of zero. However, in a real system, an integral time of zero could not be used since it would make the system unstable. In the highly simplified model of this example, the only dynamic element was the controller, but a real system always has several dynamic elements. At relatively large values of integral time, the behavior of a real system would not be unlike that found in this example; that is, the error and controller output would have forms similar to those shown in Figures 7 and 8, respectively. The error would approach zero asymptotically without changing sign. On the other hand, at relatively small values of integral time, the behavior of a real system would depart from that shown here. The corrective action of the controller is stronger and this would cause oscillatory behavior of both the error and the controller output. If the integral time were sufficiently small, the oscillations would build up and the system would be unstable.

Integral action is frequently used for flow control as in the example and in ratio control systems (1). It is also suitable for systems in which the only dynamic element is a pure time delay (2). The main problem with integral action is that it introduces a lag into the system. This was evident in Figure 8, where the controller output required time to generate a corrective effect. For this reason, integral action is usually employed in combination with proportional action.

IV. PROPORTIONAL-INTEGRAL ACTION

A. Description of Action

The initial slowness of integral action can be compensated for by using a combination of proportional and integral action. This combination is more widely used than any other and is described by the equation

$$P = K_C \left(e + \frac{1}{T_R} \int e\, dt \right) \tag{22}$$

where K_C = proportional gain, psi/psi
T_R = reset time, time units

The equation indicates that K_C/T_R is equivalent to $1/T_i$ in the case of an integral-only controller. When the proportional gain, K_C, is increased, the integral time, T_i, is decreased by the factor, $1/K_C$. This effect of K_C on T_i results in a meaningful physical interpretation of the reset time, T_R. Suppose that a step change in error occurs at time zero and this error remains constant. The change in the output pressure, P, then behaves in accordance with the following equation:

$$P = K_C e + \frac{K_C}{T_R} et \qquad (23)$$

This response is shown in Figure 9. The response of the proportional action is immediate in the amount, $K_C e$. The integral action is responsible for the linearly increasing response, $(K_C/T_R)et$. After a time period, T_R, has elapsed,

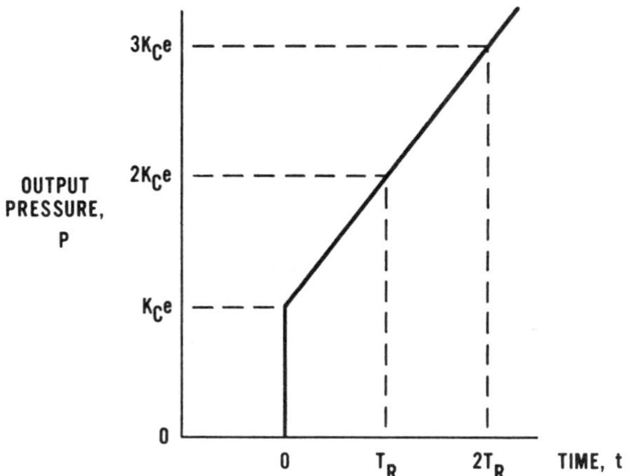

Figure 9.   Response of proportional-integral controller to step input.

the total change in controller output is $2K_C e$, so the integral action has exactly "repeated" the response from the proportional action. It is again repeated in the second time period between T_R and $2T_R$, and would continue to repeat in this manner until the controller output saturated at 15 psig. The "reset time" is then the time it takes for each repeat of the proportional action.

This repetitive property of a proportional-integral controller to a *constant* error is the basis for a commonly used method for calibrating the integral-action dial on the controller. Since T_R is the time required for one repeat of the proportional action, the reciprocal of T_R is the number of "repeats per minute" if T_R has the units of minutes. Some manufacturers calibrate the dial in repeats per minute and others calibrate it in terms of the reset time in minutes. The range of reset times for one commercial controller is 0.1 to 20 min. This range is satisfactory for most processes. However, for fast ones such as flow processes, a controller with a fast range between 0.005 and 1 min is available.

The physical interpretation of reset time with respect to a constant error is useful, but it must be emphasized that in a real, stable control system, the error will not remain constant. The controller output will act to reduce the error, so this physical interpretation would not apply exactly, but the idea is directionally correct.

B. Flow Process Example

To assess the improvement in control achieved through the addition of proportional action to integral action, we shall return again to the flow control system. Equations 13 to 15 apply and Equation 22 is used for the controller. Combining these equations, and setting $F_S = 0$, we obtain

$$F = K_C K_V K_T \left(-F - \frac{1}{T_R} \int_0^t F \, dt' \right) + K_L P_S \tag{24}$$

Since K_C/T_R is identically $1/T_i$, this equation can be written

$$F = -K_C K_V K_T F - \frac{K_V K_T}{T_i} \int_0^t F \, dt' + K_L P_S \tag{25}$$

Equation 25 is solved in the same way as Equation 17. The general solution takes the form

$$F = c_1 \exp\left[- \frac{K_V K_T}{T_i(1 + K_C K_T K_V)} t \right] \tag{26}$$

where c_1 is a constant.

The constant, c_1, depends on the initial condition of F, which is again determined by the value of F as time approaches zero through positive values. It can be seen from Equation 25 that as time approaches zero, F approaches $K_L P_S/(1 + K_C K_V K_T)$ since the integral vanishes. This same result can also be obtained on the basis of physical arguments. The desired solution then is

$$F = \frac{K_L P_S}{1 + K_C K_T K_V} \exp\left[- \frac{K_V K_T}{T_i(1 + K_C K_T K_V)} t \right] \tag{27}$$

As in the integral-only case, the deviation in the flow rate approaches zero as time approaches infinity, so there is no offset. The error, e_F, is $-F$, so Equation 27 can be used to find the integral of the error from time zero to infinity. The result of this integration is $-K_L T_i P_S/(K_V K_T)$, which is exactly the same as was found for the integral-only case. However, the introduction of proportional action does have the advantage of reducing the maximum error that occurs immediately after the step input disturbance. With just integral action, this initial error is $-K_L P_S$, but with the addition of proportional action, the initial error is reduced to $-K_L P_S/(1 + K_C K_T K_V)$. This effect is shown in Figure 10, in which typical curves are presented for the two cases. The degree of improvement depends upon the magnitude of the product, $K_C K_T K_V$. In most feedback systems, this product will be greater than one, and a "typical" value might fall between say three and five. Therefore, a fivefold improvement in the initial error might be expected.

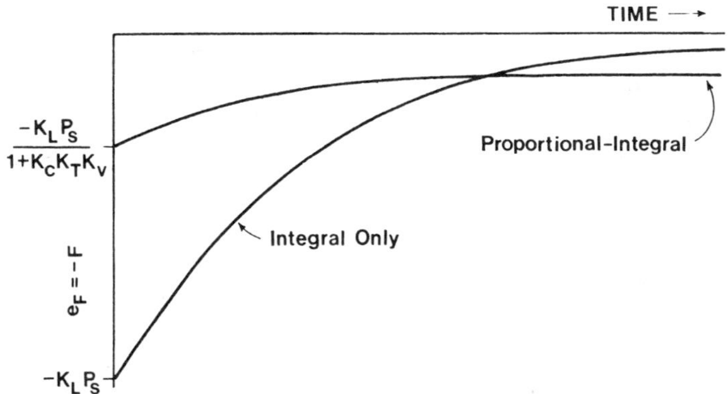

Figure 10.   Error curves for integral-only and proportional-integral actions.

Both curves display the shape of a first-order exponential decay. The time constant for integral control is $T_i/(K_V K_T)$ and that for proportional-integral control is $T_i(1 + K_C K_T K_V)/(K_T K_V)$. Therefore, the time constant for pure integral action is the smaller of the two. Since the areas beneath the curves are identical, the curves eventually cross. As the reset time, T_R and therefore T_i approach infinity, the controller action approaches purely proportional behavior. Equation 27 indicates that the corresponding response is a constant deviation of $K_L P_S/(1 + K_C K_T K_V)$, which is the offset. Also, when K_C is set equal to zero in Equation 27, the result, as expected, is Equation 20, which was obtained for integral action alone.

For the case of pure integral action, it was pointed out that the optimum integral time is zero since this results in an error integral of zero. However, for real systems, the presence of several dynamic elements in the control loop places a lower limit on the integral time below which the system is unstable. A similar stability constraint was mentioned in Chapter 5—namely, that for real systems under proportional control, there is an upper limit of controller gain beyond which the system becomes unstable. For proportional-integral control of the simple system here, again a zero integral time results in an error integral of zero. However, for real systems, there is a lower stability limit for the integral time. Furthermore, this limit is a function of the proportional gain of the controller. As this gain is increased, the lower limit of the integral time is increased, too. Therefore, a compromise is usually necessary between the proportional gain and the integral time. A large gain is desirable to reduce the magnitude of the error, while a small integral time is desirable to increase the rate at which the error is reduced.

C. Magnitude of the Error Integral

It might have appeared coincidental that the error integrals for the integral and proportional-integral systems were exactly the same. However, this is not the case. Consider the block diagram in Figure 11. Both integral and

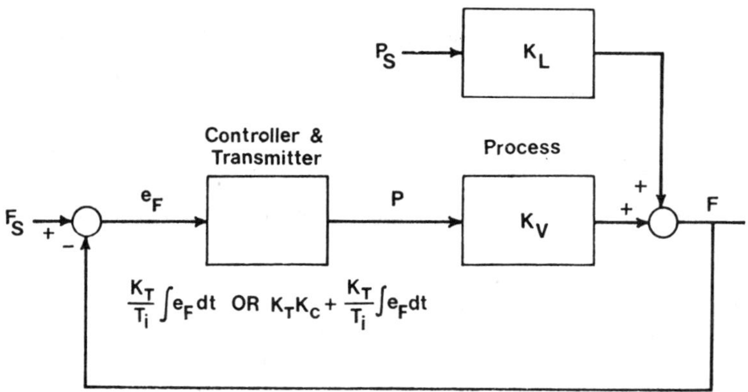

Figure 11.   Block diagram for system with either integral or proportional-integral control.

proportional-integral control action are indicated and the transmitter and controller have been combined in a single block. The integral term of the proportional-integral controller has been modified using the equality between K_c/T_R and $1/T_i$. The presence of the integral action in both controllers ultimately forces the deviation in F to zero. From the block diagram then:

$$K_V P_{\text{ult}} + K_L P_S = 0 \tag{28}$$

where P_{ult} = ultimate value of P

Now, since the ultimate value of F is zero, the ultimate error, e_F, is also zero. This means that the proportional term $K_T K_C e_F$ for the proportional-integral controller is ultimately zero. Therefore, the ultimate value of P for both controllers is

$$P_{\text{ult}} = \frac{K_T}{T_i} \int_0^\infty e_F \, dt \tag{29}$$

Substituting this expression into Equation 28 and solving for the error integral, we obtain

$$\int_0^\infty e_F \, dt = - \frac{K_L T_i P_S}{K_V K_T} \tag{30}$$

This confirms the value found by integrating the error in each case over an infinite time period.

D. Reset Windup

Although reset action solves the problem of offset, it may introduce a problem known as "reset windup" for systems subject to frequent start-ups and shutdowns. Batch chemical reactors fall into this category. Typically, the reactor is loaded up with reactants and brought up to a specified reaction temperature. At the completion of the reaction, the unit is shut down and unloaded. This cycle may be repeated several times a day. During start-up, it may be desirable to bring the reactor temperature up to the design value rapidly, but "overshoot" beyond this value may be undesirable because of detrimental side reactions.

To gain some understanding of reset windup, we shall reconsider the tank-heater process that was examined earlier in this chapter. We found that for the start-up of this process under proportional control, the temperature of the tank approached the set point asymptotically as time approached infinity, and there was no overshoot beyond the set point. However, for a real process there would be a finite maximum gain beyond which instability would occur. Furthermore, it was pointed out that there was a tendency for overshoot to occur because of the inertial effects in the process. Thus, overshoot could only be avoided by using a relatively low proportional gain, but if only proportional action were used, undesirable offset would result. Hence reset action would be necessary.

Reset action in a controller is related to the air pressure in a small bellows called the "reset bellows." At steady-state conditions, the pressure in this bellows is constant and equal to the output pressure of the controller. The only way the bellows pressure can change is if there is an error, and the change will be proportional to the integral of the error.

As in the earlier start-up example, suppose that the normal operating condition of the heater is 150°F and the normal output pressure to the valve is 9 psig. The reset windup will occur when the heater is shut down. This is best understood by referring to Figure 12 in conjunction with the following sequence of events.

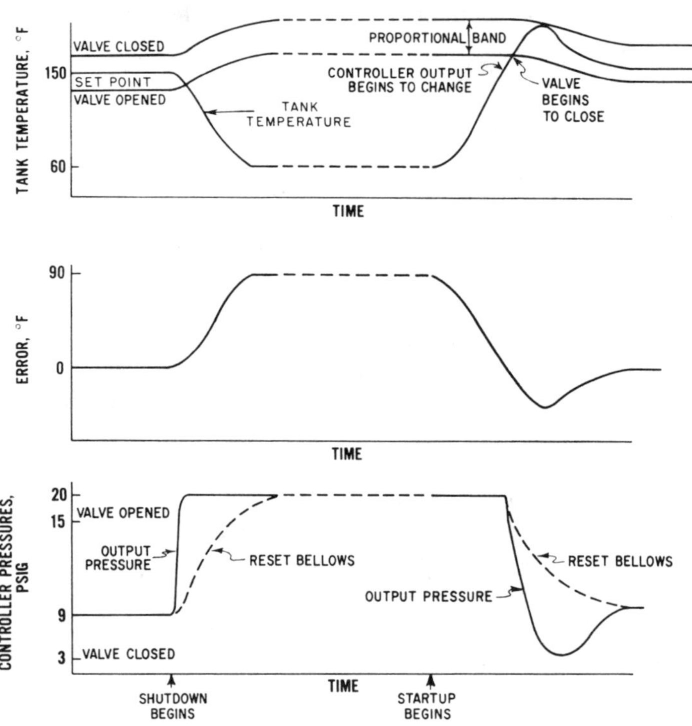

Figure 12.   Example of reset "windup."

The heater is shut down by closing a manual steam valve. The tank temperature begins to fall, causing a corresponding positive error as shown in the middle sketch of the figure. The reset action begins integrating this error, with the result that the pressure in the reset bellows begins to rise. At the same time, in accordance with the controller equation, the controller output pressure rises in a futile attempt to increase the tank temperature by opening the steam valve further. If the shutdown is sufficiently long, the pressure in the reset bellows will eventually reach the maximum output pressure of the controller. It was noted that for some controllers this could be 20 psig, and this value has been used in this example. The valve is wide open, having reached this condition when the pressure reached 15 psig. This buildup of pressure in the reset bellows is referred to as "reset windup."

The output pressure of the controller changes because of both the proportional and reset contributions; the reset bellows pressure lags behind but eventually reaches 20 psig. The output pressure rises rather rapidly in this example since the temperature is shown to fall rapidly through the proportional band. The output pressure would reach 15 psig slightly ahead of the time at which the temperature falls through the lower limit of the proportional band since there is a slight contribution of reset action to the output; however, most of the contribution is indicated to be from proportional action.

During the shutdown period, the proportional band shifts upward as the pressure in the reset bellows increases. By the time that the windup is complete—that is, when the pressure in the reset bellows has reached 20 psig— the entire proportional band lies above the set point. In fact, if the error were suddenly and instantaneously reduced to zero in some hypothetical way, the pressure to the valve would still remain at 20 psig. The actual proportional band lies a finite distance above the set point. Further insight into this shift of the band can be obtained by solving Problem 12 at the end of this chapter.

The start-up commences when the manual steam valve is opened. The set point is still at 150°F since it was never moved when the shutdown began. The tank temperature immediately begins to rise. Initially, the pressure in the reset bellows is 20 psig. Until the temperature reaches the set point of 150°F, the error is positive and this calls for output pressures in excess of 20 psig, which, of course, is impossible. However, as soon as the temperature crosses the set point, the error changes sign, and the pressure in the reset bellows begins to drop. Shortly thereafter, the controller pressure reaches 15 psig and the valve begins to close. From here on, the controller is actually "in control."

The tank temperature may gradually return to the set point, as shown in the figure, or it may return with some oscillatory behavior. It depends on the nature of the dynamic elements in the control loop and the gain and reset settings. Ultimately, the integral of the error during start-up must be sufficiently large and *negative* to change the pressure in the reset bellows from 20 psig back to the original normal pressure of 9 psig. The necessity of a negative error, of course, means that an overshoot is unavoidable. In the case of the tank heater, the overshoot is probably not troublesome, but as stated previously, if this were a batch reactor, the overshoot might be intolerable.

Windup can be a problem with any system during start-up, even though it will operate at design conditions indefinitely once it reaches them. Quite frequently, the start-up is carried out under manual control. In effect, the feedback loop is broken at the controller and the pressure to the control valve is manipulated from the controller by the operator. The situation is as shown in Figure 13. The switch may be incorporated in the controller itself and many controllers have them.

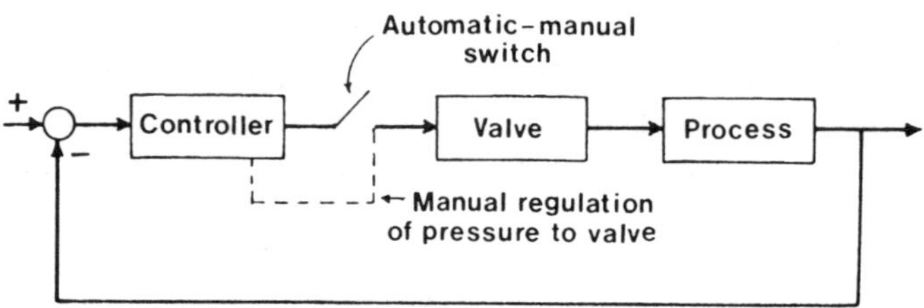

Figure 13.   Illustration of automatic-manual switch used during start-up under manual control.

The problem is that as long as the measured variable and the set point differ, the integral action tries to correct for the difference but, of course, is unable to do so since the loop is open. In the meantime, when design conditions are finally reached, the pressure to the valve is at the normal or design value. The system is now ready for transfer from manual to automatic, except that the pressure in the reset bellows is unlikely to be the same as the valve pressure. Hence, if the transfer were carried out, there would be a "bump." Various methods have been devised by manufacturers to insure "bumpless transfer." Similarly, reset windup can be avoided by using special controllers which are described in References 3 to 5i.

V. DERIVATIVE ACTION

A. Description of Action

We have seen that the addition of proportional action to integral can significantly reduce the error in a system. A third controller action, "derivative" or "rate" action, is sometimes beneficial for improving the stability of a system or for reducing overshoot. This action senses the rate of change of the error and contributes a component of the output pressure which is proportional to this rate. Therefore, for purely derivative action, the behavior is:

$$P = T'_D \frac{de}{dt} \tag{31}$$

where T_D' is a proportionality constant. One problem with purely derivative action is that when the error is constant, there is no corrective action at all. Since this is undesirable, derivative action is always combined with proportional and frequently with proportional plus integral. The corresponding mathematical descriptions of these actions are

$$P = K_C\left(e + T_D \frac{de}{dt}\right) \tag{32}$$

and

$$P = K_C\left(e + \frac{1}{T_R}\int e\,dt + T_D \frac{de}{dt}\right) \tag{33}$$

where T_D = derivative time, time units

In contrast to integral action, which tends to lag behind the input signal, derivative action tends to lead the input. This is illustrated by the response of a proportional-derivative controller to an error function in the form of a ramp:

$$e = \begin{cases} 0 & t < 0 \\ mt & t \geq 0 \end{cases}$$

where m = constant

Substituting this expression into Equation 32, the result is

$$P = K_C mt + K_C T_D m$$

The first term on the right is the contribution to the output from proportional action and the second is that from derivative action. A sketch of this result is presented in Figure 14. The proportional-plus-derivative output leads the proportional-only contribution by the derivative time, T_D. Therefore, derivative action has the effect of anticipating the input so it is sometimes referred to as "anticipatory control" or "preact control."

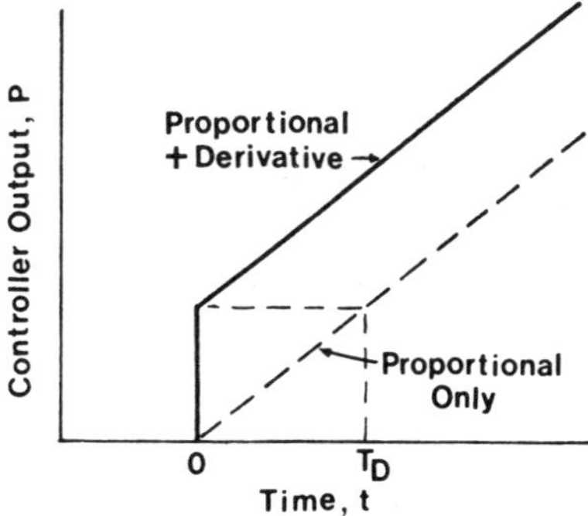

Figure 14.   Response of proportional-derivative controller to ramp input.

B. Applications

Derivative action is not as widely employed as integral and proportional actions. Integral action is frequently a necessity, to remove offset, but the advantages to be derived from derivative action are usually less obvious. It is sometimes used in controllers for batch processes to avoid overshoot of the set point (6i, 7). The idea behind this application is brought out in Problem 14 at the end of this chapter. It also is helpful in improving the response of certain types of processes since it tends to increase the stability of a control system, thus permitting the use of higher controller gains. The reasons for this will be examined in Chapter 14, which deals with frequency response.

Some control systems are not improved by the addition of derivative action. For example, if significant pure delays are present, then derivative action is unlikely to be of much benefit. It is not recommended for flow control systems. The reason for this is that flows are usually turbulent, and therefore "noisy." This noise is sensed to some extent by the flow meter and transmitter and would be amplified by derivative action, causing undue wear and tear on the control valve.

VI. PROPORTIONAL-INTEGRAL-DERIVATIVE ACTION

A. Interaction of Parameter Settings

The relationship for an "ideal" three-mode controller was given in Equation 33. On a three-mode controller, there are three dials, one for each of the parameters—proportional gain, reset time, and derivative time. In an ideal controller, the manipulation of any one of these parameters would not affect the others—in other words, there would be no interaction between the individual controller actions. While it would be possible to design such a controller, all standard three-mode controllers exhibit interacting behavior. Hence, the *effective* values of gain, reset rate, and derivative time are not the *nominal* values shown on the dials of the controller. The interactions of a particular three-mode controller can be determined from its specifications. More details concerning this subject can be found in References 5ii and 6ii.

SUMMARY

In this chapter, the actions of the most common types of process controllers have been described and illustrated. Frequently the control engineer must decide what kind of controller should be used in a system. The illustrations and discussion of this chapter have shown that proportional control results in relatively quick response but usually results in offset. Integral action is relatively slow but it can eliminate offset. Derivative action to some extent anticipates the effect of a disturbance and attempts to correct for it. To obtain more quantitative comparisons of the effects of the various controller actions in controlling a process, more complex models and methods of analysis must be used. These will be covered in the remaining chapters of this book.

A three-mode controller possesses a great deal of flexibility but this is not without its disadvantages, since the more complex a controller is, the more difficult it is to tune. Thus, as a general rule, it is best to use the simplest controller possible, consistent with the quality of control desired.

REFERENCES

1. M. Tyner and F. P. May, *Process Engineering Control*, The Ronald Press Co., New York, 1968, p. 184.
2. P. S. Buckley, *Techniques of Process Control*, John Wiley and Sons, Inc., New York, 1964, p. 75.
3. W. I. Caldwell, G. A. Coon, and L. M. Zoss, *Frequency Response for Process Control*, McGraw-Hill Book Co., Inc., New York, 1959, p. 229.
4. R. E. Clarridge, "An Improved Pneumatic Control System," *Trans. ASME* (April 1951), 297–305.
5. F. G. Shinskey, *Process-Control Systems*, McGraw-Hill Book Co., Inc., New York, 1967, (i) pp. 96–98, (ii) pp. 99–101.
6. P. Harriott, *Process Control*, McGraw-Hill Book Co., Inc., New York, 1964, (i) pp. 79–80, (ii) pp. 126–129.
7. F. G. Shinskey, "End-Point Control of Batch Processes," *Insts. and Control Systems* (February, 1972), 13.

PROBLEMS

Problem 1. A proportional controller with a chart width of 4 in. has a minimum reading of $-250°F$ and a maximum reading of $-150°F$. The proportional band is 10%.

(a)   Determine the gain, K_C, in psi/psi.
(b)   Determine the sensitivity in psi/in.

Answers: (a) 10 psi/psi
(b) 30 psi/in.

Problem 2. Reconsider Table 1 of this chapter. Suppose that the set point is at 100°F and that the measured temperature is also at this value. Since there is no error, the controller output pressure is at its normal value, which is 9 psig. For a controller gain of 1 psi/psi, reconstruct the first three columns of Table 1 and include values of the transmitter and controller output pressures for temperatures of 33.3 and 16.7°F. What is the effective proportional-action range for the transmitter-controller combination?

Answer: 150°F

Problem 3. The temperature of a continuous-flow, stirred-tank heater is controlled by an on-off controller having a differential gap of 4°F. With the set point at 150°F, the steam is turned on when the temperature falls to 148°F, and is turned off when it reaches 152°F. The steam valve has been sized so that for the normal feed rate and feed temperature, the tank temperature would reach 212°F, but without boiling, if the steam valve were to remain open indefinitely. On the other hand, if the steam valve were to remain closed indefinitely, the tank temperature would fall to the feed temperature of 60°F. The unsteady-state behavior of the process is adequately described by first-order differential equations. Unlike the illustration of Section II of this chapter, the time constants are not the same in the heating and cooling periods. The time constant during heating is 10.0 min and during cooling it is 20.0 min.

(a)   If there are no variations in the load variables, the system eventually reaches a steady-state, cyclic behavior; the valve is wide open for a period of time and then closed for a period of time, and then this cycle is repeated, over and over again. Make a sketch of this behavior.
(b)   Calculate the period of the cycle.
(c)   Will the average temperature for one of these cycles be exactly 150.00°F?

Answer: (b) 1.534 min

Problem 4. This problem is similar to Problem 3. However, the time constant during heating is 30 min and during cooling it is 60 min.

(a)   Suppose that the differential gap is 2°F. With the set point at 150°F, the steam is turned on when the temperature falls to 149°F and is turned off when it reaches 151°F. As in part *b* of problem 3, what is the period of the cycle when steady-state conditions are established?
(b)   Repeat part *a* of this problem but for a differential gap of 4°F. Theoretically, should the period be twice that found for part *a*?
(c)   What is the average temperature in the tank for a differential gap of 2°F?

Answers: (a) 2.301 min
(b) 4.603 min
(c) 150.00012°F

Problem 5. Reconsider the on-off illustration of Section II of this chapter. Suppose that the on-off controller has a differential gap of 2°F, which is centered about the set point of 150°F. The time constants for both heating and cooling are 30 min.

(a) As mentioned in part *a* of Problem 3, eventually steady-state cyclic behavior is reached. Find the period of this cycle.

(b) Suppose that the feed temperature drops from 60°F to 55°F and remains there indefinitely. Now what is the period of the cycle?

Answers: (a) 1.635 min
(b) 1.684 min

Problem 6. A pneumatic proportional-integral controller has an output pressure of 10 psi when the set point and the indicator pen point are together. The set point is suddenly moved 0.5 in. and the following data are obtained:

Time, sec	Psi
0−	10
0+	8
20	7
60	5
90	3.5

Determine the sensitivity in psi/in. and the integral time.

Problem 7.   The chart on a proportional-integral controller has a minimum value of 100°F and a maximum value of 300°F. The proportional gain is 2 psi/psi and the reset time is 1.0 min. The input pressure is fixed at 6 psig. When the set point is placed at the same temperature as that indicated by the input pen, the output pressure is 4 psig. Now suppose that the set point is instantaneously increased above the input pen by 25°F and held there for 2 min, and then returned instantaneously to the position of the input pen. Sketch the corresponding behavior of the output pressure. Note particularly the net effect of this procedure.

Problem 8. A proportional-integral controller is set for a proportional band of 20% and the chart covers the span from 0 to 200°F. Assume an air-to-open valve.

(a) When the set point and the input are both at 150°F, the output pressure is 9 psig. Find the maximum and minimum values of the proportional band in terms of temperature.

(b) Suppose that the controller is readjusted to give an output pressure of 10 psig when the set point and input are both at 150°F. (This adjustment might be carried out in the manner described in Problem 7.) Now, where does the proportional band lie in terms of temperatures if the valve is an air-to-open type? Where does it lie if the valve is an air-to-close type?

Problem 9.   The analyses in Sections III.B and IV.B led to expressions for the response of the flow control system to changes in the load variable, P_S. The results were given in Equations 20 and 27. Derive the corresponding equations for a change in set point instead. Sketch the response in each case. Also, what is the resulting offset in each case?

Problem 10.   Consider the flow control system discussed in Section IV.B in which the controller is a proportional-integral unit. The load gain, K_L, is 10 gpm/psi and the process gain, K_V, is 5 gpm/psi. The combined controller-transmitter gain, $K_C K_T$, is 2 psi/gpm and the reset time, T_R, is 10 sec. The normal flow rate is 150 gpm and the span of the flow transmitter is 100 gpm.

(a)   The supply pressure suddenly increases by 1.0 psi. What is the flow rate after 20 sec?

(b)   If the supply pressure then suddenly returns to its original value, what will the flow rate be 10 sec later? Make a sketch of the flow rate versus time including the behavior of part *a*.

(c)   Suppose that the control valve is somewhat sluggish. In other words, if a step change in pressure were made to the valve, it would take a few seconds for the valve to completely respond. Make a sketch on that made in part *b* to indicate qualitatively the effect of this slowness on the "ideal" response previously drawn.

Answers: (a) 150.148 gpm

(b) 149.693 gpm

Problem 11.   A gas-phase reactor is supplied by an upstream storage tank at pressure, P_U, as shown in Figure 15. The reactants leave the reactor at pressure, P_R, and flow through a line to a second storage tank at pressure, P_D. The reaction is quite pressure sensitive so the reactor pressure must be closely controlled. The feed flow rate, F, is the manipulated variable.

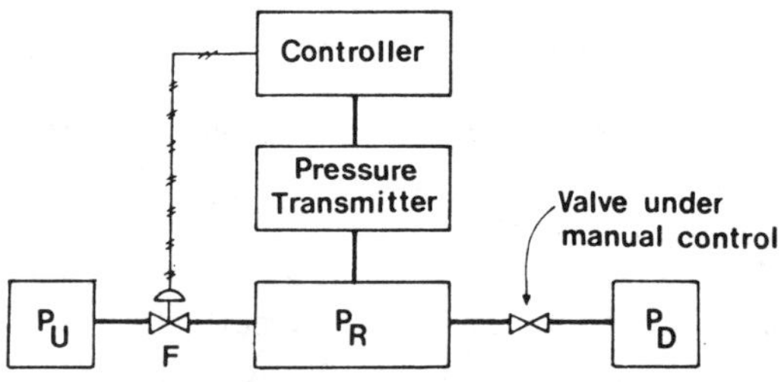

Figure 15.   Physical picture of pressure control system for Problem 11.

For the reactor alone, its unsteady-state behavior is adequately described by the following differential equation in which the variables are deviations from their normal values:

$$5\frac{dP_R}{dt} + P_R = 3P_U + 2F + 5P_D$$

where F = deviation in the feed rate from the upstream storage tank, standard cubic feed per minute (SCFM)

t = time, min

and all pressures are in psi.

The valve gain, K_V, is 1 SCFM/psi. The pressure transmitter has a span of 50 psig, which is set between 250 and 300 psig.

(a) Draw a block diagram for this system. Label the signal between each block and fill in each block with its appropriate gain or differential equation.

(b) The controller is of the proportional-integral type. However, in this part of the problem, assume that the reset time, T_R, is infinite. For a proportional gain of 0.5 psi/psi, what will the time constant for the *closed* loop be? If the supply pressure, P_U, suddenly drops by 2 psi, what will the resulting offset be? Make a sketch of the reactor pressure versus time and indicate the point at which 63.2% of the ultimate change occurs.

(c) Suppose that the reset time, T_R, is set at 4 min. Show qualitatively how the sketch of part *b* would be modified by the introduction of reset action.

(d) Suppose that the process has been operating satisfactorily for a long time and integral action is being used. The reset time is 4 min and the actual pressure to the valve is 4 psig. Suppose that P_U drops 2 psi. After a very long time, say an hour or so, and if no other disturbances are introduced in the meantime, what will the pressure to the valve be? The valve is an air-to-open type.

Answers: (b) Time constant = 4.03 min

Offset = −4.84 psi

(d) 7 psig

Problem 12.   The set point of a temperature control system is at 150°F. The span of the transmitter is 200°F and is set between 50 and 250°F. The proportional band is 10% and the reset time is 2.0 min. The normal output pressure to the valve is 9.0 psig, but during a shutdown of the process, reset windup occurs and the pressure in the reset bellows reaches 20 psig.

(a)   Where is the proportional band located after the windup is completed? Give the temperature extremes of the proportional band.

(b)   Suppose that it were physically possible to reduce the error to zero instantaneously when the start-up begins. What would the pressure to the valve do during this instantaneous change?

(c)   What instantaneous change in tank temperature would cause the valve to move?

> *Answers:* (a) 158.3 to 178.3°F
> (b) Remain constant at 20 psig
> (c) A change to 158.3°F

Problem 13.   Reconsider the stirred-tank heater discussed in Section II of this chapter and suppose that the tank temperature is under proportional-integral control. As in the illustration, the time constant is 30 min during heating and cooling. The steam valve has been sized so that for the normal feed rate and feed temperature, the tank temperature would reach 212°F, but without boiling, if the steam valve were to remain fully opened for an indefinite period of time. The combined gain of the controller and transmitter is 2.0 psi/°F and the reset time is 10.0 min.

Suppose that the heater is shut down for repairs by closing a manually operated steam valve. During the shutdown, the reset bellows windup to a pressure of 20 psig while the set point remains fixed at 150°F. The start-up is commenced at 3:30 p.m. when the manual steam valve is opened. At what time does the control valve begin to close?

> *Answer:* 3:58.1 p.m.

Problem 14.

(a)   The behavior of the controlled variable during the start-up of a process is presented in Figure 16*a*. Is the component of the controller output arising from derivative action in the same direction as that arising from proportional action? Explain how the derivative action tends to prevent overshoot of the set point.

(b)   Suppose that the controlled variable in a process suddenly begins to run away from the set point as shown in Figure 16*b*. Is the component of the controller output arising from derivative action in the same direction as that arising from proportional action? Is the derivative action advantageous in this situation?

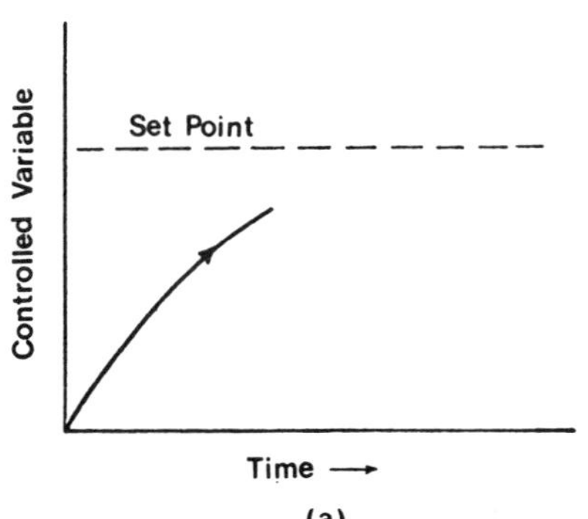

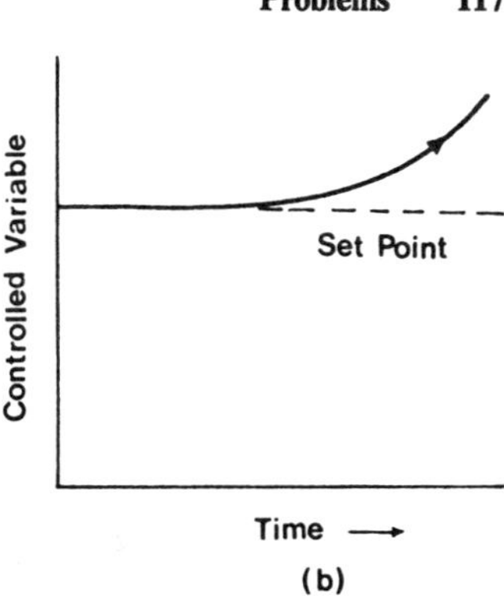

Figure 16.   Graphs for Problem 14.

CHAPTER VII

The Laplace Transformation

Actual process control systems are more complex than those that we have examined so far. Usually there are several elements that display significant dynamic behavior. These elements do not respond instantaneously to inputs and therefore they must be described by ordinary or partial differential equations. For example, suppose a stirred-tank heater is controlled by a proportional-integral controller. We have seen that a simple model for the heater itself results in a first-order differential equation. The controller action is described by a proportional plus an integral term. Furthermore, the control valve could not respond instantaneously because of its air requirements and the inertia of its moving parts. The result is that the valve might be modeled by a second- or even a third-order differential equation. Therefore, this control system is described as a whole by three differential equations plus the algebraic relationship for the comparator.

Although conventional methods could be used in the analysis of some of these complex systems, the Laplace transformation has a number of advantages over them. It may be useful for determining the response of the system to various inputs. Sometimes, the stability of a process or a system is easily ascertained using the transform. Finally, as we shall learn in Chapter 14, it can be used to determine the "frequency response" of a process or of a system.

The Laplace transformation has been the object of an immense body of mathematical research. However, the purpose here is to present just enough background so that most applications of the transform in process control can be understood. More details concerning its theory and applications can be found in References 1 and 2.

I. DEFINITION OF THE LAPLACE TRANSFORMATION

The Laplace transformation of a function, $f(t)$, is defined by the equation

$$L[f(t)] = \int_0^\infty f(t)e^{-st}\,dt = F(s) \tag{1}$$

The symbol $L[f(t)]$ indicates that the function $f(t)$ is to be transformed in accordance with Equation 1. To indicate that the resulting function depends upon s, the alternate notation, $F(s)$, is frequently used. Ordinarily, t stands for time and s is a complex parameter. Note that the transform eliminates time as a variable and introduces the new parameter, s. Not all functions are transformable, but most common functions that arise in engineering are.

As a simple first example, consider the transformation of the function:

$$f(t) = 1 \text{ for all values of time} \tag{2}$$

Substituting this into Equation 1, we obtain

$$L[1] = \int_0^\infty (1)e^{-st}\,dt = -\left.\frac{e^{-st}}{s}\right|_{t=0}^{t=\infty} = \frac{1}{s} \tag{3}$$

Since the limits of the integral run from zero to infinite time, it is apparent that the value of the function for negative values of time is of no consequence. In particular, even though $f(t)$ in Equation 2 was unity for negative times, the resulting transform is no different from the one that would be obtained for a unit step given by the following equation:

$$f(t) = \begin{cases} 0 & t < 0 \\ 1 & t \geq 0 \end{cases} \tag{4}$$

For this reason, when the transform of a function is taken, the function can be considered to be:

$$f(t) = \begin{cases} 0 & t < 0 \\ f(t) & t \geq 0 \end{cases} \tag{5}$$

The unit step function defined by Equation 4 is sometimes given the symbol, $u(t)$. Clearly, any function of time, $f(t)$, defined for both negative and positive values of time, can be converted to the form of Equation 5 by multiplication by $u(t)$.

Although the transformation disregards the behavior of the function prior to time zero, this is not a serious limitation in its use. All physical phenomena begin at some definite time—for example, at the instant a valve is opened or a change in set point is made on a controller. Prior to this time, the system can be considered to be at rest. Hence, to insure that the transform of the complete response is obtained, the time axis may be placed at or before the time of the initial disturbance.

In the first example, time did not appear in the function of Equation 2. Consider the function

$$f(t) = t \tag{6}$$

For the purposes of the transformation, this function becomes:

$$u(t)f(t) = \begin{cases} 0 & t < 0 \\ t & t \geq 0 \end{cases} \tag{7}$$

This is the ramp function of Chapter 5. As before, the transform is obtained by substituting into the defining equation:

$$L[t] = \int_0^\infty te^{-st}\, dt = \frac{1}{s^2} \tag{8}$$

The integration is accomplished using integration by parts.

II. SOME USEFUL PROPERTIES IN PROCESS CONTROL ANALYSIS

A. Frequently Used Properties

1. Linearity

Linearity is an extremely important property of the transform. In mathematical terms, this means

$$L[af_1(t) + bf_2(t)] = aL[f_1(t)] + bL[f_2(t)] \qquad (9)$$

where a and b are constants. The validity of Equation 9 is easily demonstrated using the definition of the transform and the properties of integrals.

To illustrate the use of this property, suppose that x, y, and z are all functions of time related through the equation

$$ax + by = cz \qquad (10)$$

where a, b, and c are constants. The equality is unchanged if the transform of each side is taken. The linearity property then yields

$$aX(s) + bY(s) = cZ(s) \qquad (11)$$

2. Differentiation

a. ORDINARY DIFFERENTIATION

The transform of a derivative of a function bears a simple relationship to the transform of the function itself. The key to the derivation of this result is integration by parts. The proof begins with the substitution of the derivative into the defining equation of the transform:

$$L\left[\frac{df(t)}{dt}\right] = \int_0^\infty \frac{df}{dt} e^{-st}\, dt \qquad (12)$$

The integration-by-parts equation is

$$\int u\, dv = uv - \int v\, du \qquad (13)$$

Let

$$u = e^{-st} \qquad \text{and} \qquad dv = \frac{df}{dt}\, dt \qquad (14)$$

Then

$$du = -se^{-st}\, dt \qquad \text{and} \qquad v = f(t)$$

Hence

$$\int_0^\infty \frac{df}{dt} e^{-st}\, dt = f(t)e^{-st}\Big|_0^\infty + s\int_0^\infty f(t)e^{-st}\, dt$$

$$= -f(0) + sF(s) \qquad (15)$$

Thus, the rule is

$$L\left[\frac{df(t)}{dt}\right] = sF(s) - f(0) \tag{16}$$

where $f(0)$ is the value of the function at time zero. In cases where the function is discontinuous at zero, the initial value should be evaluated as the limit of the function as time approaches zero through *positive* values. The rule given in Equation 16 can be applied to the second derivative to obtain the following result:

$$L\left[\frac{d^2f}{dt^2}\right] = s^2F(s) - sf(0) - f'(0) \tag{17}$$

where $f'(0)$ is the value of the first derivative at time zero, or if the derivative is discontinuous at zero, then the value as time approaches zero through positive values.

Higher derivatives than second can be transformed by repeated application of the above rules. For an nth-order derivative, the result is

$$L\left[\frac{d^nf}{dt^n}\right] = s^nF(s) - s^{n-1}f(0) - s^{n-2}f^{(1)}(0) - \cdots - sf^{(n-2)}(0) - f^{(n-1)}(0) \tag{18}$$

where $f^{(i)}(0)$ stands for the value of the ith derivative as time approaches zero through positive values. In particular, it should be noted that if the transform of the nth derivative of f is taken, and if the initial conditions on f and its derivatives are zero, then the transform is simply obtained by multiplying the transform of f by the nth power of s. In this sense then, the transform is similar to the D operator used in the solution of some ordinary differential equations.

b. PARTIAL DIFFERENTIATION

The Laplace transformation is useful in the solution of some partial differential equations. These arise when a variable is a function of both time and position. Some examples of these are discussed in Chapter 11.

Suppose a variable, T, is a function of position, x, and time, t, as indicated by the following notation:

$$T = T(x, t)$$

Now suppose that the relationship between T and its independent variables is given by the following partial differential equation:

$$\frac{\partial T(x, t)}{\partial t} = k\frac{\partial T(x, t)}{\partial x} \tag{19}$$

where k is a constant. The procedure for solution begins with the transformation of the equation. This merely means that the transform of both sides is taken, hence preserving the equality. To find the transform of the left-hand side, we substitute the partial derivative into Equation 1:

$$L\left[\frac{\partial T(x, t)}{\partial t}\right] = \int_0^\infty \frac{\partial T(x, t)}{\partial t} e^{-st}\, dt \tag{20}$$

The integration is with respect to t, so x is merely a parameter and has no effect on the integration. Hence, Equation 16 can be used to obtain the transformation

$$L\left[\frac{\partial T(x, t)}{\partial t}\right] = s\overline{T}(x, s) - T(x, 0) \tag{21}$$

The symbol, $\overline{T}(x, s)$, is the transform of $T(x, t)$. The initial condition, $T(x, 0)$, indicates that this is the value of the function as time approaches zero through positive values.

When the derivative on the right-hand side is substituted into Equation 1, it is advantageous if the order of integration and differentiation can be interchanged as follows:

$$L\left[\frac{\partial T(x, t)}{\partial x}\right] = \int_0^\infty \frac{\partial T(x, t)}{\partial x} e^{-st}\, dt = \frac{\partial}{\partial x}\int_0^\infty T(x, t)e^{-st}\, dt \tag{22}$$

This step is usually mathematically justifiable. The integral on the far right is the transform of $T(x, t)$, namely $\overline{T}(x, s)$. Thus, the transformation of Equation 19 can be written

$$s\overline{T}(x, s) - T(x, 0) = k\,\frac{\partial \overline{T}(x, s)}{\partial x} \tag{23}$$

There are still two independent variables, x and s. However, there are no derivatives with respect to s. Therefore, since s appears only parametrically in this equation, it is legitimate to treat the partial derivative as an ordinary derivative:

$$s\overline{T}(x, s) - T(x, 0) = k\,\frac{d\overline{T}(x, s)}{dx} \tag{24}$$

Only in the application of the boundary condition of $\overline{T}(x, s)$ is it necessary to observe the partial nature of this function.

3. Integration

The transform of an integral with respect to time is simply related to the transform of the function itself, just as was the case for a derivative. The transform is found using integration by parts and the result is

$$L\left[\int_\alpha^t f(t')\, dt'\right] = \frac{1}{s} F(s) - \frac{1}{s}\int_0^\alpha f(t')\, dt' \tag{25}$$

where t' is a dummy variable. In most problems, the lower limit will be a zero so that the second term on the right, corresponding to the "initial condition" for the integral, is zero.

B. Less Frequently Used Properties

1. Final-Value Theorem

The final value of a function of time, $f(t)$, refers to the value of the function as time approaches infinity. The final value may be obtained from the Laplace transform of the function, $F(s)$, using the final-value theorem. The theorem states that

$$\lim_{t \to \infty} \left[f(t) \right] = \lim_{s \to 0} \left[sF(s) \right] \tag{26}$$

There is one situation in which this theorem fails. If $sF(s)$ becomes infinite for any value of s whose real part is greater than or equal to zero, then $f(t)$ does not approach a limit as time approaches infinity. Consider the following function, for example:

$$F(s) = \frac{1}{s(s - 3)}$$

Since $sF(s)$ becomes infinite when the real part of s is $+3$, the final-value theorem predicts an incorrect value of the function as time approaches infinity.

2. Initial-Value Theorem

The initial value of a function refers to its value as time approaches zero through positive values. The theorem states that

$$\lim_{t \to 0} \left[f(t) \right] = \lim_{s \to \infty} \left[sF(s) \right] \tag{27}$$

3. Translation of a Function

The translation of a function refers to a displacement of it with respect to time. Although a function can be either advanced or delayed, the latter situation is more commonly the case. Consider, for example, the continuous-flow, hot water heater shown in Figure 1. Heat is supplied by an electrical heating element and the water passes the element at a constant velocity of 0.5 ft/sec. To insure that the temperature sensing element measures the cup-mixing temperature, a 10-ft mixing section follows the heating element. Thus, it takes 20 sec from the time the water leaves the element to the time at which its temperature is sensed. Such a time lag is referred to as a pure delay or dead time.

At the normal operating conditions, the inlet water temperature is 60°F and the leaving temperature is 80°F. Now suppose that at time zero,

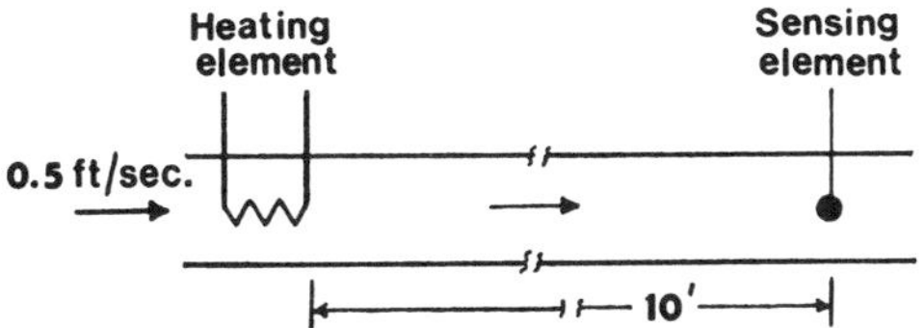

Figure 1.   Continuous-flow, hot water heater.

the electrical power supplied to the heater begins to increase linearly with time in the manner of a ramp function. The average temperature of the water at the heating element will begin to rise linearly with time. Assuming that the mixing section is well insulated, the response at the sensing element will be identical to the rise in fluid temperature at the heating element, but just displaced in time by 20 sec. These relationships are illustrated in Figure 2.

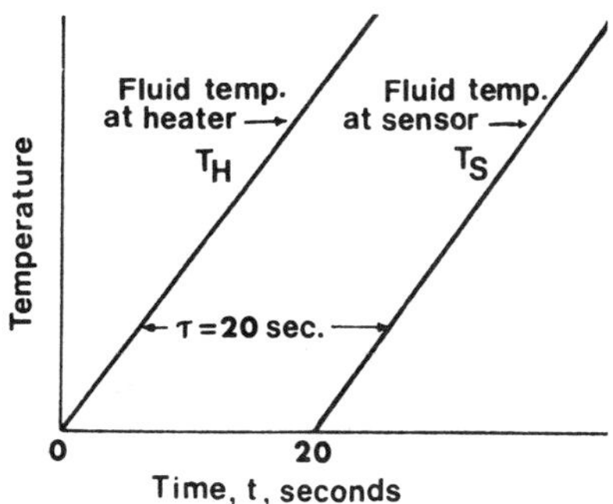

Figure 2.   Temperature of fluid at heater and sensor versus time.

More specifically, let us assume that the rate of power increase is such that the fluid temperature rises at a rate of 1°F/sec. In other words

$$T_H = t \qquad t \geq 0$$

This is a deviation equation that expresses the change in temperature at the heater from the normal value existing prior to time zero. Since there was no change prior to time zero, the initial condition on T_H is zero. Accordingly, as we saw earlier in this chapter, the transform of this ramp function is

$$T_H(s) = \frac{1}{s^2}$$

An s has been placed in parentheses only as a reminder that the transform has been taken; even this reminder is unnecessary since s appears on the right-hand side. The problem is to find the transform of the fluid temperature at the sensing element, T_S.

Looking at the figure, it can be seen that the relationship between T_S and T_H is

$$T_S(t) = T_H(t - \tau) \qquad\qquad t \geq \tau$$

In general terms, if we call $T_H(t)$ the function, $f(t)$, then the problem is to find the transform of $f(t - \tau)$. The starting point is the defining equation for the transform:

$$L[f(t - \tau)] = \int_0^\infty f(t - \tau)e^{-st}\, dt \qquad\qquad (28)$$

Nothing is changed if we replace dt by $d(t - \tau)$, since τ is a constant. Also, we can multiply and divide by $e^{-s\tau}$. Because the range of the integration must remain unchanged, the lower limit must be changed from 0 to $-\tau$. Hence, the transform is now

$$L[f(t - \tau)] = e^{-s\tau} \int_{-\tau}^\infty f(t - \tau)e^{-s(t-\tau)}\, d(t - \tau) \qquad\qquad (29)$$

Note that at the lower limit, the function-part of the integrand is $f(-2\tau)$, which is zero. It is still zero when t is zero since $f(-\tau)$ is zero; in fact, it is zero until t is equal to τ. Therefore, nothing is lost from the value of the integral by changing the lower limit to 0, so the result is

$$L[f(t - \tau)] = e^{-s\tau} \int_0^\infty f(t - \tau)e^{-s(t-\tau)}\, d(t - \tau) \qquad\qquad (30)$$

Now if we change the dummy variable by setting $t' = t - \tau$, then the integral has the form

$$L[f(t - \tau)] = e^{-s\tau} \int_0^\infty f(t')e^{-st'}\, dt' \qquad\qquad (31)$$

The integral is immediately recognized to be the definition of the transform of $f(t)$. So the result is

$$L[f(t - \tau)] = e^{-s\tau}L[f(t)] = e^{-s\tau}F(s) \qquad\qquad (32)$$

Applying this result to the heater, we see that the transform of T_S must be

$$L[T_S] = \frac{e^{-20s}}{s^2}$$

III. TRANSFORMS OF COMMON FUNCTIONS

So far in this chapter, we have been concerned with the transformation of a known function of time. Frequently, it is necessary to "invert" a given transform to find its corresponding function of time. This inversion can be carried out mathematically using some advanced techniques. Usually, however, only a table of "transform pairs" is needed. It might be expected that a very large table would be necessary, but this is not the case since most problems in process control involve *linearized* differential equations whose solutions fall into only a few categories.

We began to develop a table of transform pairs with the evaluation of the transforms for a unit step and a ramp. We turn now to the determination of the transform of a pulse, an impulse, and an exponential function.

A. Pulse Function

It was noted in Chapter 5 that in addition to the unit step and ramp functions, the pulse function is also important as a system input function. The pulse function provides a very good illustration of the use of the rule for the translation of a function.

Consider the pulse function shown in Figure 3. This is a unit pulse since

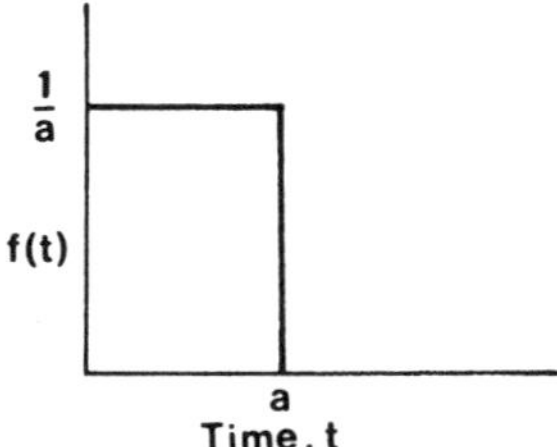

Figure 3.   Unit pulse function.

the area inclosed by the pulse is unity. This pulse can be described in two ways. One possible way is

$$f(t) = \begin{cases} 0 & t < 0 \\ 1/a & 0 < t < a \\ 0 & t > a \end{cases} \tag{33}$$

An alternative expression is obtained by noting that this function is the difference of two step functions, each of magnitude $1/a$, but with one of them delayed by the time, a. Hence

$$f(t) = \frac{1}{a} [u(t) - u(t - a)] \tag{34}$$

Using the translation rule, it can be seen that

$$F(s) = \frac{1}{as} - \frac{e^{-as}}{as} = \frac{1}{a}\left(\frac{1 - e^{-as}}{s}\right) \tag{35}$$

B. Impulse Function

Consider the limiting situation in which the pulse width, a, of a unit pulse shrinks to zero and the height of the pulse approaches infinity. This limiting function is known as a unit impulse or a delta function, and is assigned the symbol, $\delta(t)$. The transform of the delta function is obtained by taking the limit of Equation 35 as the parameter, a, approaches zero. This involves the application of L'Hôpital's rule and the result is

$$L\left[\delta(t)\right] = \lim_{a \to 0}\left(\frac{1 - e^{-as}}{as}\right) = 1 \tag{36}$$

C. Exponential Function

When simple systems are disturbed by step, ramp, and impulse inputs, they usually respond in terms of exponentials or sinusoids of time. We have already partially observed this fact in the examples of Chapters 5 and 6. It must be emphasized that this was a consequence of the simplicity of the models chosen and the linearization of the equations. Furthermore, one may recall from a study of elementary differential equations that the solutions to ordinary linear differential equations of higher order than one are still additive combinations of constants, exponentials, and sinusoids when the inputs are relatively simple. Since these are the kinds of equations we shall be dealing with in control, the foundations of a table of transform pairs are essentially complete upon the determination of the transform of a general exponential function.

The general exponential function has the following form:

$$f(t) = e^{-at} \tag{37}$$

where the constant, a, can be a complex number. Since the transformation only involves values of a function for positive values of time, the exponential function has the form shown in Figure 4 assuming that the constant, a, is a positive real number. The transform is:

$$L[u(t)e^{-at}] = \int_0^\infty e^{-(s+a)t}\,dt = -\frac{1}{s+a}\,e^{-(s+a)t}\Big|_0^\infty = \frac{1}{s+a} \tag{38}$$

This integral converges if the real part of $s + a$ is positive.

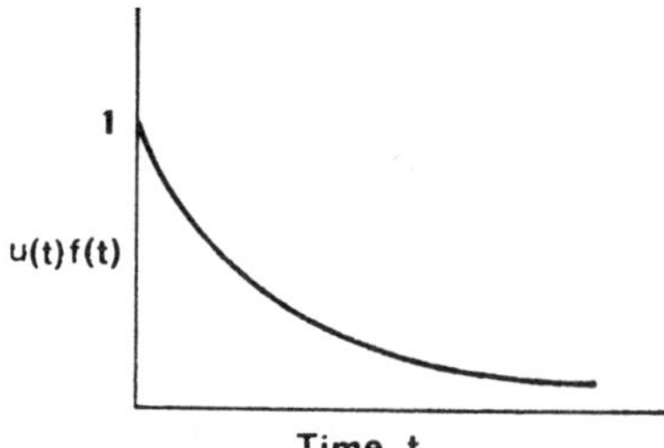

Figure 4.   Decaying exponential function.

   This result can be conveniently used to obtain the transform of the following sinusoidal function:

$$f(t) = \sin \omega t \tag{39}$$

where ω is a constant. The sine function can be expressed in terms of imaginary exponential functions:

$$\sin \omega t = \frac{e^{i\omega t} - e^{-i\omega t}}{2i} \tag{40}$$

where $i = \sqrt{-1}$

Using Equation 38, we obtain:

$$L[u(t) \sin \omega t] = \frac{1}{2i}\left[\frac{1}{s - i\omega} - \frac{1}{s + i\omega}\right] = \frac{\omega}{s^2 + \omega^2} \tag{41}$$

IV. TABLE OF TRANSFORMS

In a process control problem, it is frequently necessary to transform several functions of time. After a few algebraic manipulations of the resulting transforms, a single transform is obtained that must then be inverted from the "s domain" back into the "time domain." The original transformations are seldom a problem, but the inversion sometimes is. The usual procedure is to consult a table of transform pairs. If the desired form is found, then the inversion can be made immediately.

   For the problems in this text, a fairly limited table is adequate. In Table 1, the transform operations previously discussed are summarized. The transform pairs are presented in Table 2. To be perfectly correct, each of the time functions on the right should be multiplied by the unit step

TABLE 1. SUMMARY TABLE OF TRANSFORM OPERATIONS

Function of Time	Transform
$f(t)$	$F(s)$
$af_1(t) + bf_2(t)$	$aF_1(s) + bF_2(s)$
$\dfrac{df(t)}{dt}$	$sF(s) - f(0)$
$\displaystyle\int_\alpha^t f(t')\,dt'$	$\dfrac{1}{s}F(s) - \dfrac{1}{s}\displaystyle\int_0^\alpha f(t')\,dt'$
$\displaystyle\lim_{t \to \infty}\,[f(t)]$	$\displaystyle\lim_{s \to 0}\,[sF(s)]$
$\displaystyle\lim_{t \to 0}\,[f(t)]$	$\displaystyle\lim_{s \to \infty}\,[sF(s)]$
$f(t - \tau)$	$e^{-s\tau}F(s)$

TABLE 2. TRANSFORM PAIRS

Number	Transform	Time Function
1	1	$\delta(t)$
2	$\dfrac{1}{s}$	1
3	$\dfrac{1}{s^2}$	t
4	$\dfrac{1}{s^n}$	$\dfrac{1}{(n-1)!}\,t^{n-1}$ $n = $ positive integer
5	$\dfrac{1}{s+a}$	e^{-at}
6	$\dfrac{1}{Ts+1}$	$\dfrac{1}{T}\,e^{-t/T}$
7	$\dfrac{\omega}{s^2+\omega^2}$	$\sin \omega t$
8	$\dfrac{s}{s^2+\omega^2}$	$\cos \omega t$
9	$\dfrac{n!}{(s+a)^{n+1}}$	$e^{-at}t^n \quad n \geq 0,\ \text{integer}$
10	$\dfrac{\omega}{(s+a)^2+\omega^2}$	$e^{-at}\sin \omega t$
11	$\dfrac{s+a}{(s+a)^2+\omega^2}$	$e^{-at}\cos \omega t$
12	$\dfrac{1}{1+2\zeta s/\omega + s^2/\omega^2}$	$\dfrac{\omega}{\sqrt{1-\zeta^2}}\,e^{-\zeta\omega t}\sin \omega\sqrt{1-\zeta^2}\,t$

TABLE 2. (*continued*)

Number	Transform	Time Function
13	$\dfrac{1}{(1 + Ts)^n}$	$\dfrac{1}{T^n(n - 1)!}\, t^{n-1}e^{-t/T}$
14	$\dfrac{1}{(1 + Ts)(1 + 2\zeta s/\omega + s^2/\omega^2)}$	$\dfrac{T\omega^2 e^{-t/T}}{1 - 2\zeta T\omega + T^2\omega^2}$ $+ \dfrac{\omega e^{-\zeta\omega t}\sin(\omega\sqrt{1 - \zeta^2}\,t - \psi)}{[(1 - \zeta^2)(1 - 2\zeta T\omega + T^2\omega^2)]^{1/2}}$ $\text{where } \psi = \tan^{-1}\dfrac{T\omega\sqrt{1 - \zeta^2}}{1 - T\zeta\omega}$
15	$\dfrac{1}{(1 + T_1 s)(1 + T_2 s)}$	$\dfrac{1}{T_1 - T_2}(e^{-t/T_1} - e^{-t/T_2})$
16	$\dfrac{1}{s(1 + 2\zeta s/\omega + s^2/\omega^2)}$	$1 + \dfrac{1}{\sqrt{1 - \zeta^2}}$ $\times\, e^{-\zeta\omega t}\sin(\omega\sqrt{1 - \zeta^2}\,t - \psi)$ $\text{where } \psi = \tan^{-1}\dfrac{\sqrt{1 - \zeta^2}}{-\zeta}$
17	$\dfrac{1}{s(1 + Ts)}$	$1 - e^{-t/T}$
18	$\dfrac{1}{s(1 + Ts)(1 + 2\zeta s/\omega + s^2/\omega^2)}$	$1 - \dfrac{T^2\omega^2}{1 - 2T\zeta\omega + T^2\omega^2}e^{-t/T}$ $+ \dfrac{e^{-\zeta\omega t}\sin(\omega\sqrt{1 - \zeta^2}\,t - \psi)}{\sqrt{1 - \zeta^2}(1 - 2\zeta T\omega + T^2\omega^2)^{1/2}}$ $\text{where } \psi = \tan^{-1}\dfrac{\sqrt{1 - \zeta^2}}{-\zeta}$ $+ \tan^{-1}\dfrac{T\omega\sqrt{1 - \zeta^2}}{1 - T\zeta\omega}$
19	$\dfrac{1}{s(1 + T_1 s)(1 + T_2 s)}$	$1 + \dfrac{1}{T_2 - T_1}(T_1 e^{-t/T_1} - T_2 e^{-t/T_2})$

function, $u(t)$, to emphasize that each function is zero prior to time zero. Most tables omit this multiplier since it is implied.

In many cases, the exact form of the transform to be inverted does not appear in tables. Sometimes the transform can be rearranged into simpler forms that do appear in a table. For example, suppose that the following transform is to be inverted:

$$F(s) = \frac{1}{s(1 + Ts)} \tag{42}$$

Using a technique known as "partial fractions," this transform can be expressed in the following way:

$$F(s) = \frac{1}{s(1 + Ts)} = \frac{1}{s} - \frac{1}{s + 1/T} \tag{43}$$

Each of the simple terms on the right-hand side can be inverted immediately since they would appear in even a short table. The method of partial fractions is frequently successful in reducing the complexity of a transform and this method will be discussed in detail in this chapter.

When relatively complex transforms are encountered, comprehensive tables should be consulted. The following sources are suggested:

1. "Handbook of Mathematical Functions," Ed. by M. Abramowitz and I. A. Stegun, Nat. Bur. of Standards (1964).
2. "Handbook of Laplace Transformation—Fundamentals, Applications, Tables, and Examples," by F. E. Nixon, Prentice-Hall, Inc., Englewood Cliffs, N.J. (Second Edition, 1965).
3. "Principles of Automatic Controls," by F. E. Nixon, Prentice-Hall, Inc., Englewood Cliffs, N.J. (1953).

The first is a U.S. Government publication that contains a wealth of mathematical tables including a very long list of transforms. The second carries a very large set of tables with transforms of the type frequently encountered in control. The last contains an excellent, but abridged form of the tables in the second reference.

V. USE OF THE TRANSFORM IN SOLVING DIFFERENTIAL EQUATIONS

A. A Simple Example

The use of the transform in the solution of a differential equation is best illustrated by an example. Consider the following equation:

$$T\frac{dy}{dt} + y = 1 \quad \text{and} \quad y(0) = \tfrac{1}{2} \tag{44}$$

where T is the time constant. The equality is preserved if the transform of both sides is taken. Using the first three operations of Table 1, the result is:

$$T(sY - \tfrac{1}{2}) + Y = \frac{1}{s} \tag{45}$$

The function, Y, is $Y(s)$, but no reminder is needed since the equation contains s. We wish to find $y(t)$, so Equation 45 is algebraically solved for Y:

$$Y = \frac{1}{s(Ts + 1)} + \frac{T/2}{Ts + 1} \tag{46}$$

Referring to Table 2, the first term on the right is the same as transform pair No. 17 and the second is the same as No. 6. Hence, $Y(s)$ can be inverted to $y(t)$ using the corresponding time functions:

$$y = (1 - e^{-t/T}) + \frac{T}{2}\frac{1}{T}e^{-t/T} \tag{47}$$

This expression can be reduced to

$$y = 1 - \tfrac{1}{2}e^{-t/T} \tag{48}$$

As t approaches zero, y approaches $\tfrac{1}{2}$, so the initial condition is satisfied. Substitution of y into Equation 44 shows that it is the desired solution.

B. Solution of Simultaneous Ordinary Differential Equations

The transform is particularly convenient in cases where there is a set of simultaneous differential equations to be solved. The continuous-flow, stirred-tank heater will be used to illustrate this. In previous discussions of this process, it was assumed that the steam control valve would respond instantaneously to changes in air pressure from the controller. However, if the valve were large, the steam flow rate might be approximated by the following first-order differential equation:

$$\tau_V \frac{dW_S}{dt} + W_S = K_V P_V \tag{49}$$

where τ_V = time constant of valve
W_S = change in flow rate of steam
t = time
K_V = valve gain
P_V = change in pressure to the valve

The variables, W_S and P_V, are changes from original steady-state values. Hence, at time zero, W_S is zero. The input disturbance is the change in pressure to the valve, P_V. To simplify the problem, assume that at time zero, a step change in pressure is made of such magnitude that the product, $K_V P_V$, is unity. Hence, Equation 49 may be written:

$$\tau_V \frac{dW_S}{dt} + W_S = 1 \tag{50}$$

We have seen that according to one possible model for a tank heater, the temperature responds to a change in steam flow in accordance with a first-order differential equation:

$$\tau_P \frac{dT}{dt} + T = K_P W_S \tag{51}$$

where τ_P = time constant of tank heater
T = change in temperature of fluid in tank
K_P = process gain

Again, assume that at time zero, the change in tank temperature is zero.

The problem is to find the response in tank temperature to the disturbance, $K_V P_V$, which is equal to unity. The solution by Laplace transforms is purely algebraic. Equations 50 and 51 are transformed and the transformation is especially simple since the initial conditions on W_S and T are both zero. The result is the following set of transformed equations:

$$\tau_V s W_S + W_S = \frac{1}{s} \tag{52}$$

$$\tau_P s T + T = K_P W_S \tag{53}$$

These two equations can be rewritten as follows:

$$(\tau_V s + 1)W_S = \frac{1}{s} \tag{54}$$

$$(\tau_P s + 1)T = K_P W_S \tag{55}$$

Although the symbol, W_S, appears in both Equations 50 and 52, it stands for $W_S(t)$ in Equation 50 and for $W_s(s)$ in Equation 52. This is also the case for the symbol, T, in Equations 51 and 53. Use of the same symbol is convenient and should result in no ambiguity.

Solving Equation 54 for W_S, substituting the result into Equation 55, and solving for T, we obtain

$$T = \frac{K_P}{s(\tau_P s + 1)(\tau_V s + 1)} \tag{56}$$

This is in the form of transform pair No. 19 in Table 2. The solution is therefore

$$T = K_P \left[1 + \frac{1}{\tau_P - \tau_V} (\tau_V e^{-t/\tau_V} - \tau_P e^{-t/\tau_P}) \right] \tag{57}$$

Equation 56 was the result of a special case in which a unit step change in the quantity, $K_V P_V$, occurred at time zero. However, a very interesting result occurs if we generalize Equation 56 to permit P_V to be some unspecified function. For this general situation, Equation 52 becomes

$$\tau_V s W_S + W_S = K_V P_V \tag{58}$$

or

$$(\tau_V s + 1) W_S = K_V P_V \tag{59}$$

This result, combined with Equation 55 yields

$$T = \left(\frac{K_V}{\tau_V s + 1} \right) \left(\frac{K_P}{\tau_P s + 1} \right) P_V \tag{60}$$

When $K_V P_V$ is a unit step, then Equation 60 reduces to Equation 56. The form of Equation 60 is especially suggestive. It can be seen from Equation 59 that the first factor on the right of Equation 60 is W_S/P_V, and from Equation 55 that the second factor is T/W_S. In other words, not surprisingly,

$$T = \frac{W_S}{P_V} \frac{T}{W_S} P_V \tag{61}$$

This suggests an extension of the block diagram concept as illustrated in Figure 5. The two diagrams are essentially identical, but the lower one indicates that each component was modeled by a first-order linear differential equation, and the corresponding time constants and gain factors are indicated. Each block then represents the ratio of the transform of the response of the dependent or output variable to the transform of the input or forcing function, and this ratio is referred to as the "transfer function." The transfer function completely characterizes the dynamics of a linear system described by a linear differential equation. It may be thought of as an "operator" in

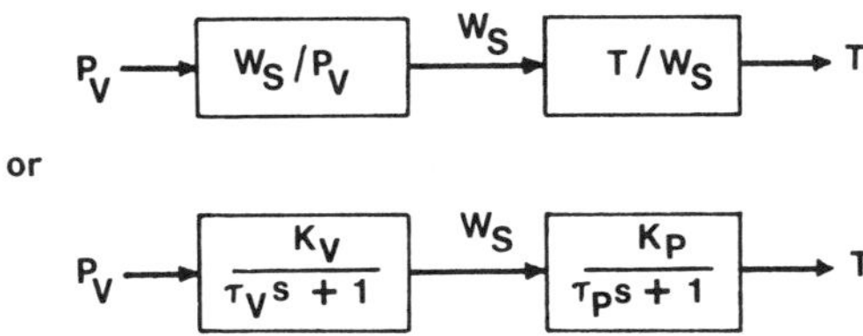

Figure 5. Transfer functions are multiplied in block diagrams.

the sense that it "operates" on the input function to produce the corresponding response of the output function.

In the prior example, the initial conditions on W_S and T were assumed to be zero. If the initial condition on W_S had been 0.5, then the transform of Equation 50 would have been

$$\tau_V(sW_S - 0.5) + W_S = K_V P_V \tag{62}$$

When the transfer function, W_S/P_V, is solved for, two terms result:

$$\frac{W_S}{P_V} = \frac{K_V}{\tau_V s + 1} + \frac{0.5\tau_V}{P_V(\tau_V s + 1)} \tag{63}$$

The second term on the right results from the initial condition. As indicated by the block diagram, when this transfer function is multiplied by the transfer function, T/W_S, two terms will again result. And if there had also been an initial condition on T, there would have been two terms for the transfer function, T/W_S. Thus, the presence of initial conditions complicates the algebra, and in some applications of the transform would lead to rather cumbersome manipulations. When dealing with deviation equations, it is not only convenient but logical to assume zero initial conditions and transfer functions are always defined on this basis.

VI. PARTIAL FRACTIONS

It was noted earlier that the method of partial fractions can sometimes be employed to reduce some transforms to simpler forms. There will be very little need for this technique in solving problems in this book. However, some knowledge of it is important here for the insight that it provides regarding the qualitative nature of the solution of some types of differential equations. More specifically, it is often possible to determine rather easily whether the solution will consist of decaying exponentials of time, oscillatory functions, and unbounded functions.

The method will be explained using the following general differential equation with constant coefficients:

$$a_n \frac{d^n y}{dt^n} + a_{n-1} \frac{d^{n-1} y}{dt^{n-1}} + \cdots + a_1 \frac{dy}{dt} + a_0 y$$

$$= b_m \frac{d^m x}{dt^m} + b_{m-1} \frac{d^{m-1} x}{dt^{m-1}} + \cdots + b_1 \frac{dx}{dt} + b_0 x \tag{64}$$

As pointed out before, y represents the response of the system and x represents the excitation or input. In most cases of interest, the right-hand side is simpler than the left, but Equation 64 illustrates the possible complexity of the most general case. For a physically realizable system, m must be less than or equal to n, as proven in Reference 3.

When the Laplace transform of both sides is taken, initial conditions are introduced. The resulting transform can then be solved for the transform of the response, $Y(s)$:

$$Y = \frac{c_m s^m + c_{m-1} s^{m-1} + \cdots + c_1 s + c_0}{s^n + d_{n-1} s^{n-1} + \cdots + d_1 s + d_0} = \frac{N(s)}{D(s)} \tag{65}$$

where $N(s)$ and $D(s)$ represent the numerator and denominator polynomials, respectively. Note particularly that the numerator and denominator have been divided by a common factor so that the coefficient of s^n is unity. Since the denominator is a polynomial in s, it is possible to factor it. Hence, Equation 65 may be rewritten as

$$Y = \frac{N(s)}{(s - s_1)(s - s_2) \cdots (s - s_n)} \tag{66}$$

where $s_1, s_2, \ldots,$ and s_n are the n roots of the polynomial. These roots may be either real or complex, and as we shall see, it is necessary to locate where they lie in order to obtain a qualitative description of the solution of the equation. If the denominator polynomial, $D(s)$, is of high order, it can be rather difficult to determine their exact locations, although it is often relatively easy to determine the regions of the complex plane in which they are located. If the denominator is first order, the root is evident by inspection, and if second order, the quadratic formula enables a direct calculation of the roots. For third and higher orders, synthetic division or trial-and-error methods may be satisfactory.

Having determined the roots, the next step depends upon their nature. Three cases must be considered. The simplest is that in which each of the roots is distinct and real. In the second case, some of the roots are complex. The third and rarest case in process control arises when at least one of the roots is repeated. Each of these three cases is discussed in some detail in the paragraphs that follow.

A. Real and Distinct Roots

When the roots are real and distinct, Equation 66 can be expressed as a sum of partial fractions of the following form:

$$Y = \frac{N(s)}{D(s)} = \frac{A_1}{s - s_1} + \frac{A_2}{s - s_2} + \cdots + \frac{A_n}{s - s_n} \tag{67}$$

The expansion contains n terms, one for each of the factors of $D(s)$. The coefficients, $A_1, A_2, \ldots, A_n$, of these partial fractions are evaluated in the following way. Suppose, for example, that we wish to find A_2. Both sides of Equation 67 are multiplied by $(s - s_2)$ to yield

$$(s - s_2)\frac{N(s)}{D(s)} = \frac{A_1(s - s_2)}{s - s_1} + A_2 + \cdots + \frac{A_n(s - s_2)}{s - s_n} \qquad (68)$$

Since $D(s)$ contains $(s - s_2)$ as a factor, cancellation effectively occurs on the left-hand side. Since s is a variable, it can be assigned any value. If s is given the value of s_2, then all terms on the right-hand side vanish with the exception of A_2. Therefore

$$A_2 = \left[(s - s_2)\frac{N(s)}{D(s)} \right]_{s=s_2} = \frac{N(s_2)}{(s_2 - s_1)(s_2 - s_3) \cdots (s_2 - s_n)} \qquad (69)$$

The same method is used for determining the coefficients of the other terms.

Having found each of the n coefficients for Equation 67, the inversion into the time domain is carried out term by term using transform pair No. 5 of Table 2. Hence, the complete inverse transform of Equation 67 is

$$L^{-1}[Y(s)] = A_1 e^{+s_1 t} + A_2 e^{+s_2 t} + \cdots + A_n e^{+s_n t} \qquad (70)$$

where the symbol, $L^{-1}[Y(s)]$, indicates that the transform, $Y(s)$, is to be inverted to the time domain.

An example will be helpful in illustrating the discussion thus far. Consider the following differential equation:

$$\frac{d^3 y}{dt^3} + 2\frac{d^2 y}{dt^2} - \frac{dy}{dt} - 2y = x \qquad (71)$$

with the following initial conditions:

$$y(0) = 1$$
$$y'(0) = 2$$
$$y''(0) = 3$$

where $y'(0)$ and $y''(0)$ are the initial values of the first and second derivatives, respectively. Transforming, we obtain

$$(s^3 Y - s^2 - 2s - 3) + 2(s^2 Y - s - 2) - (sY - 1) - 2Y = X \qquad (72)$$

Capital letters have been used to indicate the transformed variables. This equation can be rearranged with factors of Y appearing only on the left-hand side:

$$(s^3 + 2s^2 - s - 2)Y = X + s^2 + 4s + 6 \qquad (73)$$

In order to develop the problem further, it is necessary to specify the nature of the input function. For convenience, assume a unit step that has the transform, $1/s$; substituting this for X and solving for Y, we obtain

$$Y = \frac{s^3 + 4s^2 + 6s + 1}{s(s^3 + 2s^2 - s - 2)} \tag{74}$$

A transform of the form of Equation 74 would not be found in tables but it is easily simplified using partial fractions. The first step is to factor the denominator. It is known that at least one root of a third-order polynomial must be real, and therefore it can be located fairly quickly by trial and error. Having located this root, the remaining two can be found by the quadratic formula. The roots are 1, -1, and -2, so the expansion takes the form

$$Y = \frac{s^3 + 4s^2 + 6s + 1}{s(s - 1)(s + 1)(s + 2)} = \frac{A_1}{s - 1} + \frac{A_2}{s + 1} + \frac{A_3}{s + 2} + \frac{A_4}{s} \tag{75}$$

A_1 is then found by multiplying through by $s - 1$ and setting s in the result equal to $+1$. The result is that A_1 is 2. In like fashion, A_2 is found by multiplying through by the factor, $s + 1$, and setting s in the result equal to -1; A_2 is found to be -1. The same procedure yields a value of $\frac{1}{2}$ for A_3 and a value of $-\frac{1}{2}$ for A_4. Therefore

$$Y = \frac{s^3 + 4s^2 + 6s + 1}{s(s - 1)(s + 1)(s + 2)} = \frac{2}{s - 1} + \frac{-1}{s + 1} + \frac{\frac{1}{2}}{s + 2} + \frac{-\frac{1}{2}}{s} \tag{76}$$

The final step is the term-by-term inversion of the right-hand side of Equation 76, using transform pairs Nos. 2 and 5. Hence, the solution of the differential equation is

$$y = 2e^t - e^{-t} + \tfrac{1}{2}e^{-2t} - \tfrac{1}{2} \tag{77}$$

B. Complex Roots

It is not unusual for complex roots to occur. In order to understand their implications, we must begin by returning to Equation 64. The function, y, originates from one or more differential equations that describe the behavior of real physical processes and systems. This means that not only y is real but also all of the coefficients, $a_1, a_2, \ldots, a_n, b_1, b_2, \ldots, b_m$. It follows that the function, Y, must necessarily be a real function of s; in other words, for any real value of s, Y must also be real.

Suppose now that the denominator of Equation 65 is a third-order polynomial. Therefore, there are three roots, one of which must be real. The other two can either be real or complex, but if they are complex, they must be complex conjugates. Assuming that two of the roots are complex and have the values of $-b - ic$ and $-b + ic$, then the partial fraction expansion takes the following form:

$$Y = \frac{A_1}{s - a} + \frac{A_2 + iB_2}{s + b + ic} + \frac{A_3 + iB_3}{s + b - ic} \tag{78}$$

The coefficients of the complex terms are generally complex themselves and therefore have been expressed as $A_2 + iB_2$ and $A_3 + iB_3$.

Since the left-hand side will be real for all real values of s, the sum of the two complex terms on the right-hand side must be real. Since a real number can only result from the addition of conjugate pairs, the last two terms must be conjugates of each other. Furthermore, since their denominators are already conjugates, it can be shown that their numerators must also be conjugates. Therefore, $A_2 = A_3$ and $B_2 = -B_3$. Hence, Equation 78 can be written

$$Y = \frac{A_1}{s - a} + \frac{A_2 + iB_2}{s + b + ic} + \frac{A_2 - iB_2}{s + b - ic} \tag{79}$$

Each term on the right is in the form of transform pair No. 5. The first term gives the simple exponential $A_1 e^{at}$. The complex pairs invert to

$$(A_2 + iB_2)e^{-(b+ic)t} + (A_2 - iB_2)e^{-(b-ic)t}$$

The following trigonometric identity can be used to simplify this expression:

$$e^{x+iy} = e^x (\cos y + i \sin y)$$

The result is

$$2e^{-bt}(A_2 \cos ct + B_2 \sin ct)$$

The latter factor involving the cosine and sine of ct can be combined into a single sine function of ct plus a phase angle giving the final form:

$$2\sqrt{A_2{}^2 + B_2{}^2}\, e^{-bt} \sin (ct + \phi)$$

where

$$\phi = \tan^{-1} \frac{A_2}{B_2}$$

Hence, the complete inversion of Equation 78 can be written in either of the following two forms:

$$y(t) = A_1 e^{+at} + 2e^{-bt}(A_2 \cos ct + B_2 \sin ct) \tag{80}$$

or

$$y(t) = A_1 e^{+at} + 2\sqrt{A_2{}^2 + B_2{}^2}\, e^{-bt} \sin (ct + \phi) \tag{81}$$

C. Multiple Roots

When repeated roots occur, the procedure for expansion is slightly modified. Consider, for example, the following differential equation:

$$\frac{d^3y}{dt^3} - 3\frac{dy}{dt} + 2y = 1 \tag{82}$$

with the following initial conditions:

$$y''(0) = y'(0) = y(0) = 0$$

The input function in this case is a unit step. Upon Laplace transformation and rearrangement, we obtain

$$Y(s) = \frac{1}{s(s^3 - 3s + 2)} \tag{83}$$

With reference to the cubic polynomial, trial and error rapidly locates the roots at $+1$ and -2, but the root at $+1$ is a "double root." Hence:

$$Y(s) = \frac{1}{s(s - 1)^2(s + 2)} \tag{84}$$

Although only three factors appear in the denominator of Equation 84, four terms are necessary for the partial fraction expansion:

$$Y(s) = \frac{A}{s} + \frac{B}{s + 2} + \frac{C_1}{(s - 1)^2} + \frac{C_2}{s - 1} = \frac{1}{s(s - 1)^2(s + 2)} \tag{85}$$

In the general situation involving an nth-order root, n terms are required, representing the nth and lower powers of the factor.

No problems are encountered in obtaining the coefficients for the first two terms of Equation 85. The coefficient, A, is found by multiplying through by s and setting s equal to zero with the result that A is $\frac{1}{2}$. Similarly, B is found by multiplying through by $(s + 2)$ and setting s equal to -2 with the result that B is $-\frac{1}{18}$. The same routine is followed to find C_1. When both sides are multiplied by $(s - 1)^2$, the result is

$$\frac{1}{s(s + 2)} = \frac{A(s - 1)^2}{s} + \frac{B(s - 1)^2}{s + 2} + C_1 + C_2(s - 1) \tag{86}$$

If s is set equal to $+1$, C_1 is found to be $\frac{1}{3}$.

The only problem is to find the coefficient, C_2. The "usual" routine would call for multiplication by $(s - 1)$, but C_2 becomes indeterminant when s is set equal to $+1$. However, Equation 86 is differentiable with respect to s.

When the derivative of both sides is taken, C_1 vanishes and the following equation results:

$$-2\frac{s+1}{s^2(s+2)^2} = A(s-1)\frac{(s+1)}{s^2} + B(s-1)\frac{(s+5)}{(s+2)^2} + C_2 \tag{87}$$

The first two terms on the right-hand side have been purposely written in this form to show that $(s-1)$ is a factor. Hence, when s is set equal to $+1$, these terms drop out; only C_2 remains on the right-hand side and the left-hand side is $-\frac{4}{9}$. The complete expansion then is

$$Y(s) = \frac{1}{s(s-1)^2(s+2)} = \frac{\frac{1}{2}}{s} + \frac{-\frac{1}{18}}{s+2} + \frac{\frac{1}{3}}{(s-1)^2} + \frac{-\frac{4}{9}}{(s-1)} \tag{88}$$

The inversion of Equation 88 into the time domain yields

$$y(t) = \tfrac{1}{2} - \tfrac{1}{18}e^{-2t} + \tfrac{1}{3}te^t - \tfrac{4}{9}e^t \tag{89}$$

When a root is repeated three or even more times, the procedure follows along the same lines. In the case of three, for example, the third coefficient is found by differentiating a second time. In the general case of n repeated roots, $n-1$ differentiations are required. A further variant would be the presence of several sets of repeated roots. This would not present any new theoretical problems, but just make the algebra more lengthy.

VII. CHARACTERISTIC EQUATION AND THE NATURE OF THE SOLUTION

Frequently, a detailed solution of a linear differential equation is not required. Perhaps one is only interested in the stability of the system. We recall that for a *linear* system, the system is stable if it remains bounded for all bounded inputs. The answer to the stability question lies in the denominator of the transform of the output. Although this can be shown generally, it is perhaps best illustrated by an example. The general proof would follow exactly the same lines of reasoning.

For convenience, we shall reconsider a previous example:

$$\frac{d^3y}{dt^3} + 2\frac{d^2y}{dt^2} - \frac{dy}{dt} - 2y = x \tag{71}$$

with the following initial conditions:

$$y(0) = 1$$
$$y'(0) = 2$$
$$y''(0) = 3$$

Transformation yields

$$(s^3 Y - s^2 - 2s - 3) + 2(s^2 Y - s - 2) - (sY - 1) - 2Y = X$$

$$(72)$$

Solving for Y, we obtain

$$Y = \frac{\overbrace{X}^{\substack{\text{Input}\\\text{Function}}} + \overbrace{s^2 + 4s + 6}^{\substack{\text{Initial Condition}\\\text{Terms}}}}{\underbrace{s^3 + 2s^2 - s - 2}_{\text{System Factors}}}$$

$$(90)$$

This equation is in the general form of Equation 65. Note particularly that the input function and initial conditions determine the numerator; on the other hand, the denominator is determined by the nature of the system itself—in effect, how the physical system combines the output and its derivatives.

We already know the roots of the denominator of Equation 90 and can therefore indicate the nature of a partial fraction expansion at this point:

$$Y = \frac{X + s^2 + 4s + 6}{s^3 + 2s^2 - s - 2} = \frac{A_1'}{s - 1} + \frac{A_2'}{s + 1} + \frac{A_3'}{s + 2} + \sum \psi(X) \qquad (91)$$

The summation, $\sum \psi(X)$, on the right-hand side consists of one or more terms, determined by the nature of the input function itself. The constants, A_1', A_2', and A_3', are functions of the input function and the initial conditions. They have been primed here to indicate that they are not the same as the constants of the previous example unless the input function is the same.

Inversion of Equation 91 yields the following result:

$$y = A_1' e^t + A_2' e^{-t} + A_3' e^{-2t} + L^{-1} \left[\sum \psi(X) \right] \qquad (92)$$

If the input were zero, then the solution would consist of the sum of the three exponential terms shown. Recalling the classical method of solution of linear differential equations, we recognize that these terms are the same exponentials as those that arise in the complementary solution. They characterize the *inherent* nature of the system and in particular, whether it is stable or unstable. The important point is that they arise from the roots of the denominator polynomial of the transform of the output, Y. Because of their significance, these roots are referred to as the "characteristics" of the polynomial and the equation formed when the polynomial is set equal to zero is referred to as the "characteristic equation." Referring to Equation 92, the presence of the positive exponential automatically indicates that the system is unstable, since it is unbounded.

To illustrate the previous statements, let us consider the solution for three different input functions; for brevity, only the results are presented here.

Case 1: $x = 0$ No input.

$$y = \tfrac{11}{6}e^t - \tfrac{3}{2}e^{-t} + \tfrac{2}{3}e^{-2t}$$

Case 2: $x = u(t)$ unit step. This case was covered in detail in the earlier example.

$$y = 2e^t - e^{-t} + \tfrac{1}{2}e^{-2t} - \tfrac{1}{2}$$

Case 3: $x = e^{-3t}$.

$$y = \tfrac{15}{8}e^t - \tfrac{7}{4}e^{-t} + e^{-2t} - \tfrac{1}{8}e^{-3t}$$

Comparing the three cases, we see that changing the bounded input only changes the values of the coefficients of the terms. In Cases 2 and 3, the presence of an input has only added an extra term and these added terms are bounded because the inputs were bounded.

The preceding discussion suggests that if only the stability of a system is to be determined, the nature of the input can be ignored, other than that it is presumed to be bounded; furthermore, the initial conditions have no relevance as far as the stability is concerned. In short, only the location of the roots of the characteristic equation must be ascertained.

In the most general case, the roots will be complex. As previously mentioned, complex roots always occur in pairs and lead to a sinusoid multiplied by an exponential. The real part of the root determines the magnitude and sign of the exponent of the exponential and the imaginary part determines the frequency of the oscillation. If the real part of the root is positive, then the exponential will grow without bound and the system is unstable; if the real part is negative, the exponential will decay with time. For stability, there must be no roots with positive real parts. As shown in Figure 6, any root to the right of the imaginary axis indicates an unstable system.

The occurrence of repeated roots only assumes importance if the root is purely imaginary. To see this, consider the effect of a repeated root first to the left of the imaginary axis, then to the right of the axis, and finally on the imaginary axis. If the root is to the left, then it will result in a decaying exponential multiplied by a function of t^{m-1}, where m stands for the number of times the root occurs at the point in question. This fact can be seen by looking at transform pair No. 9, for example. The decaying exponential factor dominates over the power of t, thus making the term a stable one. If the root is to the right of the imaginary axis, then we have a function of

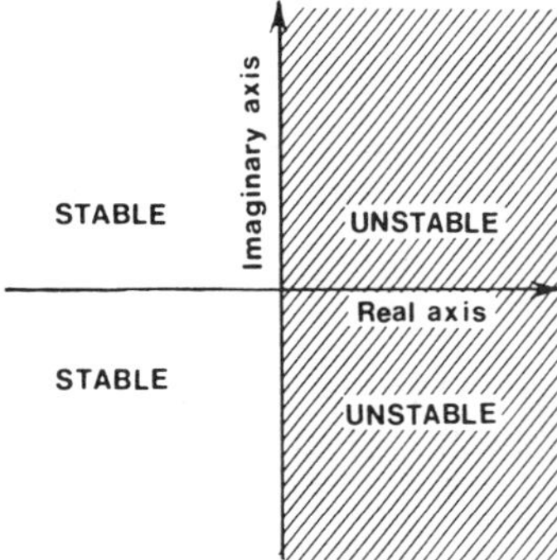

Figure 6.   Stable and unstable regions for roots in the complex plane.

t^{m-1}, multiplying a positive exponential, which, of course, would be unstable even without the power of t multiplying it. Finally, looking at a pair of purely imaginary roots, we see that if they are of multiplicity one, they invert to a sinusoid of constant amplitude. Some authors classify this as a case of limited stability. However, if the imaginary roots are of multiplicity greater than one, then the sinusoid will be multiplied by a function of t^{m-1}, causing the amplitude of the sinusoid to grow without bound, and therefore the system is unstable.

Although the locations of the roots of the characteristic equation do provide sufficient information to determine the stability of a system, finding their locations may present some problems. While this is easily accomplished for quadratic equations, and fairly easily accomplished for cubics, the complexity compounds rapidly for higher orders. Actually, the exact locations of the roots are not really important; it is only necessary to know if there are *any* roots with positive real parts. This fact was recognized at least as early as 1868 when Maxwell was studying flyball governors. He obtained a characteristic equation of fifth order and sought a method for determining sufficient conditions that all roots of a polynomial have negative real parts. His interest in this problem ultimately led Routh to develop a criterion several years later, and in 1893, Hurwitz formulated a similar criterion.

Since these early developments, other methods have been presented. Roughly, they fall into two classes. Some are purely algebraic, as for example, the Routh and Hurwitz methods. Others fall into the nonalgebraic class. Among these are the Nyquist criterion, the Bode criterion, and the root locus method. The Routh criterion is presented in Section VIII of this chapter, and the Bode criterion will be discussed and developed in Chapter 14. The Hurwitz and Nyquist criteria and the root locus method are discussed in a number of the references cited in this text.

VIII. THE ROUTH TEST

The Routh test begins with the ordering of the characteristic equation into the form

$$a_0 s^n + a_1 s^{n-1} + a_2 s^{n-2} + \cdots + a_{n-1} s + a_n = 0 \qquad (93)$$

where the a's are the constant coefficients. There are two necessary conditions for stability that can be quickly checked. The first is that all of the coefficients, a_1 through a_n, must be present, and the second is that all of the coefficients must have the same sign. These requirements are not sufficient conditions for stability since all of the coefficients may be present and have the same sign, yet the system may still be unstable.

In most problems in process control, the characteristic polynomial will not exceed four or five terms. However, the method is better illustrated if a few more terms are assumed to be present, say eight. The test involves a "Routh array," the first two rows of which are taken directly from the characteristic equation. The remaining terms in the array are calculated in accordance with the procedure to be described here. In general, the array consists of a total of $n + 1$ rows. For the case here in which Equation 93 consists of eight terms, n is 7 and there will be eight rows in the array as shown below:

$$
\begin{array}{cccc}
a_0 & a_2 & a_4 & a_6 \\
a_1 & a_3 & a_5 & a_7 \\
b_1 & b_3 & b_5 & \\
c_1 & c_3 & c_5 & \\
d_1 & d_3 & & \\
e_1 & e_3 & & \\
f_1 & & & \\
g_1 & & &
\end{array}
$$

If n is an even number, then the first row has one more member than the second. The elements of the third through the eighth rows are calculated successively by a routine that somewhat resembles the evaluation of a second-order determinant. For example, the calculation of the first two elements of the third and fourth rows are given by the equations:

$$b_1 = \frac{a_1 a_2 - a_0 a_3}{a_1} \qquad b_3 = \frac{a_1 a_4 - a_0 a_5}{a_1}$$

$$c_1 = \frac{b_1 a_3 - a_1 b_3}{b_1} \qquad c_3 = \frac{b_1 a_5 - a_1 b_5}{b_1} \qquad (94)$$

The routine for all remaining elements is apparent by inspection. With the exception of the first two rows of the array, each row is evaluated using the previous two rows.

The Routh criterion states that the system is stable if all of the terms in the first column of the array have the same sign. This means that there are no roots on or to the right of the imaginary axis. If they do not have the same sign, then the number of roots with positive real parts is equal to the number of changes of sign in the first column. In setting up the characteristic equation, it is standard practice to arrange it so that the sign of a_0 is positive. The number of sign changes in the first column of the array is determined by examining the elements in order—a_0, a_1, b_1, c_1, ... Thus, if b_1 has a different sign than a_1, then this counts as one sign change; if c_1 has a different sign than b_1, then this is an additional change, and so forth.

There are several unusual situations that call for special alterations of the Routh procedure. The first is that in which a zero is obtained for an element in the first column of an array but at least one element in the same row differs from zero. The second is that in which all elements of a row are zero. These two cases are discussed and illustrated in Reference 4.

Although the Routh method reveals the number of roots having positive real parts, one of its limitations is that it gives no indication of their exact locations. Sometimes this information is needed for the design of a control system, and if so, the root locus method can be employed. The disadvantage of the root locus procedure is that it usually requires more work than the Routh method. A second limitation of the Routh method is that it can only be used when the characteristic equation is a polynomial. If pure time delays are present in a system, the characteristic equation does not take this form and stability must be determined by some other method, such as the Bode criterion.

We conclude this chapter with two examples of the Routh method. For the first, consider the following differential equation:

$$\frac{d^4y}{dt^4} - 4\frac{d^3y}{dt^3} + 5\frac{d^2y}{dt^2} + 98\frac{dy}{dt} - 200y = 1 \tag{95}$$

For simplicity, assume all initial conditions are zero. The transform of Y reduces to:

$$Y(s) = \frac{1}{s(s^4 - 4s^3 + 5s^2 + 98s - 200)} \tag{96}$$

The factor, s, in the denominator arises from the unit step forcing function. The factor in the parentheses is the characteristic polynomial, and therefore the characteristic equation is

$$s^4 - 4s^3 + 5s^2 + 98s - 200 = 0 \tag{97}$$

Since all of the coefficients do not have the same sign, the system must be unstable. There will be five rows in the final array, since n is four. The first two rows of the Routh array are:

$$
\begin{array}{ccc}
1 & 5 & -200 \\
-4 & 98 &
\end{array}
$$

We then calculate the b-row members:

$$b_1 = \frac{(-4)(5) - (1)(98)}{-4} = 29.5$$

$$b_3 = \frac{(-4)(-200) - (1)(0)}{-4} = -200$$

The array is then:

$$
\begin{array}{ccc}
1 & 5 & -200 \\
-4 & 98 & \\
29.5 & -200 &
\end{array}
$$

The c-row members are:

$$c_1 = \frac{(29.5)(98) - (-4)(-200)}{29.5} = 70.88$$

$$c_3 = 0$$

Now the array is:

$$
\begin{array}{ccc}
1 & 5 & -200 \\
-4 & 98 & \\
29.5 & -200 & \\
70.88 & 0 &
\end{array}
$$

The d-row members are:

$$d_1 = \frac{(70.88)(-200) - (29.5)(0)}{70.88} = -200$$

$$d_3 = 0$$

Hence, the final Routh array is:

$$
\begin{array}{ccc}
1 & 5 & -200 \\
-4 & 98 & \\
29.5 & -200 & \\
70.88 & 0 & \\
-200 & 0 &
\end{array}
$$

There are three reversals of sign for the members of the first column, which means that there are three roots with positive real parts.

This example is quite easily checked. Although the characteristic equation is quartic, and therefore could have two sets of complex roots, trial and error rather quickly locates a root at $+2$, and a second root is found at -4. Hence, division by the factors $(s - 2)$ and $(s + 4)$ yields the remaining quadratic polynomial, $s^2 - 6s + 25$. From the quadratic formula, the remaining two roots are found to be $3 + i4$ and $3 - i4$. The three roots with positive real parts confirm the result of the Routh test.

An interesting example arises from a closed-loop system under proportional control where there are three elements, each described by a first-order linear differential equation. It turns out that the characteristic equation for this combination results from the following form:

$$K_o + (T_1 s + 1)(T_2 s + 1)(T_3 s + 1) = 0 \tag{98}$$

or

$$T_1 T_2 T_3 s^3 + (T_1 T_2 + T_1 T_3 + T_2 T_3)s^2 + (T_1 + T_2 + T_3)s$$
$$+ (1 + K_o) = 0 \tag{99}$$

where T_1, T_2, and T_3 are the time constants of the three elements and K_o is the overall gain of the system. In this case, the only adjustable parameter is the controller gain, K_C, which is a factor in the overall gain, K_o. The problem is to find out if there are any values of the overall gain, and hence the controller gain, which will make the system unstable.

The characteristic equation differs from the preceding example in that one coefficient in the polynomial is a variable. The time constants will be assumed to be positive here, so all of the coefficients are positive. Furthermore, none of them is missing so the two necessary conditions for stability are met. The first two rows of the Routh array are as follows:

$$T_1 T_2 T_3 \qquad\qquad\qquad T_1 + T_2 + T_3$$

$$T_1 T_2 + T_1 T_3 + T_2 T_3 \qquad\qquad\qquad 1 + K_o$$

Since the elements of the first column are positive, the system will be unstable if the first calculated element, b_1, can be negative; in other words, we must answer the following question:

$$b_1 = \frac{(T_1 T_2 + T_1 T_3 + T_2 T_3)(T_1 + T_2 + T_3) - (T_1 T_2 T_3)(1 + K_o)}{T_1 T_2 + T_1 T_3 + T_2 T_3} \overset{?}{<} 0$$

Rearranging, we find that b_1 can be less than zero if

$$1 + K_o > \left(\frac{1}{T_1} + \frac{1}{T_2} + \frac{1}{T_3}\right)(T_1 + T_2 + T_3) \tag{100}$$

Multiplying and dividing the right-hand side by T_1 and defining:

$$R_2 = \frac{T_2}{T_1}$$

and   (101)

$$R_3 = \frac{T_3}{T_1}$$

we obtain the result that

$$1 + K_o > \left(1 + \frac{1}{R_2} + \frac{1}{R_3}\right)(1 + R_2 + R_3) \tag{102}$$

and therefore, the system will be unstable if:

$$K_o > \left(1 + \frac{1}{R_2} + \frac{1}{R_3}\right)(1 + R_2 + R_3) - 1 \tag{103}$$

It is interesting at this point to see what Equation 103 predicts for the maximum overall gain. Suppose that all three time constants of the system are equal. Then $R_2 = R_3 = 1$, and the maximum gain is eight. On the other hand, if one of the time constants is very small compared to the others, then the maximum gain increases. In the limit as the time constant approaches zero, the maximum gain approaches infinity.

Significantly, Equation 103 shows that the maximum gain is only dependent on the *ratios* of the time constants, and not upon their actual values.

SUMMARY

A considerable number of the properties of the Laplace transformation have been covered in this chapter. The emphasis has been on the applications of the transform rather than upon the mathematical theory behind it. The only extension of the material here will be concerned with the use of the transform in determining the frequency response of a process or system. This aspect will be covered in Chapter 14.

REFERENCES

1. S. C. Gupta, *Transform and State Variable Methods in Linear Systems*, John Wiley and Sons, Inc., New York, 1966.
2. R. V. Churchill, *Operational Mathematics*, Second Ed., McGraw-Hill Book Co., Inc., New York, 1958.

3. L. B. Koppel, *Introduction to Control Theory with Applications to Process Control*, Prentice-Hall, Inc., Englewood Cliffs, N.J., 1968, pp. 3–6.
4. G. J. Murphy, *Control Engineering*, D. Van Nostrand Co., Inc., New York, 1959, pp. 65–67.

PROBLEMS

Problem 1.   The differential equation describing a first-order system is given by

$$\tau \frac{dy}{dt} + y = x$$

Show that the response of this system is the same for the following two cases:

(1) $y(0) = 1,\ x = 0$

(2) $y(0) = 0,\ x = \tau\delta(t)$

Problem 2.   A system is described by the following first-order differential equation:

$$2\frac{dy}{dt} + 5y = 10x$$

(a)   What is the time constant of this system?
(b)   What is the gain factor?
(c)   Find the Laplace transformation of the response if the initial value of y is 1.
(d)   Suppose that the input disturbance is a unit step. Now what is the Laplace transformation of the response?
(e)   Apply the initial-value theorem to part d of this problem to verify that the initial condition is satisfied.
(f)   Apply the final-value theorem to part d.
(g)   Utilizing only transform pairs Nos. 2 and 5, find the time response of y.

Problem 3.   The response of a process to an input, x, is described by the following differential equation:

$$\frac{d^2y}{dt^2} + 2\frac{dy}{dt} - 8y = 5x$$

with the following initial conditions:

$$y(0) = y'(0) = 0$$

(a)   For a unit step input, solve this equation using Laplace transforms.
(b)   Locate the roots of the characteristic equation in the complex plane. Is the response stable?

Problem 4.   Given the differential equation

$$\frac{d^2 y}{dt^2} + 4\frac{dy}{dt} + 13y = 1$$

with the following initial conditions:

$$y(0) = y'(0) = 0$$

(a)   Find the solution of this equation by Laplace transformation. If any oscillatory solutions exist, express them in terms of a sinusoidal function of time plus a phase angle.

(b)   Check to see that the solution meets both of the initial conditions and that it satisfies the differential equation.

(c)   Apply the Routh criterion to this problem.

$$\textit{Answer: (a)}\quad y = \tfrac{1}{13} + \tfrac{1}{13}\sqrt{1.444}\ e^{-2t} \sin\,(3t + 4.12 \text{ rad})$$

Problem 5.   Given the differential equation

$$\frac{d^3 y}{dt^3} + \frac{d^2 y}{dt^2} + 2\frac{dy}{dt} - 4y = 1 + t$$

with the following initial conditions:

$$y(0) = y'(0) = y''(0) = 0$$

Solve this equation by Laplace transformation.

   Hint: One root of the characteristic equation is at $+1$.

$$\textit{Answer:}\quad y = -\tfrac{1}{4}t - \tfrac{3}{8} + \tfrac{2}{7}e^t + \frac{\sqrt{28}}{56}\,e^{-t}\sin\,(\sqrt{3}\,t + 1.24 \text{ rad})$$

Problem 6.   Do Problem 9 of Chapter 6 using Laplace transforms.

Problem 7.   The roots of the characteristic equation for 10 systems are presented below.

       System   1 $+2i, -2i$
                2    $-4, -4$
                3    $3 + 4i;\ 3 - 4i$
                4   -5
                5    $-3 - 6i, -3 + 6i, 3 + 4i, 3 - 4i$
                6    $-4 + 5i, -4 + 5i, -4 - 5i, -4 - 5i$
                7    $-3, +7, -2$
                8    $-6, +7, +3 + 2i, +3 - 2i$
                9    $-3 - 2i, -3 - 2i$
               10    $-8, -10$

For each system, characterize its behavior in terms of one or more of the properties in the following list:

1. Stable.
2. Unstable.
3. Limited stability.
4. Oscillatory.
5. Decaying exponential.
6. Growing exponential.
7. Impossible for a real physical system.

Problem 8.   Consider a system whose dynamic response, y, to a disturbance, x, is described by a first-order linear differential equation:

$$\tau \frac{dy}{dt} + y = x$$

Furthermore, suppose that the initial condition on y is zero. Using Laplace transforms, solve the following:

(a) If the input is a unit step, determine the response and sketch it. Check to see that the solution meets the given initial condition.
(b) If the input is a pulse function as given by Equation 33, determine the response and sketch it. Check to see that the solution meets the initial condition.
(c) Now suppose that the input is a unit impulse. As demonstrated in the text, this function is a limiting case of the pulse function. Using the fact that the transform of the unit impulse is unity, determine the response and check to see that the solution meets the initial condition. Explain the unexpected (or is it expected?) result of this check. A restudy of part *b* may prove helpful.

Answers: (a) $y = (1 - e^{-t/\tau})\, \mathbf{u}(t)$

$$(b)\ \ y = \frac{1}{a}\, \mathbf{u}(t) \left\{ 1 - e^{-t/\tau} \right\} - \frac{1}{a}\, u(t - a) \left\{ 1 - e^{-(t-a)/\tau} \right\}$$

Problem 9.   A feedback control system is composed of two first-order elements and is under proportional-integral control. For a unit change in set point, the transform of the output, $Y(s)$, is given by the expression:

$$Y(s) = \frac{1}{s} \cdot \frac{\dfrac{K_o(T_R s + 1)}{T_R s(T_1 s + 1)(T_2 s + 1)}}{1 + \dfrac{K_o(T_R s + 1)}{T_R s(T_1 s + 1)(T_2 s + 1)}}$$

Using the Routh test, show that for stability, the following condition must be met:

$$T_R > \frac{K_o}{1 + K_o} \frac{T_1 T_2}{T_1 + T_2}$$

Problem 10.   Given the partial differential equation

$$\frac{\partial T}{\partial t} = - \frac{\partial T}{\partial x}$$

where $T = T(x, t)$, and the following initial and boundary conditions:

$$T(x, 0) = 0$$

$$T(0, t) = u(t)$$

find $T(x, t)$ using Laplace transforms.

Answer: $T(x, t) = u(t - x)$

PART II

Analysis of Simple Processes

In Part I, very simple models of processes were used to illustrate the behavior of some systems under various types of controller actions. For example, a continuous-flow, stirred-tank heater was modeled by a first-order differential equation, and the flow process in Chapter 6 was simply described by a pure gain factor. In some situations, such simple models may be satisfactory. However, usually more exact analyses are desired, and these call for more complex models than those previously used. In Part II, many different processes will be studied and a number of models will be developed. Since there is such a large variety of processes, it might seem that "process dynamics" could scarcely be introduced in just a few brief chapters. However, this is fortunately not the case.

As chemical engineering has developed, the concepts of "transport phenomena" and "unit operations" have been extremely helpful in unifying seemingly unrelated complex processes. The basic idea of these two concepts is that analogies do exist between processes. These analogies can be used to adapt the solutions of problems in one field to problems in another field. Important also, these concepts indicate useful ways of placing numerous types of processes into a few categories. For example, the category of unit operations places the processes of distillation and absorption under the general heading of "staged operations." The category of transport phenomena draws an analogy between heat transfer in a boundary layer and mass transfer in a boundary layer.

In a similar way, the unsteady-state analysis of a number of processes has indicated that they fall into well-defined groups. For example, the response of a temperature bulb falls into the same category as the response of one model of a continuous-flow, stirred-tank heater. In the analysis of control systems, the recognition and utilization of such analogies cannot be overemphasized.

While there are many analogies among processes, there are also analogies among electrical, mechanical, and chemical process systems. These arise because the electrical concepts of resistance, capacitance, and inductance exist in mechanical and chemical process systems as well. These analogies will be pointed out as the various types of processes are discussed.

Returning to the categories of unsteady-state behavior of processes, it is useful to adopt the following breakdown:

    I.   Pure Delay
   II.   Pure Capacitance
 III.   First Order
 IV.   Second Order
   V.   Distributed

Each of these types of elements will be discussed in some detail, mostly by looking at specific examples. Clearly, it will not be possible to cover all types of processes, but with the techniques to be presented, the reader should be able to investigate many others.

The topics of pure delay and pure capacitance are covered quite readily. First-order processes are so numerous and varied in character that these will be covered in two chapters. Second-order processes, which are to some extent an extension of the ideas developed for first-order processes, are covered in a single chapter. Distributed processes are basically the most common type, but they are usually approximated by simpler models. Hence, only one chapter is devoted to them.

CHAPTER VIII

Pure Delay, Capacitance,
and First-Order Processes
Based on Mass Balances

I. ANALOGIES BETWEEN CHEMICAL PROCESSES AND ELECTRICAL CIRCUITS

There are many analogies between the physical components and behavior of chemical processes and those of electrical circuits. These analogies are useful in illustrating the relationships between physically unrelated problems. Furthermore, they provide an additional way of looking at the dynamics of a process, and this frequently leads to greater insight and understanding. Only a rudimentary knowledge of electrical engineering is necessary in order to utilize them. In the chapters that follow, electrical analogies will be illustrated for a number of processes.

A. Two Classes of Elements

1. Active Elements

All electrical circuits contain one or more energy sources that are called "active" elements. Two common examples of these are batteries and power generators. The remaining components of a circuit are "passive" elements since they are not sources of energy. The transient analysis of an electrical circuit concerns the determination of the current through and voltage across one or more of the passive elements in the circuit.

The situation is basically the same in chemical processes—that is, there are active elements such as pumps and compressors that supply mass or energy to passive elements such as pipes and tanks. Consider, for example, a storage tank as illustrated in Figure 1. A flow source supplies liquid to the tank at a variable rate. The tank is a passive element and we might be interested in the response of the tank in terms of the level, h, and leaving flow rate, F_o, to variations in the entering flow, F_i.

As we consider various processes in this and succeeding chapters, it will be found that active elements fall into either one of two categories, flow sources or potential sources. Similarly, electrical circuits are commonly

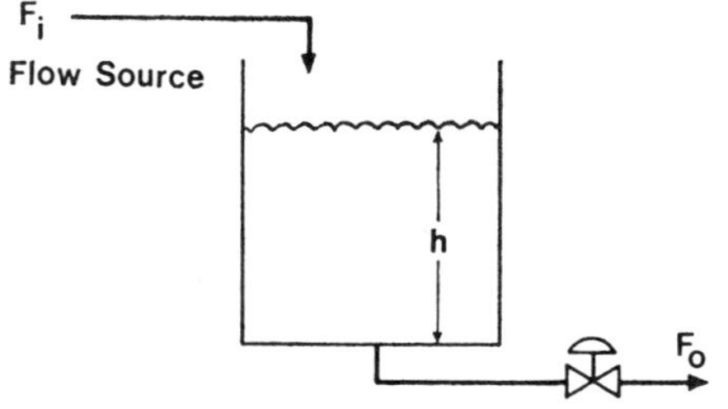

Figure 1. Storage tank with variable feed rate.

spoken of as "current-driven" or "voltage-driven." The distinctions between these two are best understood from the examples to be presented later, but for purposes of future classification, the analogies for the two cases are presented in Table 1. Most circuits discussed in electrical engineering textbooks involve voltage sources rather than current sources. On the other hand, as we shall see shortly, current-source and voltage-source situations are both common in chemical processes.

TABLE 1. ACTIVE ELEMENT ANALOGIES

"Current" Sources			
Physical Situation	Examples	"Current"	Typical Units
Electrical circuits	Current generator	Current	Amperes (coulombs/sec)
Fluid flow	Pump or compressor	Flow rate of fluid	ft^3/sec
Heat transfer	Electrical heating	Heat transfer rate	Btu/sec
Mechanical translation	Momentum source	Velocity	ft/sec

"Voltage" sources			
Physical Situation	Examples	"Voltage"	Typical Units
Electrical circuits	Battery	Voltage	Volts
Fluid flow	Hydraulic pressure or pneumatic pressure	Pressure	lb_f/ft^2
Heat transfer	Temperature difference in a heat exchanger	Temperature	°F
Mechanical translation	Gravitational force	Force	lb_f

2. Passive Elements

The passive elements of electrical circuits are resistors, capacitors, and inductors. Capacitors and inductors store electrical energy, while resistors dissipate it. Basically, each of these passive elements differs in the way in which the driving force across the element is related to the flow rate through it. For example, a resistor has the property that the driving force across it is proportional to the flow rate. This relationship, and those for a capacitor and an inductor, are summarized in Table 2. Inductors fall in the more general category of "inertances," and this terminology is used in the table.

TABLE 2. DEFINITIONS OF ELECTRICAL CIRCUIT ELEMENTS

Element	Relationship
Resistance	$R = \dfrac{\text{Driving force}}{\text{Flow rate}}$
Capacitance	$C = \dfrac{\text{Flow rate}}{\text{Rate of change of driving force}}$
Inertance	$I = \dfrac{\text{Driving force}}{\text{Rate of change of flow rate}}$

In accordance with these definitions, the analogies for passive elements are summarized for the several physical situations in Table 3. Notice that there are two equivalent expressions for capacitance. It is the ratio of a flow rate to the rate of change of a driving force, but since these are both rates with respect to time, this ratio is the same as the ratio of a stored quantity to a driving force. Interestingly, there is no analogy for inertance in heat transfer. For a more extensive discussion of the analogies, the reader is referred to Chapter 3 of the textbook by Tyner and May (1).

Slight variations in these definitions have been used by other authors. For example, resistance in fluid flow is sometimes defined as the ratio of the driving force to the velocity of the fluid instead of to its volumetric flow rate. Even when the definitions given in Table 2 are used, units other than those indicated in Table 3 for driving forces and flow rates can just as well be employed.

TABLE 3. PASSIVE ELEMENT ANALOGIES

	Resistance	
Physical Situation	Examples	Typical Units Driving Force/Flow Rate
Electrical circuits	Resistor	$\text{Volt/ampere} = \text{ohm}$ $= \text{volt, sec/coulomb}$
Fluid flow	Fluid friction in lines, valves, and fittings	$\dfrac{\text{lb}_f}{\text{ft}^2}\bigg/\dfrac{\text{ft}^3}{\text{sec}} = \dfrac{\text{lb}_f,\ \text{sec}}{\text{ft}^5}$
Heat transfer	Heat transfer resistance	$^\circ\text{F}\bigg/\dfrac{\text{Btu}}{\text{sec}}$
Mechanical translation	Dash pot	$\text{lb}_f\bigg/\dfrac{\text{ft}}{\text{sec}}$

	Capacitance	
Physical Situation	Examples	Typical Units Flow Rate/Rate of Change of Driving Force or Stored Quantity/Driving Force
Electrical circuits	Capacitor	$\text{Ampere}\bigg/\dfrac{\text{volt}}{\text{sec}} = \dfrac{\text{coulomb}}{\text{sec}}\bigg/\dfrac{\text{volt}}{\text{sec}}$ $= \dfrac{\text{coulomb}}{\text{volt}} = \text{farad}$
Fluid flow	Volume	$\dfrac{\text{ft}^3}{\text{sec}}\bigg/\dfrac{\text{lb}_f}{\text{ft}^2}\bigg/\text{sec} = \dfrac{\text{ft}^3}{\text{lb}_f/\text{ft}^2}$
Heat transfer	Heat capacitance	$\dfrac{\text{Btu}}{\text{sec}}\bigg/{}^\circ\text{F/sec} = \dfrac{\text{Btu}}{{}^\circ\text{F}}$
Mechanical translation	Spring, elasticity	$\dfrac{\text{ft}}{\text{sec}}\bigg/\dfrac{\text{lb}_f}{\text{sec}} = \dfrac{\text{ft}}{\text{lb}_f}$

	Inertance	
Physical Situation	Examples	Typical Units Driving Force/ Rate of Change of Flow Rate
Electrical circuits	Inductor	$\text{volt}\bigg/\dfrac{\text{ampere}}{\text{sec}} = \dfrac{\text{volt, sec}^2}{\text{coulomb}}$ $= \text{henry}$
Fluid flow	Mass	$\dfrac{\text{lb}_f}{\text{ft}^2}\bigg/\dfrac{\text{ft}^3}{\text{sec}^2} = \dfrac{\text{lb}_f,\ \text{sec}^2}{\text{ft}^5}$
Heat transfer	None	
Mechanical translation	Mass	$\text{lb}_f\bigg/\dfrac{\text{ft}}{\text{sec}^2} = \dfrac{\text{lb}_f,\ \text{sec}^2}{\text{ft}}$

Most processes in chemical engineering can be characterized by combinations of resistances and capacitances, or of resistances and inertances. Only rarely does one encounter a process that must be described by a combination of all three of the passive elements. In this and the next several chapters these properties will be studied, but in the contexts of the processes in which they occur. Purely resistive situations are common, but since resistances do not alter the time behavior of flows, there is no need for a detailed study of their behavior. Purely capacitive processes do exist, although their occurrence is rare. Several will be suggested but they are mainly of interest because they are limiting cases of resistance-capacitance combinations. Processes possessing only inertance never occur since there is also always some resistance present.

Before turning to purely capacitive processes, we shall begin with a review of a rather special element—namely, a pure time delay. This element is very important in process control, but is usually of little significance in electrical systems since electrical signals travel with the speed of light.

II. PURE DELAY

We have already met some examples of a pure delay in control systems. The first arose in Chapter 1 where one occurred in the example of the shower system. The control valves were some distance from the shower head so a pure delay resulted. Another example was that discussed in Chapter 7 where a water heater in a pipe was some distance upstream of the sensing element. Pure delays arise where there is some transportation time for a signal between two points in a control system. For this reason, terms such as "transport lag," "distance/velocity lag," and "dead time" are synonymous with "pure delay."

In Chapter 7, the transfer function for a pure delay was obtained, although not labeled as such. For completeness, the development of the transfer function will be reviewed here. Shinskey (2) illustrates a pure delay by a conveyer belt as shown in Figure 2a. Solids are fed from a hopper at a controlled rate to a conveyer belt. They fall on the belt at x and are weighed continuously at y. The delay between x and y is the time, τ_D. A pure delay arises in line blending as shown in Figure 2b. Stream A and Stream B are combined in a mixing-T at x and the composition is analyzed at y. The transportation time between x and y is τ_D.

In either of these examples, the relationship between x and y is

$$y(t) = x(t - \tau_D) \tag{1}$$

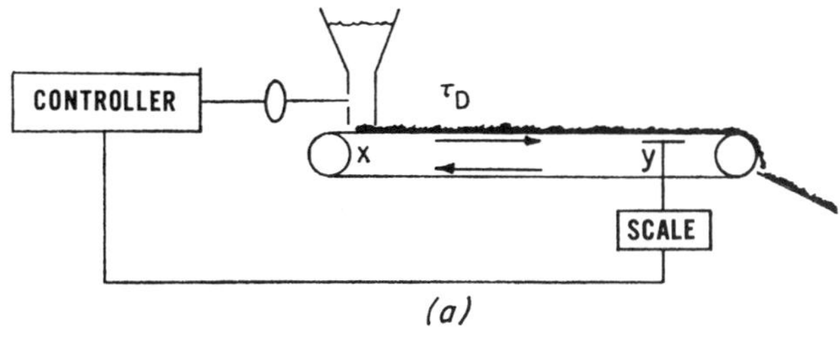

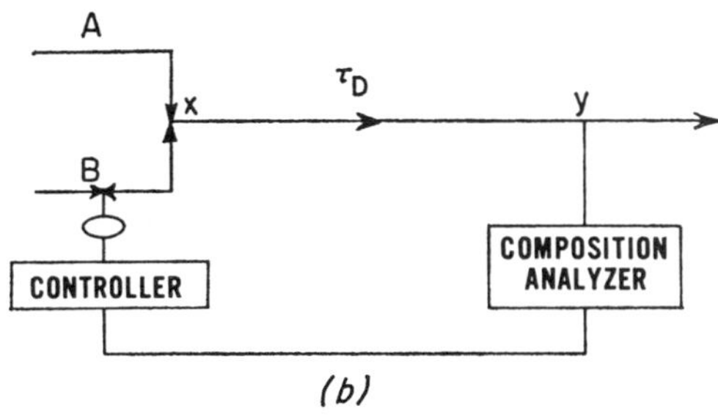

Figure 2. Illustrations of pure time delays. (*a*) Conveyor belt transport delay. (*b*) Pipeline delay.

As in Chapter 5, the variables can be defined in terms of an original steady-state value plus a deviation:

$$x(t) = \bar{x} + \Delta x(t) \tag{2}$$

$$y(t) = \bar{y} + \Delta y(t) \tag{3}$$

The steady-state values, $\bar{x}$ and $\bar{y}$, are intentionally not indicated to be functions of time. Furthermore, it is physically obvious that

$$\bar{x} = \bar{y} \tag{4}$$

Substituting Equations 2 and 3 into Equation 1, and subtracting Equation 4 from the result, we obtain

$$\Delta y(t) = \Delta x(t - \tau_D) \tag{5}$$

Prior to time zero, there are no deviations. As shown in Section II.B.3 of Chapter 7, the Laplace transformation is

$$Y(s) = e^{-\tau_D s} X(s) \tag{6}$$

where $Y(s)$ and $X(s)$ are the transforms of the deviations. Therefore, the transfer function is

$$\frac{Y(s)}{X(s)} = e^{-\tau_D s} \tag{7}$$

Normally, pure delays do not occur in such very obvious ways as in the conveyer-belt and pipeline examples here. However, as will be explained in Chapter 13, they are *effectively* present to some extent in most systems.

III. PURELY CAPACITIVE PROCESSES

A capacitor acts as a storage unit for either mass or energy. Since this energy must usually overcome a resistance in reaching the capacitor, purely capacitive processes are rather rare. Hence, only one example is given here for a mass-balance situation.

A. Liquid Storage Tank with a Constant-Displacement Effluent Pump

Consider the storage tank shown in Figure 3. Liquid enters the tank at a variable rate, F_i ft^3/sec, but is removed at a *constant* rate, F_o ft^3/sec, by a constant-displacement pump. The liquid level in the tank is h ft. For purposes of discussion, we shall assume that the level will be the measured variable for a control system that will stop the feed if the level reaches the top and will stop the pump if the level reaches the bottom. Thus, the problem is to determine the response of h to changes in F_i.

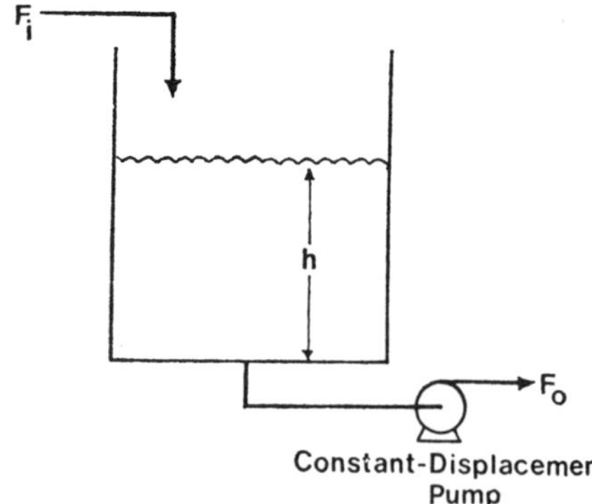

Figure 3.   Purely capacitive storage tank.

An unsteady-state material balance begins the analysis:

$$\text{input} - \text{output} = \text{accumulation}$$

$$F_i \quad - \quad F_o \quad = \frac{dV}{dt} = A\,\frac{dh}{dt} \tag{8}$$

where $V =$ volume of liquid in the tank, ft^3
$\quad A =$ cross-sectional area of tank, ft^2

In terms of steady-state and deviation variables, we have:

$$F_i = \bar{F}_i + \Delta F_i$$

$$h = \bar{h} + \Delta h$$

$$F_o = \bar{F}_o$$

The leaving flow rate exhibits no deviation since the speed of the constant-displacement pump is not manipulated. Substitution of these expressions into Equation 8 yields

$$\bar{F}_i + \Delta F_i - \bar{F}_o = A\,\frac{d(\bar{h} + \Delta h)}{dt} \tag{9}$$

The derivative, $d\bar{h}/dt$, is zero since $\bar{h}$ is a constant. The steady-state equation is

$$\bar{F}_i - \bar{F}_o = 0 \tag{10}$$

Subtracting this equation from Equation 9 results in

$$\Delta F_i = A\,\frac{d\,\Delta h}{dt} \tag{11}$$

The deviations at time zero are zero. Laplace transformation yields the transfer function

$$\frac{h}{F_i} = \frac{1}{As} \tag{12}$$

The Δ notation has been dropped for convenience and the symbols h and F_i stand for the transforms of these quantities.

B. Electrical Analogy

Referring to Equation 11, this equation can be rearranged to the form:

$$A = \frac{\Delta F_i}{\dfrac{d\,\Delta h}{dt}} \tag{11a}$$

Thus, the cross-sectional area is the ratio of the change in flow rate to the rate of change of the change in liquid level. The liquid level actually represents the driving force for flow leaving the tank, even though the withdrawal rate is kept constant by the pump. Thus, the area is the ratio of the change in flow rate to the rate of change of the change in the driving force. Referring to Table 2, it can be seen that this ratio basically conforms with the definition and behavior of a capacitance. It is a somewhat broader and more general definition, since it is defined in terms of *changes* in the variables:

$$C \equiv A = \frac{\textit{change} \text{ in flow rate}}{\text{rate of change of the \textit{change} in driving force}} \tag{13}$$

If the capacitance is constant, then this definition gives the same value of capacitance as that in Table 2. The broader definition of Equation 13 is necessary in chemical processes since the capacitance is frequently not constant. In spite of this, the analogy to electrical capacitance is still valid and proves to be very useful.

The units of capacitance in this example are ft^2 which do not agree with those in Table 3. The discrepancy arises because the driving force for flow has been expressed in terms of feet of fluid flowing rather than in terms of pressure in lb_f/ft^2.

When the cross-sectional area of the tank is constant, the capacitance is constant and independent of the level. An electrical capacitor also has this property of constancy over a specified range of voltages. As noted before, the capacitance can be thought of as the ratio of a stored quantity to a driving force. In terms of the tank, this means that the capacitance is the ratio of the volume of liquid stored to the driving force, which is the level of the liquid; this ratio is the cross-sectional area of the tank. On the other hand, if the cross-sectional area is variable, then the capacitance is not the ratio of the holdup to the level, and the following more general definition must be employed:

$$\text{tank capacitance} = \frac{dV}{dh} = A \tag{14}$$

where A is a function of h

As may be inferred from the transfer function presented in Equation 12, the transfer function relating voltage to current for a capacitor is $1/Cs$, since h is analogous to voltage and F_i is analogous to current. The example treated here illustrates a flow- or current-source situation, supplying only a capacitor. The equivalent electrical circuit is shown in Figure 4, in which F_i represents the change or deviation in inlet flow and h represents the change in liquid head in the tank. The rationale for this circuit can be seen from

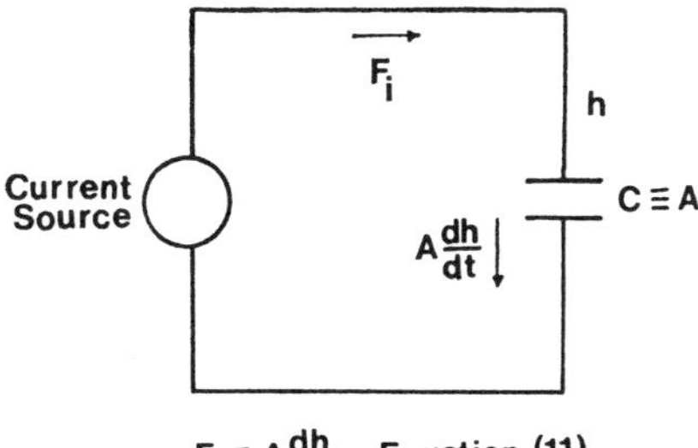

Figure 4. Electrical analogy of storage tank driven by a flow source.

Equation 11, which is repeated in the figure. The Δ's have been omitted from the equation for convenience.

IV. FIRST-ORDER PROCESSES BASED ON MASS BALANCES

A. Definition of a First-Order Process

A first-order process is one that is satisfactorily modeled by a linear, first-order differential equation. As we have seen in Chapter 5, this equation has the general form

$$\tau \frac{dy}{dt} + y = Kx \tag{15}$$

where y = output or response variable
x = input or forcing function
τ = time constant
t = time
K = gain factor

The transfer function is obtained by Laplace transformation with a zero initial condition:

$$\frac{Y}{X} = \frac{K}{\tau s + 1} \tag{16}$$

When the input is a unit step, the response is given by the equation

$$y = K(1 - e^{-t/\tau}) \tag{17}$$

This response is shown graphically in Figure 5.

In this chapter, the word "process" is used loosely to include such components of the system as valves and sensing elements, as well as the

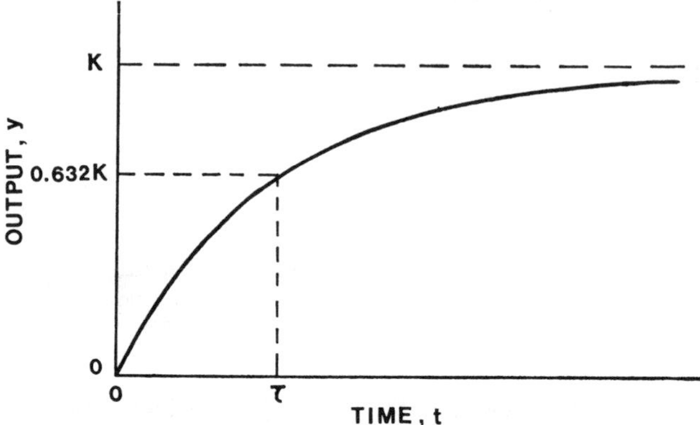

Figure 5.   First-order response to a unit step input.

process being controlled. Therefore, the complete control system will likely consist of a number of "processes." It will be apparent as we proceed through this and succeeding chapters that few, if any, real processes are truly first order, especially over wide ranges of operating conditions. However, for purposes of control system design and analysis, first-order models may be perfectly satisfactory. Furthermore, they are the logical starting point for the development and understanding of more complex models.

In the development of a transfer function, an unsteady-state balance is required—either a momentum, a mass, or an energy balance. Only a few problems in control involve momentum balances. One is treated at the end of Chapter 9 and a second more involved case is discussed in Chapter 11. The bulk of the problems involve mass and energy balances. Because of the large number of processes to be covered, those concerned with mass balances will be treated in this chapter and those concerned with energy and momentum balances will be discussed in the next.

B. Liquid Storage Tanks

The dynamic behavior of the storage tank of Figure 1 will now be examined in detail. Liquid enters the tank at F_i ft^3/sec and leaves at F_o ft^3/sec. Unlike the pure capacitance example just discussed, the leaving flow rate depends on the level, h, which is the driving force for flow through the exit line and valve.

1. Identification of the Desired Transfer Functions

As shown in Figure 1, the valve might be a pneumatic control valve, or it could be a simple hand-operated one. The flow rate from the tank is a function of both the valve position and the liquid level. On the other hand, the entering rate, F_i, is some arbitrary function of time.

For the sake of discussion, we shall assume that the level is to be controlled by manipulation of the stem position of the valve. Then the feed rate becomes a load variable. In broadest terms, the block diagram takes the form shown in Figure 6. The pressure to the valve, P_V, determines the stem position, which in turn determines the area for flow. From the diagram, it can be seen that the transfer functions of interest will be h/P_V for Process 1 and h/F_i for Process 2.

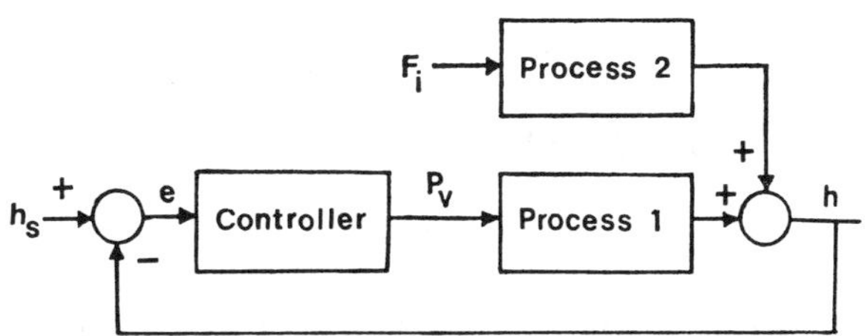

Figure 6.   Block diagram for level control of storage tank.

Broadly speaking, level control falls into one of two categories. In the first, the objective is to maintain the level itself under close control. For example, in a continuous-flow, stirred-tank reactor, it is important to maintain the liquid level constant so that the reactor holdup time and area for heat transfer are constant. As a second example, the reboiler of a distillation column is often under level control to insure that the boiler tubes are always submerged.

The second category of level control arises in the case of surge tanks. A surge tank serves as a storage element between two processing units. It is used to permit a relatively constant feed rate to the downstream unit even though the feed rate to the tank varies over a rather wide range. The level is controlled, rather than the leaving flow rate itself, to avoid the possibility of having the tank overflow or run dry.

2. Derivation of Transfer Functions

a. UNSTEADY-STATE MATERIAL BALANCE

The analysis begins with the unsteady-state mass balance for the tank:

$$F_i - F_o = A \frac{dh}{dt} \tag{18}$$

These variables can be written in terms of the normal value plus a deviation:

$$F_i = \bar{F}_i + \Delta F_i \tag{19}$$

$$F_o = \bar{F}_o + \Delta F_o \tag{20}$$

$$h = \bar{h} + \Delta h \tag{21}$$

The steady-state equation is

$$\bar{F}_i - \bar{F}_o = 0 \tag{22}$$

Substituting Equations 19 to 21 into Equation 18, and subtracting Equation 22, we obtain

$$\Delta F_i - \Delta F_o = A \frac{d\,\Delta h}{dt} \tag{23}$$

or, dropping the Δ notation for convenience:

$$F_i - F_o = A \frac{dh}{dt} \tag{24}$$

where the variables are now deviations. Because of the linearity of the original total differential equation given in Equation 18, Equation 24 is identical with it.

b. EFFECT OF THE CONTROL VALVE ON LEAVING FLOW RATE

Valves are essentially square-root devices; that is, the flow rate through a valve is proportional to the square root of the driving force across the valve. This same behavior is exhibited by orifice and venturi meters, and also by turbulent flow through pipes and fittings over small ranges of flow rate. Thus, for *fixed* valve position, the flow rate through the valve is related to the tank level as shown in Figure 7. The proportionality constant, c_{P_V}, is dependent on the valve pressure, P_V. When the variations in pressure to the valve are taken into account, a family of curves results, one curve for each valve pressure. This is illustrated in Figure 8 for an air-to-open valve.

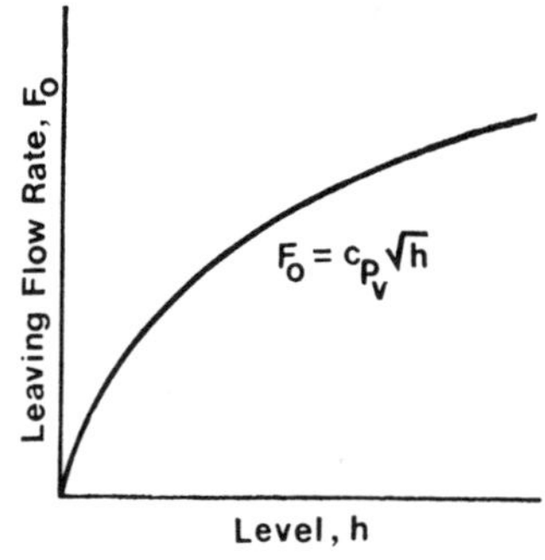

Figure 7. Flow rate versus liquid level for fixed valve position.

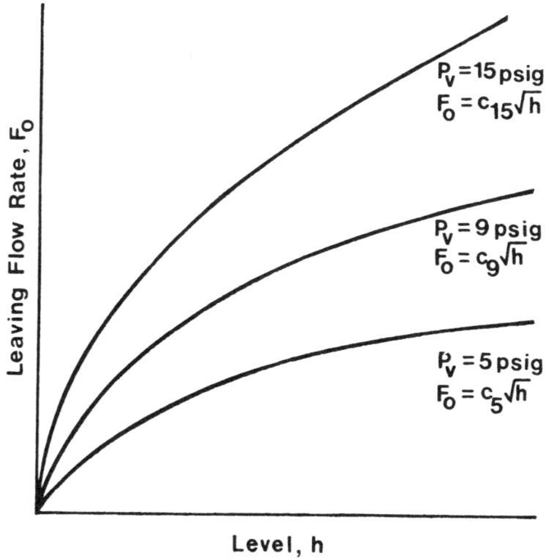

Figure 8.   Flow rate versus liquid level with valve pressure as a parameter.

Since F_o is a function of the two variables, P_V and h, changes in it can be expressed by a Taylor expansion in terms of these two variables. If it is assumed that deviations in P_V and h from the normal values will be small, then the change in leaving flow rate can be approximated by the linear terms of the expansion

$$F_o = \underbrace{\frac{\partial F_o}{\partial P_V} P_V}_{F_{o_1}} + \underbrace{\frac{\partial F_o}{\partial h} h}_{F_{o_2}} \qquad (25)$$

In this equation, F_o, P_V, and h are deviations from their normal values and the partial derivatives are evaluated at the normal valve pressure and normal tank level. The two terms on the right have been identified by subscripts to indicate that one is the change in leaving flow rate resulting from the change in valve pressure, and the other is the change resulting from the change in level.

At a fixed level, the change in flow with valve pressure is determined by the valve trim and the flow resistances in the line; this sensitivity of the flow rate to the valve pressure is expressed mathematically by the partial derivative, $\partial F_o/\partial P_V$, which is the valve gain, K_V:

$$K_V = \frac{\partial F_o}{\partial P_V} \qquad (26)$$

At a fixed valve position, the change in flow with level can be calculated in the following way. The flow rate is proportional to the square root of the level:

$$F_o = c_{P_V}\sqrt{h} \tag{27}$$

where c_{P_V} is the valve coefficient at the normal valve pressure. Taking the partial derivative of F_o with respect to h, we obtain:

$$\frac{\partial F_o}{\partial h} = \frac{1}{2}\frac{c_{P_V}}{\sqrt{h}}$$

Note that this derivative has been evaluated at the normal level, $\bar{h}$. But from Equation 27

$$c_{P_V} = \frac{\bar{F}_o}{\sqrt{\bar{h}}}$$

Therefore

$$\frac{\partial F_o}{\partial h} = \frac{\bar{F}_o}{2\bar{h}} \tag{28}$$

C. FORMULATION OF TRANSFER FUNCTIONS

Substitution of Equation 25 into Equation 24 yields

$$F_i - \frac{\partial F_o}{\partial P_V}P_V - \frac{\partial F_o}{\partial h}h = A\frac{dh}{dt} \tag{29}$$

This can be rearranged into the standard form:

$$\frac{A}{\partial F_o/\partial h}\frac{dh}{dt} + h = \frac{\partial h}{\partial F_o}F_i - \frac{\partial F_o/\partial P_V}{\partial F_o/\partial h}P_V \tag{30}$$

The time constant, load gain factor, and process gain factor are apparent; that is:

$$\tau = \frac{A}{\partial F_o/\partial h} \tag{31}$$

$$K_P = -\frac{\partial F_o/\partial P_V}{\partial F_o/\partial h} \tag{32}$$

$$K_L = \frac{\partial h}{\partial F_o} \tag{33}$$

Substitution of these factors and transformation of Equation 30 gives

$$(\tau s + 1)h = K_L F_i + K_P P_V \tag{34}$$

The transfer function, h/P_V, is found by setting F_i equal to zero. The result is

$$\frac{h}{P_V} = \frac{K_P}{\tau s + 1} \tag{35}$$

Similarly, the transfer function, h/F_i, is found by setting P_V equal to zero. Hence:

$$\frac{h}{F_i} = \frac{K_L}{\tau s + 1} \tag{36}$$

Both of these transfer functions have the standard first-order form given in Equation 16. Note that the time constant for changes in P_V is the same as that for changes in F_i.

3. The Block Diagram

The block diagram can now be detailed as shown in Figure 9.

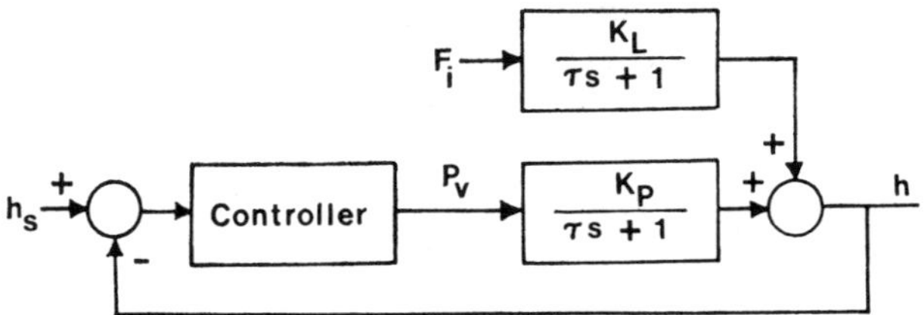

Figure 9.   Detailed block diagram for level control system.

In the design of this control system, the chief considerations will be the cross-sectional area of the tank, its height, and the characteristics of the valve. The process gain was given in Equation 32:

$$K_P = -\frac{\partial F_o/\partial P_V}{\partial F_o/\partial h} \tag{37}$$

As noted in Equation 26, the numerator is the valve gain, K_V. For purposes of analysis and understanding, the process transfer function can be separated into two parts, as indicated in Figure 10. One purpose of making this split

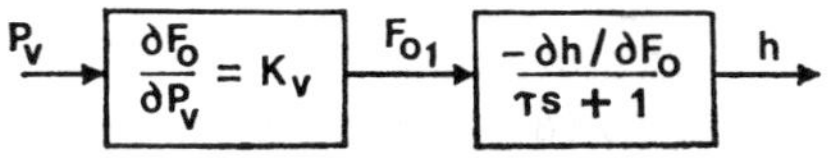

Figure 10.   Two-block representation of process.

is to emphasize that one part of the process gain is the valve gain, K_V, which is closely related to the valve trim. Another purpose is to point out that the variable between the two blocks is F_{o_1}, which is the contribution to the change in leaving flow caused by the change in pressure to the valve. The gain factor in the second block, $-\partial h/\partial F_o$, might be referred to as the "tank gain, K_P'" and is obtained from Equation 28 if the flow is turbulent.

4. Electrical Analogy of the Process

First-order behavior is often indicative of the presence of a combination of capacitance and resistance, and this is the case here. The capacitance is readily identified as the tank and the equation for this was given in Equation 14:

$$C = \frac{dV, \text{ft}^3}{dh, \text{ft}} = A \tag{14}$$

There is resistance to flow in both the exit line and the valve, but most of the resistance occurs at the valve. As shown in Table 3, the resistance is defined as the ratio of the driving force to the flow rate. For consistency with the definition of the tank capacitance, the driving force is expressed in the units of feet of fluid flowing. Therefore, the resistance is

$$R = \text{resistance to fluid flow} = \frac{\text{change in driving force for flow, } h, \text{ ft}}{\text{change in flow rate, } F_o, \text{ ft}^3/\text{min}}$$

$$= \frac{\partial h}{\partial F_o} \tag{38}$$

As in the case of capacitance, the resistance analogy is not defined as the ratio of the driving force to the flow rate, but instead as the ratio of the changes in these quantities. From Equation 28, we see that the resistance is $2\bar{h}/\bar{F}_o$, which could vary appreciably depending on the normal operating conditions.

The time constant and various gains can be written in terms of these analogies. Thus, using Equations 14, 26, and 38, Equations 31 to 33 respectively become:

$$\tau = RC \tag{39}$$

$$K_P = -K_V R \tag{40}$$

$$K_L = R \tag{41}$$

The development of the electrical analogy for this process begins with the material balance given by Equation 29. This equation can be expressed in terms of the resistance, capacitance, and valve gain, using Equations 14, 26, and 38. The result is

$$F_i - K_V P_V - \frac{h}{R} = C \frac{dh}{dt} \tag{42}$$

The first term on the left, F_i, is a variable flow rate that originates from a flow source; in an electrical network, this would be a current source. The second term is the change in flow rate caused by a change in valve pressure. The last term on the left represents the change in flow rate through the valve caused by a change in level. This change in level is analogous to a change in voltage in an electrical circuit. The electrical analogy is shown in Figure 11. Equation 42 is repeated in the figure as an aid in understanding the analogy.

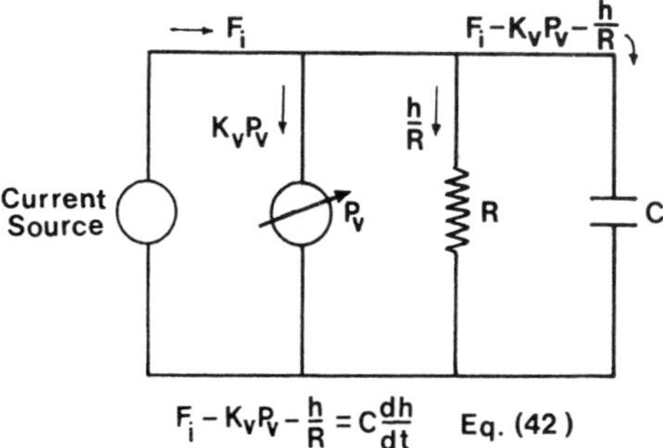

Figure 11.   Electrical analogy for a storage tank.

As just mentioned, the flow labeled $K_V P_V$ is the change caused by a change in pressure to the valve. If this pressure were constant, then there would be no change, and that leg of the circuit would be removed. In effect, this leg acts as a variable shunt across the tank capacitance. Actually, it is not a resistance since the flow through it is not dependent on the head, h, but upon the value of P_V. The symbol of a circle with an arrow has been used to indicate the variable nature of this leg.

A significant feature of this combination of a tank and a valve is that all of the capacitance of the process is associated with the tank while most of the resistance is associated with the valve. Actually, there is some resistance to flow in the lines and a small entrance loss as the fluid enters the line from the tank. It is apparent from the physical picture of the process that these resistive losses are in series with each other, so from a conceptual point of view, all of them can be considered to occur at the valve. In effect, the capacitance and resistance are "lumped" at specific locations rather than distributed along the process. Such a process is referred to as a "lumped parameter" process. Cases of distributed resistance, capacitance, and inertance are common and usually are difficult to handle analytically. Chapter 11 is devoted to some of these.

5. *Example 1*

A cylindrical tank, 10 ft in diameter and 15 ft high, is acting as a surge tank between two processing units. The level in the tank is under proportional control by manipulating the flow leaving the tank. The normal level is 7 ft and the normal flow rate is 55 ft^3/min. The span of the level transmitter is 15 ft, which is set between 0 and 15 ft. The normal pressure to the valve, $\bar{P}_V$, is 9 psig, the valve trim is linear, and the valve is an air-to-close type. The controller gain is 5 psi/psi.

(a)   Draw the block diagram of the system showing all transfer functions.
(b)   Calculate the overall gain, K_o, of the system and check its dimensions.
(c)   If F_i changes from 55 ft^3/min to 58 ft^3/min, calculate the ultimate change in tank level using the gains that have been determined.
(d)   In view of the application of this tank as a surge unit, would a high or low controller gain be advisable? Explain.

Solution:

(a)

$$\frac{\partial F_o}{\partial h} = \frac{\bar{F}_o}{2\bar{h}} = \tfrac{55}{14} = 3.93 \text{ ft}^3/\text{min/ft}$$

Since the valve is closed at 15 psig and the flow is 55 ft^3/min at 9 psig, the valve gain is:

$$K_V = \frac{\partial F_o}{\partial P_V} = \frac{-55\text{ft}^3/\text{min}}{(15-9)\text{ psi}} = -9.17\frac{\text{ft}^3/\text{min}}{\text{psi}}$$

$$A = 25\pi = 78.6 \text{ ft}^2$$

$$\tau = \frac{A}{\partial F_o/\partial h} = \frac{78.6 \text{ ft}^2}{3.93 \text{ ft}^3/\text{min/ft}} = 20 \text{ min} \qquad \text{(Equation 31)}$$

$$K_P = \frac{\partial F_o/\partial P_V}{\partial F_o/\partial h} = \frac{-9.17\,\text{ft}^3/\text{min/psi}}{3.93 \text{ ft}^3/\text{min/ft}} = -2.33\,\text{ft/psi} \qquad \text{(Equation 32)}$$

$$K_L = \frac{\partial h}{\partial F_o} = \frac{1}{3.93 \text{ ft}^3/\text{min/ft}} = 0.255 \frac{\text{ft}}{\text{ft}^3/\text{min}} \qquad \text{(Equation 33)}$$

$$K_T = \frac{(15-3)\text{ psi}}{(15-0)\text{ ft}} = 0.8 \text{ psi/ft}$$

$$K_C = 5 \text{ psi/psi} \quad \text{(given)}$$

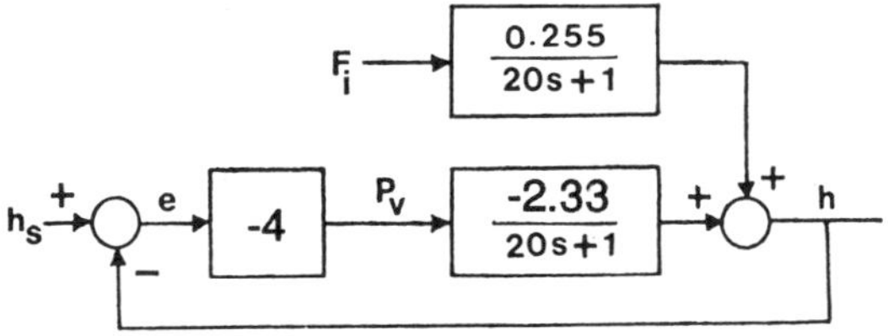

Figure 12.   Block diagram for Example 1.

The complete block diagram is presented in Figure 12. In the preceding calculation of K_P using Equation 32, the negative sign has been disregarded since the controller action is set for negative feedback, as indicated on the block diagram. The controller action can always be set for either positive or negative action to achieve negative feedback, and therefore all gains in the loop are taken as positive numbers.

(b) $$K_o = K_C K_T K_P = 5 \text{ psi/psi} \times 0.8 \text{ psi/ft} \times 2.33 \text{ ft/psi}$$
$$= 9.32 \text{ dimensionless}$$

(c) The change in level can be found by analogy to Equation 37 of Chapter 4:

$$h = \frac{K_L}{1 + K_o} F_i$$

F_i here is the change in feed rate, which is 3 ft³/min, and h is the ultimate change in level. Substituting, we obtain

$$h = \frac{0.255}{1 + 9.32} \, 3 = 0.0741 \text{ ft}$$

(d) The purpose of a surge tank is to minimize the effect of disruptions in the flow from the upstream unit on the flow to the downstream unit. Rapid changes in F_o would presumably have a detrimental effect on the operation of the following unit. Therefore, the valve should change slowly as the level changes, and a low controller gain is required.

6. Self-Regulation

A tank that empties through a valved line, or even an open line, displays a property commonly referred to as "self-regulation." This behavior is not exhibited by a tank that empties through a constant-displacement pump. To illustrate this property, consider a tank with a valved line. Suppose there is a sudden step increase in feed rate that persists for an indefinite time. The level will begin to increase in accordance with the typical first-order behavior shown in Figure 5. As the level rises, the leaving flow rate also increases and eventually a new level is reached that is sufficient to maintain the leaving flow rate equal to the feed rate. Thus, the tank possesses self-regulation that prevents it from overflowing if the step change is not too large.

Turning now to the tank with the constant-displacement pump, assume that prior to the step change, the entering and leaving flow rates are equal, so the level is constant. After the change, the level will continue to rise linearly with time and eventually the tank will overflow. Thus, this purely capacitive process displays no self-regulation.

In the case of the first-order process, we see that the resistance of the valve has the effect of providing some negative feedback that tends to "control" the level, even though there is no actual process controller present. This point is more fully developed in Problem 5 at the end of this chapter.

7. Closed Tank with Liquid Under Constant Pressure

The tank considered in Figure 1 is open to the atmosphere and empties through a line also open to the atmosphere. Quite frequently, a tank is closed to the atmosphere so the pressure in the tank is different from that on the outlet side of the valve. In the usual situation, the tank pressure is greater than the outlet pressure.

Consider the flow characteristic of a valve shown in Figure 13a. The flow has been presumed to be turbulent, so that the flow rate is proportional

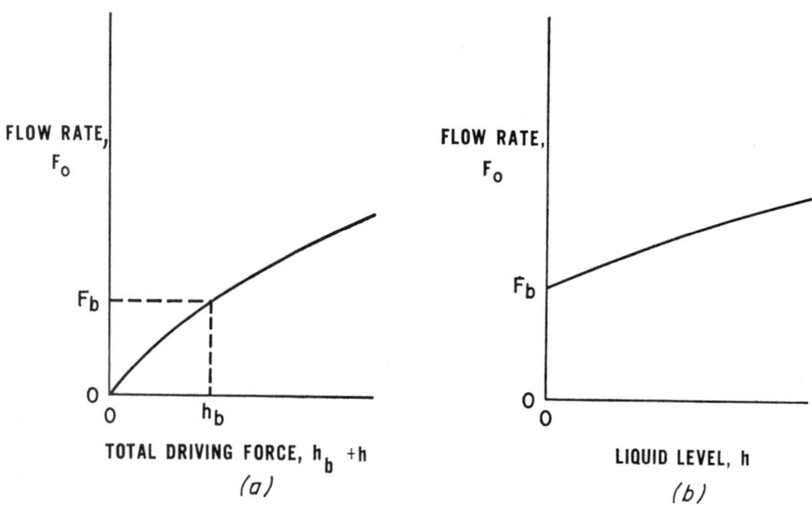

Figure 13.   Flow rate characteristics of a valve. (a) Flow rate versus total driving force across valve. (b) Flow rate versus liquid level.

to the square root of the level. The abscissa represents the total driving force for flow, which is the sum of the liquid head in the tank plus the pressure head from the gas. If the outlet pressure is atmospheric, then the pressure head of the gas is proportional to the gage pressure:

$$h_b = \frac{P_b g_c}{\rho g} \tag{43}$$

where h_b = pressure head, ft of fluid flowing
  P_b = gage pressure, lb_f/ft^2
  g = acceleration of gravity, 32.17 ft/sec^2
  g_c = Newton's law conversion factor, 32.17 lb_m, ft/lb_f, sec^2
  ρ = liquid density, lb_m/ft^3

In Figure 13a, a specific value of h_b is shown. In effect, this gas pressure biases the flow rate so that it can never be less than F_b. Assuming that h_b is constant, a correlation of flow rate versus liquid level takes the form shown in Figure 13b. Equation 28 can be used for the calculation of $\partial F_o/\partial h$, but the total head, $h_b + \bar{h}$, is used rather than the liquid level alone; thus Equation 28 is modified to the following form:

$$\frac{\partial F_o}{\partial h} = \frac{\bar{F}_o}{2(h_b + \bar{h})} \tag{44}$$

8. Multiple Outlets

Consider the tank shown in Figure 14 in which there are two outlet lines. For generality, suppose that the outlet pressures are also variables. These have been shown as the pressure heads, h_{o_1} and h_{o_2}. Valve No. 1 is pneumatically operated, but Valve No. 2 is hand-operated and is assumed to remain in a fixed position. The flow rates through the two valves are linear in the driving

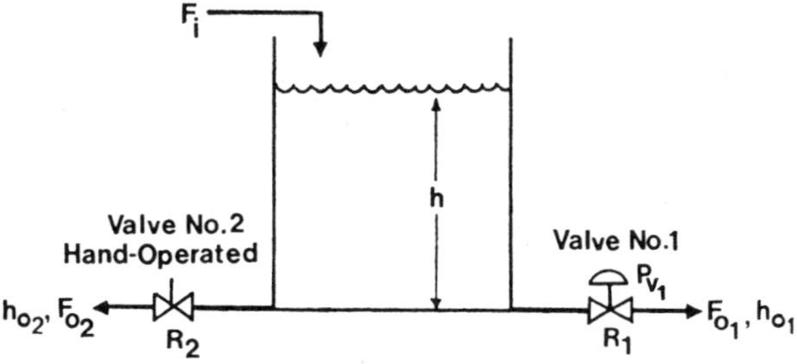

Figure 14.   Storage tank with two outlets.

forces, $h - h_{o_1}$ and $h - h_{o_2}$, for small changes. Thus the material balance can be written in terms of the deviations in F_i, h, h_{o_1}, h_{o_2}, and P_{V_1} as follows:

$$F_i - \frac{h - h_{o_1}}{R_1} - \frac{h - h_{o_2}}{R_2} - K_V P_{V_1} = C\frac{dh}{dt} \tag{45}$$

No term is included for the effect of a change in valve position of Valve No. 2 since it is fixed.

The utility of an electrical analogy is nicely illustrated by the case in which the outlet pressures, h_{o_1}, and h_{o_2}, are constant. Then the two deviation terms corresponding to these pressures drop out of Equation 45 to yield the following equation:

$$F_i - \frac{h}{R_1} - \frac{h}{R_2} - K_V P_{V_1} = C\frac{dh}{dt} \tag{45a}$$

This equation and its electrical analogy are presented in Figure 15. The two resistances are in parallel. It may be recalled from heat transfer theory or

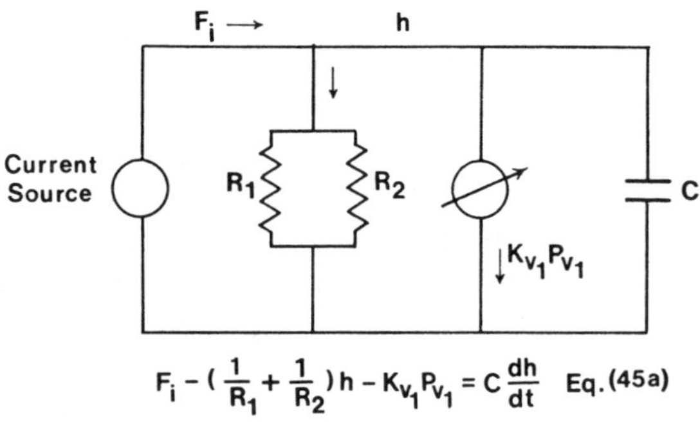

Figure 15.   Electrical analogy of Figure 14.

from electrical theory that the equivalent resistance, R_e, to a parallel combination can be calculated from the following relation:

$$\frac{1}{R_e} = \frac{1}{R_1} + \frac{1}{R_2} \quad \text{or} \quad R_e = \frac{R_1 R_2}{R_1 + R_2} \tag{46}$$

Hence, the transfer functions, h/F_i and h/P_{V_1}, could be found from Equations 39 to 41 with the substitution of R_e for R.

9. *Example 2*

Suppose the objective of the control system for Example 1 is to maintain the level at 7 ft in spite of flow variations between 35 and 75 ft^3/min. The normal flow rate is 55 ft^3/min.

(a) What kind of valve trim will maintain the gain of the process constant over this range of flow rates?
(b) Calculate the time constant and holdup time at each of the three flow rates of 35, 55, and 75 ft^3/min.
(c) Suppose the tank were enclosed and contained an inert gas having a constant pressure of 26 psig. If the flow rate is 55 ft^3/min of water and the level is 7 ft, calculate the tank gain, the time constant, and the holdup time.

Solution:

(a) The tank gain, K'_P, is given by Equation 28 when there is no biasing gas pressure above the liquid.

$$\text{at} \quad F_o = 35 \qquad K'_P = \frac{2\bar{h}}{\bar{F}_o} = \frac{2(7)}{35} = 0.4 \text{ ft/ft}^3/\text{min}$$

$$\text{at} \quad F_o = 55 \qquad K'_P = \frac{2(7)}{55} = 0.255 \text{ ft/ft}^3/\text{min}$$

$$\text{at} \quad F_o = 75 \qquad K'_P = \frac{2(7)}{75} = 0.187 \text{ ft/ft}^3/\text{min}$$

The tank gain is therefore inversely proportional to the flow rate; hence, if a valve trim can be chosen so that the valve gain will be proportional to the flow rate, the required constancy of the process gain will be met. Basically, this is the property of an equal percentage valve, as mentioned in Chapter 4:

$$K_V = \frac{dA_v}{dP_V} = kA_v$$

where A_v = area for flow through the valve port
  k = constant

For constant pressure drop across the valve, the flow rate is proportional to the area for flow. There are situations in which a constant process gain is not desirable, so this example only serves to demonstrate how a constant gain could be achieved if required.

It is important to recognize that the flow characteristics of the valve are not only a function of the valve trim, but also of the characteristics of the upstream and downstream lines connected to the valve. For example, suppose that the valve opening increases slightly. This will result in an increase in flow that will be accompanied by an increase in pressure drop through these lines. If the total pressure drop across the lines and valve is constant, then this means that the pressure drop across the valve must decrease. Therefore, the effect of the increase in opening is counteracted to some extent by the decrease in pressure drop across the valve. Thus, although the *inherent characteristic* of the valve is equal percentage, the *effective characteristic* will only be approximately equal percentage. A quantitative treatment of this subject can be found in Reference 3.

(b)   The time constant is given by Equation 31:

$$\tau = \frac{A}{\partial F_o / \partial h} = A K_P'$$

at $\quad F_o = 35 \text{ ft}^3/\text{min}$ $\qquad \tau = 78.6(0.4) = 31.4 \text{ min}$
at $\quad F_o = 55 \text{ ft}^3/\text{min}$ $\qquad \tau = 78.6(0.255) = 20.0 \text{ min}$
at $\quad F_o = 75 \text{ ft}^3/\text{min}$ $\qquad \tau = 78.6(0.187) = 14.7 \text{ min}$

The holdup time, T_H, is the ratio of the holdup of liquid to the throughput rate, F_o. Therefore

$$T_H = \frac{V}{F_o} = \frac{Ah}{F_o}.$$

at $\quad F_o = 35 \text{ ft}^3/\text{min}$ $\qquad T_H = \frac{78.6(7)}{35} = 15.7 \text{ min}$

at $\quad F_o = 55 \text{ ft}^3/\text{min}$ $\qquad T_H = \frac{78.6(7)}{55} = 10.0 \text{ min}$

at $\quad F_o = 75 \text{ ft}^3/\text{min}$ $\qquad T_H = \frac{78.6(7)}{75} = 7.33 \text{ min}$

Two interesting points come out of this. First, the time constant varies with the feed rate. Although this is perhaps no surprise, it is significant in that the response of the system to changes in feed rate can be expected to vary with the value of the feed rate. Second, for any given feed rate, the time constant is twice the holdup time. This is not a coincidence, as may be shown through the use of Equations 28 and 31.

(c)   The bias pressure, h_b, is given by Equation 43:

$$h_b = \frac{26(144)}{62.4} = 60.0 \text{ ft}$$

The tank gain, K_P', is given by Equation 44, since the liquid is now under gas pressure.

$$K_P' = \frac{2(\bar{h} + h_b)}{F_o} = \frac{2(7 + 60)}{55} = 2.44 \ \text{ft/ft}^3/\text{min}$$

$$\tau = K_P' A = 2.44(78.6) = 192 \ \text{min}$$

$$T_H = \frac{78.6(7)}{55} = 10.0 \ \text{min}$$

The ratio of the time constant to the holdup time is about $19:1$ here, whereas it was $2:1$ in part *b*. The presence of the inert gas greatly decreases the speed of response of the process. This sluggishness results from the greatly decreased property of self-regulation. Suppose there is an increase in feed rate. The level immediately begins to rise, but this rise is accompanied by a relatively small initial increase in flow out of the tank because the slope of the flow-versus-level curve is so flat in the region of operation. On the other hand, had there been no inert gas present, the initial rise in level would have been accompanied by a relatively rapid initial increase in flow because of the steepness of the curve.

C. Gas Storage Tanks and Pressurized Vessels

Processes involving gas flows and gas storage are very common in chemical plants. In some ways they are quite similar to liquid systems but important differences arise because gases are compressible while liquids are relatively incompressible.

The pressure in a tank is analogous to the liquid level in a tank. Pressure control systems fall into the same two categories that liquid level systems do. The pressure in a process may be important for its own sake, as for example in a gaseous reactor. On the other hand, the tank may be used for surge capacity, in which case the control of pressure is of secondary importance in comparison with the control of the flow rate of gas leaving the tank.

1. Some Complexities

In this section, we shall investigate the response of pressure in a tank to each of several disturbances. The development will be carried out in a way that is completely analogous to that used for liquids. Although most of the ideas presented for liquid systems carry over to gaseous systems, the actual details of the adaptation present some problems, which are anticipated by examining Figure 16. The tank pressure, P_2, is assumed to be under control by manipulating the position of either Valve 1 or Valve 2. The flow through the first valve is a function of both P_1 and P_2, and the flow through the second is a function of both P_2 and P_3.

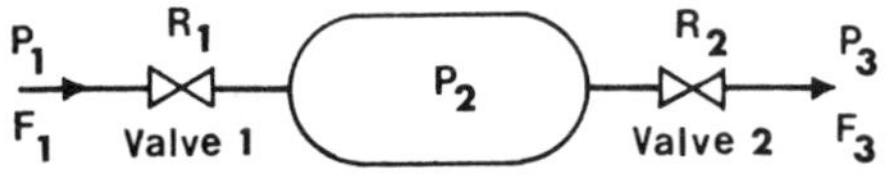

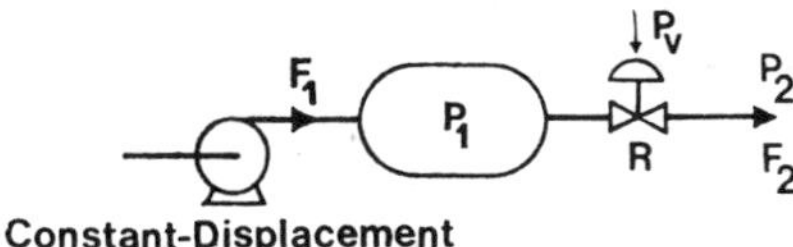

Figure 16.   Simple gas storage tank.

As in liquid flows, gas flows through lines and valves are usually turbulent. However, with gases there is the possibility of critical flow. Furthermore, gas flows can fall anywhere between isothermal and adiabatic. Relatively rigorous treatments of the behavior of gases in storage and flow systems can be found in References 4 to 6. For the purposes of a control analysis, more approximate methods are usually satisfactory.

2. Flow-Source, Single-Exit Example

We shall begin by examining a simple problem, closely analogous to the liquid-level example previously treated. Referring to Figure 17, a tank is fed with a gas from a constant-displacement compressor and the gas leaves the tank through a valve. The goal is to develop the transfer functions, P_1/P_2, P_1/F_1, and P_1/P_V.

Constant-Displacement
Pump

Figure 17.   Flow-source, single-exit gas storage tank.

a.   UNSTEADY-STATE MATERIAL BALANCE

As with liquid storage processes, the unsteady-state analysis for gas storage processes begins with the unsteady-state mass balance:

Rate of Mass Flow In $-$ Rate of Mass Flow Out $=$ Accumulation

$$(47)$$

Gas flow rates can be expressed in terms of a number of units. Although moles per minute or pounds per minute could be used, the units of standard cubic feet per minute, abbreviated "SCFM," are usually used. The conditions of pressure and temperature for defining a standard cubic foot are arbitrary.

One author (7) defines it in terms of the normal temperature and pressure of the vessel, and another (8) bases it on the normal temperature and one atmosphere. The two definitions are simply related through the ratios of the absolute pressures if ideal gas behavior is assumed. The basis of the normal temperature and one atmosphere will be used in this book. Thus, in a manner analogous to that of the liquid storage process, the unsteady-state material balance is

$$F_1 - F_2 = C\frac{dP_1}{dt} \tag{48}$$

where F_1 = flow in, SCFM
F_2 = flow out, SCFM
C = tank capacitance, SCF/psi
t = time, min

b. EFFECT OF THE VALVE ON THE LEAVING FLOW RATE

The leaving flow rate, F_2, depends on the upstream pressure, P_1, the downstream pressure, P_2, and on the valve position or pressure, P_V. As in Equation 25 for liquid flow through a valve, the change in F_2 can be approximated by the linear terms of a Taylor expansion in the three variables:

$$\Delta F_2 = \underbrace{\frac{\partial F_2}{\partial P_1}\Delta P_1}_{\Delta F_{2_1}} + \underbrace{\frac{\partial F_2}{\partial P_2}\Delta P_2}_{\Delta F_{2_2}} + \underbrace{\frac{\partial F_2}{\partial P_V}\Delta P_V}_{\Delta F_{2_3}} \tag{49}$$

The three partial derivatives are evaluated at the normal pressures, $\bar{P}_1$, $\bar{P}_2$, and $\bar{P}_V$.

The first partial derivative, $\partial F_2/\partial P_1$, is the slope of a plot of F_2 versus P_1 with P_2 and P_V at their normal values. A typical curve is shown in Figure 18 with the partial derivative taken at the normal tank pressure, $\bar{P}_1$. The flow drops to zero when the tank pressure is the same as the outlet pressure, $\bar{P}_2$. The region of sonic flow will be discussed later and it will be assumed here that the flow is below the sonic region.

As a means of obtaining the slope shown in Figure 18, the flow rate curve could be obtained from experimental measurements. However, this is not necessary if the analysis is confined to small changes in P_1. For liquids, it was pointed out earlier that the flow rate is proportional to the square root of the pressure drop. This relationship will approximately apply to gases, in spite of their compressibility, if the pressure in the tank does not change very much. Thus, in the region of the normal pressure drop, we can use a relationship similar to Equation 27, which applied to liquids:

$$F_2 = k_2\sqrt{P_1 - P_2} \tag{50}$$

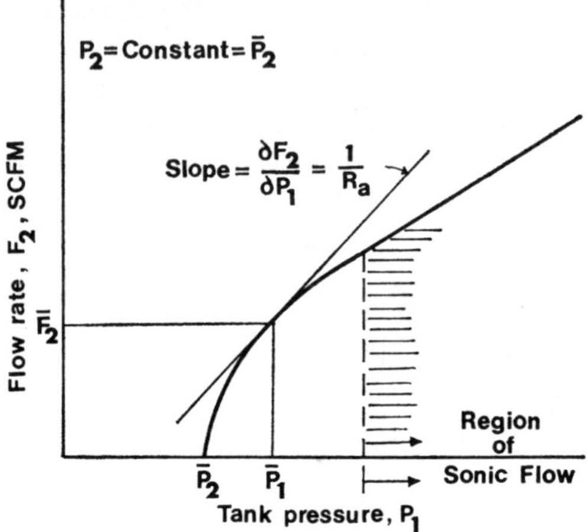

Figure 18.   Flow variation with tank pressure.

where k_2 = a constant for a given valve position

$$\frac{\partial F_2}{\partial P_1} = \frac{k_2}{2}\frac{1}{\sqrt{P_1 - P_2}}$$

But

$$k_2 = \frac{\bar{F}_2}{\sqrt{\bar{P}_1 - \bar{P}_2}} \tag{51}$$

Therefore

$$\frac{\partial F_2}{\partial P_1} = \frac{\bar{F}_2}{2(\bar{P}_1 - \bar{P}_2)} = \frac{1}{R_a} \tag{52}$$

This equation is analogous to Equation 28 for liquids. Since the partial derivative is the ratio of the change in flow to the change in upstream pressure, it is the analog of an inverse resistance which we shall call R_a.

The second partial derivative in Equation 49, $\partial F_2/\partial P_2$, is the slope of a plot of F_2 versus P_2 with P_1 and P_V at their normal values. This curve takes the form shown in Figure 19. Again, assuming that the flow is below sonic velocity, the slope at the normal pressure, $\bar{P}_2$, can be found using steps analogous to those followed in Equations 50 to 52; the result is

$$\frac{\partial F_2}{\partial P_2} = -\frac{\bar{F}_2}{2(\bar{P}_1 - \bar{P}_2)} = -\frac{1}{R_b} \tag{53}$$

This partial derivative is also in the nature of an inverse resistance, which is labeled $-R_b$. As can be seen in Figure 19, the slope is negative; hence, with a minus sign, R_b will be a positive quantity as resistances usually are.

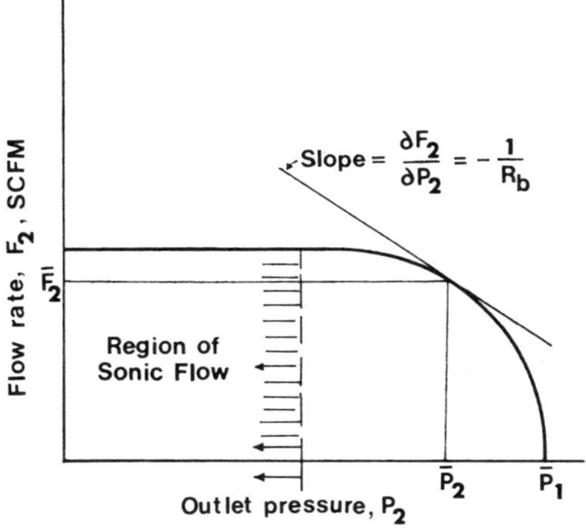

Figure 19.   Flow rate variation with outlet pressure.

Finally, the last partial derivative in Equation 49, $\partial F_2/\partial P_V$, is not an inverse resistance, because P_V is the pressure to the valve motor and not a driving force for flow; in fact, this derivative is the valve gain, K_V. Thus, using Equations 52 and 53, Equation 49 can be expressed as

$$\Delta F_2 = \frac{\Delta P_1}{R_a} - \frac{\Delta P_2}{R_b} + K_V \, \Delta P_V \tag{54}$$

The approximation sign of Equation 49 has been dropped for convenience, bearing in mind that the equation is an approximation.

From Equations 52 and 53 it can be seen that Equation 49 can be expressed as

$$\Delta F_2 = \frac{\overline{F}_2}{2\,\overline{\Delta P}} \Delta P_1 - \frac{\overline{F}_2}{2\,\overline{\Delta P}} \Delta P_2 + K_V \, \Delta P_V \tag{55}$$

where

$$\overline{\Delta P} = \overline{P}_1 - \overline{P}_2$$

and

$$K_V = \frac{\partial F_2}{\partial P_V}$$

or

$$\Delta F_2 = \frac{\Delta P_1 - \Delta P_2}{\dfrac{2\,\overline{\Delta P}}{\overline{F}_2}} + K_V \, \Delta P_V \tag{56}$$

A comparison of Equations 52 and 53 shows that R_a and R_b are equal, so that Equation 49 can be written

$$\Delta F_2 = \frac{\Delta P_1 - \Delta P_2}{R} + K_V \, \Delta P_V \tag{57}$$

where

$$R = R_a = R_b$$

This equivalence of the resistances, R_a and R_b, occurred because the flow was turbulent and subsonic. The resistances would also be equal if the flow were laminar, although this flow regime is physically very unlikely.

When the flow is in the sonic region, the resistances are no longer equal. Referring to Figure 19, the flow rate is zero when the outlet pressure is equal to the pressure in the tank. As the outlet pressure is reduced, the flow rate increases until sonic velocity is reached in the valve. Any further reduction in outlet pressure results in no further increase in flow rate. The slope of the partial derivative, $\partial F_2/\partial P_2$, in the sonic region is zero and the resistance, R_b, is infinite:

$$R_b = \infty \text{ in sonic flow} \tag{58}$$

The situation is somewhat different when the outlet pressure is held constant while the tank pressure increases. Turning to Figure 18, again the flow rate is zero when the tank and outlet pressures are equal. As the tank pressure is increased, the flow rate increases until sonic velocity occurs in the valve. Further increases in tank pressure result in nearly proportionate increases in flow rate *in terms of SCF*. This proportionality occurs because the upstream gas density increases with increasing tank pressure, so the number of SCF passing through the line can still increase in spite of the constancy of sonic velocity in the valve. The slope in the sonic region is a constant, and furthermore, if extrapolated to zero tank pressure, passes through the origin. Thus, the resistance in sonic flow, R_s, is given by the expression

$$R_s = \frac{\bar{P}_1}{\bar{F}_2} \quad \text{in sonic flow} \tag{59}$$

where $\bar{P}_1$ is the *absolute* pressure. Thus, when the flow is sonic, Equation 54 takes the form

$$\Delta F_2 = \frac{\Delta P_1}{R_s} + K_V \, \Delta P_V \quad \text{in sonic flow} \tag{60}$$

Sonic velocity occurs in a valve when the critical pressure ratio is reached. This ratio is given by the equation

$$\frac{P_c}{P_1} = \left(\frac{2}{\delta + 1} \right)^{\delta/(\delta-1)} \tag{61}$$

where δ is the ratio of the heat capacity at constant pressure to that at constant volume. The critical pressure ratio for monatomic gases is 0.49, 0.53 for diatomic gases, and slightly higher for more complex gases. For saturated steam, the ratio is about 0.58 and about 0.55 for moderately superheated steam.

c. CAPACITANCE OF TANK

For a liquid storage tank of constant cross-sectional area, the capacitance can be thought of as the ratio of the volumetric holdup to the level in the tank. In an analogous way, the capacitance of a gas tank can be thought of as the ratio of the volume of the tank to the pressure, providing that the temperature of the tank remains constant. Usually, unsteady-state behavior in a gas storage tank falls between isothermal and adiabatic. The isothermal case is the simpler of the two and will be taken up first.

(1) *Isothermal Behavior*

Suppose that the normal temperature of a tank is 70°F. Then a SCF, as we have defined it, is the amount of gas contained in a cubic foot at 70°F and 14.7 psia. If the tank is cubical with sides 10 ft long, it will hold 1000 SCF when the pressure is 14.7 psia. For isothermal conditions, the capacitance of the tank is the number of these standard cubic feet it can hold per psi, namely 1000/14.7 or 68.0 SCF/psi.

(2) *Adiabatic Behavior*

When pressure changes occur in the tank under adiabatic conditions, the capacitance is not the simple ratio of the volume to the standard pressure of 14.7 psia. Adiabatic behavior will occur if the tank is well insulated, or if fairly rapid changes in pressure occur.

The capacitance of a tank under adiabatic conditions would be expected to be less than that under isothermal conditions. This can be seen as follows. Suppose there is a slight increase in pressure. This will be accompanied by a slight increase in temperature as a result of the adiabatic compression. At this slightly above normal temperature, the tank cannot hold as many SCF as it could at its normal temperature, so its capacitance is less than that for isothermal behavior.

The derivation of the adiabatic capacitance proceeds as follows. The capacitance is given by the general capacitance equation

$$C = \frac{d(\text{mass})}{d(\text{driving force})} = \frac{d(\rho V)}{dP_1} \tag{62}$$

where V = volume of tank, ft^3

ρ = gas density in SCF/ft^3 of tank volume where a SCF is based on the normal temperature and 1 atmosphere

An adiabatic expansion or compression is commonly described by

$$\frac{P_1}{\rho^n} = \text{constant}, \beta \tag{63}$$

where n = constant (polytropic exponent)

Therefore

$$\frac{P_1 V^n}{\rho^n} = \beta V^n \tag{64}$$

It follows that

$$\rho V = \frac{V P_1^{1/n}}{\beta^{1/n}} \tag{65}$$

Substitution of Equation 65 into Equation 62 yields

$$C = \frac{V\rho}{nP_1} \tag{66}$$

But for ideal gas behavior at the normal temperature, the density is proportional to the pressure

$$\rho = \frac{\text{number of SCF}}{\text{ft}^3 \text{ of tank volume}} = \frac{P_1, \text{psi}}{14.7, \text{psi (ft}^3/\text{SCF)}}$$

Therefore, Equation 66 can be rewritten as

$$C = \frac{V}{n(14.7)} \tag{67}$$

When the expansion or compression is isothermal, $n = 1$, so this result reduces to the relationship previously deduced. When the process is adiabatic and reversible, then it is said to be isentropic and n is the ratio of the heat capacity at constant pressure to that at constant volume. Eckman (9) points out that many tests have shown that the polytropic exponent falls between 1.0 and 1.2 for uninsulated metal vessels at common pressures and temperatures. Isothermal behavior will be assumed in all illustrations in this text.

d. FORMULATION OF TRANSFER FUNCTIONS

Having considered the details concerning the valve and the tank, we return now to Equation 48 to make the appropriate substitutions. The variables are first written in terms of the normal value plus a deviation:

$$\bar{F}_1 + \Delta F_1 - \bar{F}_2 - \Delta F_2 = C\frac{d(\bar{P}_1 + \Delta P_1)}{dt} \tag{68}$$

But

$$\bar{F}_1 - \bar{F}_2 = 0 \qquad \text{(original steady-state equation)} \tag{69}$$

and

$$\frac{d\bar{P}_1}{dt} = 0$$

Subtraction of the steady-state equation results in

$$\Delta F_1 - \Delta F_2 = C\frac{d\,\Delta P_1}{dt} \tag{70}$$

If the flow through the valve is subsonic, substitution of Equation 57 into Equation 70 yields

$$\Delta F_1 - \frac{\Delta P_1 - \Delta P_2}{R} - K_V\,\Delta P_V = C\frac{d\,\Delta P_1}{dt} \tag{71}$$

No ambiguity will result at this point if the Δ's are omitted for convenience; the variables now are all deviations from the original steady-state values. Rearrangement to the standard form gives

$$RC\frac{dP_1}{dt} + P_1 = P_2 + RF_1 - K_V RP_V \tag{72}$$

Laplace transformation of Equation 72 yields the three transfer functions:

$$\frac{P_1}{P_2} = \frac{1}{\tau s + 1} \qquad \text{gain factor} = 1 \tag{73}$$

$$\frac{P_1}{F_1} = \frac{R}{\tau s + 1} \qquad \text{gain factor} = R \tag{74}$$

$$\frac{P_1}{P_V} = \frac{-K_V R}{\tau s + 1} \qquad \text{gain factor} = -K_V R \tag{75}$$

where $\tau = RC$

Suppose, for example, that the tank is a reactor whose reactants are a gas mixture fed at a fairly constant rate from a previous process or storage tank. Pressure control is achieved by manipulation of the valve position. The corresponding block diagram is shown in Figure 20. The main load variable is the feed rate, F_1. If P_2 is atmospheric pressure, this load variable will probably not be significant. Furthermore, we have seen that if flow through the line is sonic, then P_2 will cease to be a load variable at all.

e. ELECTRICAL ANALOGY

The electrical analogy for this case is only slightly different from that presented for a liquid storage tank with a valve in the exit line. It is most easily

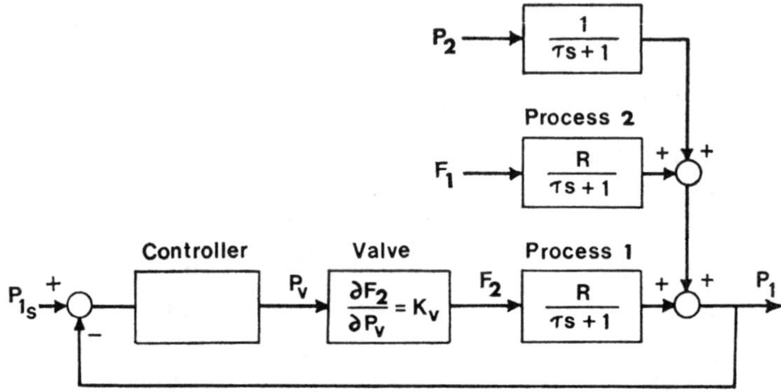

Figure 20. Block diagram of pressure control system for gaseous reactor.

drawn by looking at the unsteady-state material balance in the form where each term represents a flow rate. For subsonic flow, Equation 71 is in this form. The electrical analogy is presented in Figure 21 and the equation is shown as an aid in understanding the analogy. The Δ's have been omitted from the variables for convenience.

Although this process was introduced as a flow-source example, the pressure, P_2, is analogous to a voltage as can be seen by referring back to Table 1. Thus, P_2 appears in Figure 21 as a voltage source and this circuit

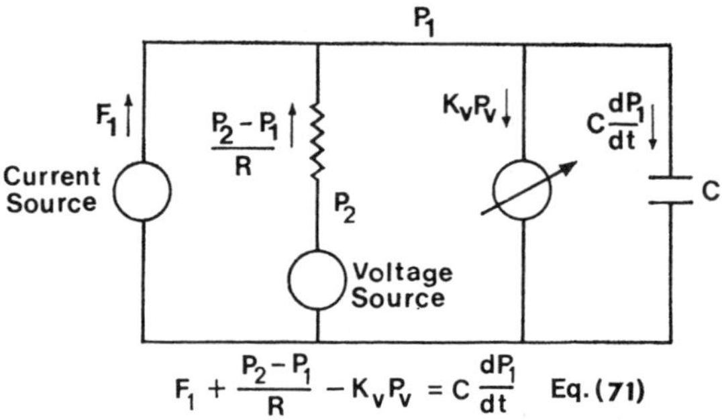

Figure 21. Electrical analogy of Figure 17 for subsonic flow.

has both kinds of active elements. If there are no changes in P_2, then this source is omitted, although the resistance, R, remains in this leg; on the other hand, if there are no changes in F_1, then the current-source leg is omitted.

3. Pressure-Source, Multiple-Exit Example

In the previous example, the feed gas was supplied by a constant-displacement pump, which is a flow source. Quite frequently, the feed gas is throttled through a valve from a high-pressure gas source. Since the feed rate is determined by the pressure of the source, this source is analogous to a voltage source rather than to a current source.

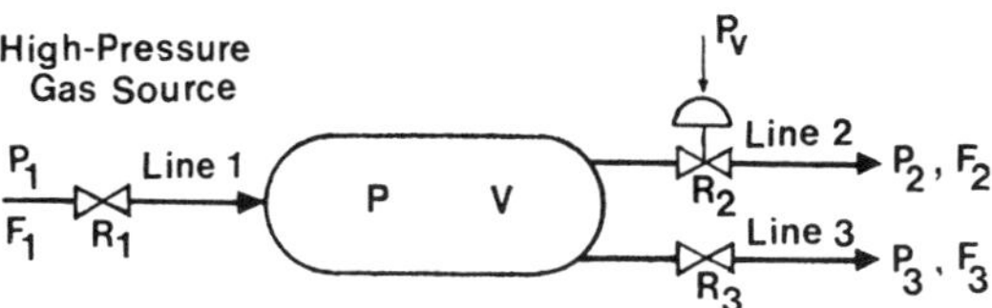

Figure 22.   Pressure-source, multiple-exit gas storage tank.

Consider the process shown in Figure 22. There is one source and two outlet lines. The material balance is:

$$F_1 - F_2 - F_3 = C\frac{dP}{dt} \tag{76}$$

where F_1, F_2, F_3, and P represent deviations from normal values. Each of these flows is then written in terms of pressure changes and resistances in the same manner as Equation 54:

$$\left(\frac{P_1}{R_{1a}} - \frac{P}{R_{1b}}\right) - \left(\frac{P}{R_{2a}} - \frac{P_2}{R_{2b}}\right) - \left(\frac{P}{R_{3a}} - \frac{P_3}{R_{3b}}\right) - K_V P_V = C\frac{dP}{dt} \tag{77}$$

The valve in Line 2 is a pneumatic control valve and the last term on the left accounts for changes in flow caused by changes in the valve pressure, P_v. Similar terms are not required for Lines 1 and 3 since the valves in these lines are hand operated and their stem positions are assumed to remain fixed. If the flow is subsonic in the feed line, then

$$R_{1a} = R_{1b} = \frac{2(\bar{P}_1 - \bar{P})}{\bar{F}_1}$$

Analogous statements apply to the other streams, providing the flow in each of them is subsonic. On the other hand, if the flow in the feed line is sonic, then R_{1b} is infinite and the term containing it drops out; R_{1a} is simply the ratio, $\bar{P}_1/\bar{F}_1$. Similar statements again apply to the other two streams if their flows are sonic.

For the sake of discussion, we shall assume for this example that flow is subsonic in all of the lines. Therefore, Equation 77 can be rewritten as

$$C \frac{dP}{dt} = \left(\frac{P_1 - P}{R_1} \right) - \left(\frac{P - P_2}{R_2} \right) - \left(\frac{P - P_3}{R_3} \right) - K_V P_V \tag{78}$$

Rearrangement to the standard form yields

$$\frac{C}{\frac{1}{R_1} + \frac{1}{R_2} + \frac{1}{R_3}} \frac{dP}{dt} + P = \frac{P_1}{R_1 \left(\frac{1}{R_1} + \frac{1}{R_2} + \frac{1}{R_3} \right)} +$$

$$\frac{P_2}{R_2 \left(\frac{1}{R_1} + \frac{1}{R_2} + \frac{1}{R_3} \right)} + \frac{P_3}{R_3 \left(\frac{1}{R_1} + \frac{1}{R_2} + \frac{1}{R_3} \right)} + \frac{-K_V P_V}{\frac{1}{R_1} + \frac{1}{R_2} + \frac{1}{R_3}} \tag{79}$$

The presence of sonic flow presents no difficulty. For example, if the flow in Line 3 were sonic, then R_{3b} in Equation 77 is infinite. Therefore, the gain factor for changes in P_3 will be zero; however, the resistance, R_{3a}, will enter into the expression for the time constant and the gain factors for the other variables. As an example, consider the following problem.

a. EXAMPLE 3

A gas storage tank supplies two processes, as shown in Figure 23. The first process is connected to the tank through a manual valve and operates at a normal pressure of 40 psig; the normal flow rate is 600 SCFM. The second process is connected to the tank through a control valve and operates at a normal pressure of 10 psig. The pressure in the tank is under proportional

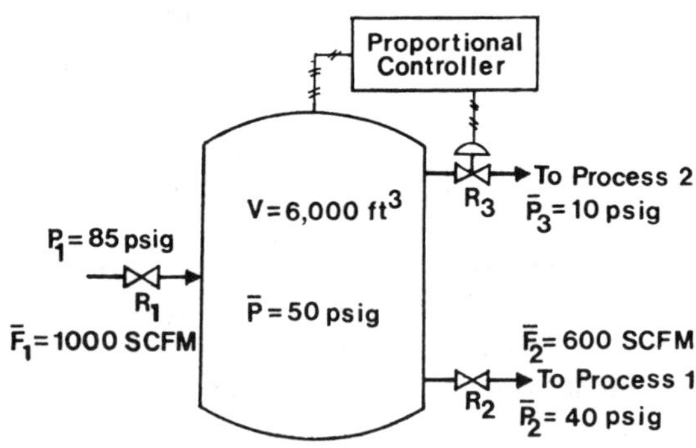

Figure 23.   Physical picture of system for Example 3.

control, the manipulated variable being the flow to the second process. The storage tank is supplied through a valved line by a process normally at 85 psig and at a rate of 1000 SCFM. The storage tank has a volume of 6000 ft^3.

The pressure transmitter has a span of 50 psi and the controller sensitivity is set at a value of 10.0 psi/in. A 5%, equal percentage valve is used.

(a) Draw a block diagram of the system with the transfer functions in the blocks.

(b) Suppose there is a step increase in P_2 of 0.5 psi. Calculate the ultimate change in P.

Solution:

(a) First calculate the pressure ratios in terms of *absolute* pressures to see if there is sonic flow in any of the lines:

$$\frac{\bar{P}}{\bar{P}_1} = \frac{64.7}{99.7} = 0.649 \qquad \text{subsonic}$$

$$\frac{\bar{P}_2}{\bar{P}} = \frac{54.7}{64.7} = 0.845 \qquad \text{subsonic}$$

$$\frac{\bar{P}_3}{\bar{P}} = \frac{24.7}{64.7} = 0.382 \qquad \text{sonic}$$

The critical pressure ratio for most gases falls between 0.49 and 0.58. Now calculate the resistances:

$$\left.\begin{array}{l} R_1 = \dfrac{2(85 - 50)}{1000} = 0.070 \dfrac{\text{psi}}{\text{SCFM}} \\[2em] R_2 = \dfrac{2(50 - 40)}{600} = 0.0333 \dfrac{\text{psi}}{\text{SCFM}} \end{array}\right\} \text{Using Equation 52}$$

$$\bar{F}_3 = \bar{F}_1 - \bar{F}_2 = 1000 - 600 = 400 \text{ SCFM}$$

$$R_3 = \frac{50 + 14.7}{400} = 0.162 \frac{\text{psi}}{\text{SCFM}} \qquad\qquad \text{Equation 59}$$

From Equation 79, it can be seen that the time constant is

$$\tau = \frac{C}{\dfrac{1}{R_1} + \dfrac{1}{R_2} + \dfrac{1}{R_3}} = \frac{6000/14.7}{\dfrac{1}{0.07} + \dfrac{1}{0.0333} + \dfrac{1}{0.162}} = \frac{6000/14.7}{50.5}$$

$$= 8.09 \text{ min}$$

The gain factors for the several load variables are

$$K_{P_1} = \cfrac{1}{R_1\left(\cfrac{1}{R_1} + \cfrac{1}{R_2} + \cfrac{1}{R_3}\right)} = \frac{1}{0.07(50.5)} = 0.283 \text{ unitless}$$

$$K_{P_2} = \cfrac{1}{R_2\left(\cfrac{1}{R_1} + \cfrac{1}{R_2} + \cfrac{1}{R_3}\right)} = \frac{1}{0.0333(50.5)} = 0.595 \text{ unitless}$$

$K_{P_3} = 0$ since the flow is sonic and small changes in downstream pressure have no effect on the flow

Referring to Equation 79, the process gain is

$$K_P = \cfrac{-K_V}{\cfrac{1}{R_1} + \cfrac{1}{R_2} + \cfrac{1}{R_3}}$$

As in the case of liquids, this gain is considered to be the product of the valve gain, K_V, and the process gain, K_P'. Thus,

$$K_P' = \cfrac{1}{\cfrac{1}{R_1} + \cfrac{1}{R_2} + \cfrac{1}{R_3}} = 0.0198 \text{ psi/SCFM}$$

The valve is a 5%, equal percentage type. This is the same type as used in the example in Section 1.B of Chapter 4. It was pointed out there that the percent change in flow is based on the actual flow rate at the point under consideration, but that the percent change in pressure is based on the pressure range of the valve motor. Thus

$$\frac{\Delta F_3}{\bar{F}_3} \bigg/ \frac{\Delta P_V}{15 - 3} = 5$$

The valve gain is therefore

$$K_V \equiv \frac{\Delta F_3}{\Delta P_V} = 5\frac{\bar{F}_3}{12} = \frac{5(400)}{12} = 167 \text{ SCFM/psi}$$

The transmitter span is 50 psi so a 50-psi change in the tank pressure will cause 4 in. of pen travel. Therefore, since the controller gain is 10 psi/in., the combined controller and transmitter gain is

$$K_C K_T = 10.0 \frac{\text{psi change in output}}{\text{inch of pen travel}} \times \frac{4 \text{ in. of pen travel}}{50\text{-psi change in pressure}}$$

$$= 0.80 \text{ psi/psi}$$

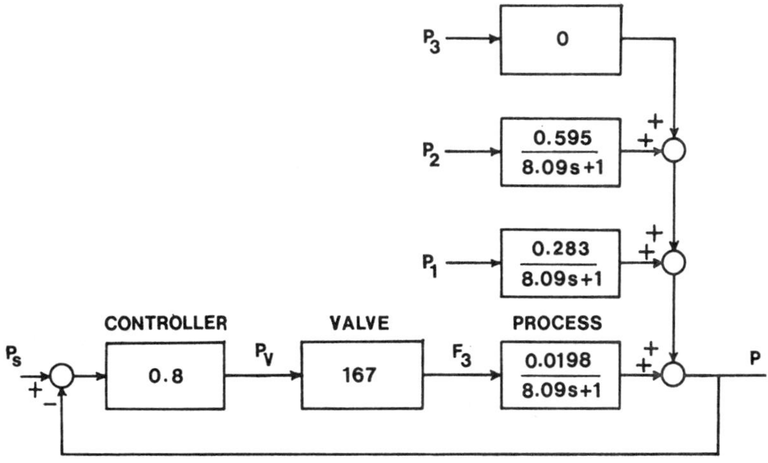

Figure 24. Block diagram for Example 3.

The block diagram can now be constructed and is shown in Figure 24. The negative sign in the expression for K_p has been disregarded since the controller action is set for negative feedback as indicated on the block diagram. The controller action can always be set for either positive or negative action to achieve negative feedback, and therefore all gains in the loop are taken as positive numbers.

(b) The ultimate change in pressure is found by analogy to Equation 37 of Chapter 4:

$$P = \frac{K_L}{1 + K_o} P_2$$

$$K_L \equiv K_{P_2} = 0.595$$

$$K_o = 0.8 \, \frac{\text{psi}}{\text{psi}} \times 167 \, \frac{\text{SCFM}}{\text{psi}} \times 0.0198 \, \frac{\text{psi}}{\text{SCFM}} = 2.64 \text{ unitless}$$

$$P = \frac{0.595}{1 + 2.64} (0.5) = 0.0817 \text{ psi}$$

4. Dead-Ended Tanks and Bellows

A dead-ended tank is a special case of those just discussed in that it has but one port. Sometimes such tanks are expandable in the form of a bellows, a balloon, or a cylinder. Bellows are extensively used in the mechanisms of pneumatic controllers.

The capacitance of a bellows will be derived here and those for cylinders and balloons can be found in Reference 7. Consider the bellows shown in

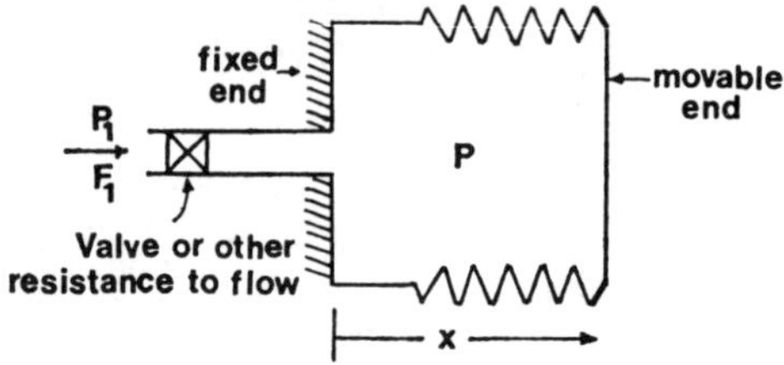

Figure 25. Bellows.

Figure 25. The walls are pleated like those of the air-holder of an accordian. Now, suppose that the pressure is greater outside the bellows than inside so that the flow is in the direction indicated in the figure. The mass balance is

$$\text{input} = \text{accumulation}$$

$$F_1 = \frac{d(\rho V)}{dt} \tag{80}$$

where F_1 = flow in, SCFM
ρ = density of gas, SCF/ft³
V = volume of bellows, ft³

Referring to Table 2, we see that capacitance is defined as the ratio of the flow rate to the rate of change of the driving force. The left-hand side of the mass balance is the flow rate; the only problem then is to relate the right-hand side to the rate of change of the driving force, namely dP/dt.

We begin by expanding the right-hand side into two terms:

$$F_1 = \rho \frac{dV}{dt} + V \frac{d\rho}{dt} \tag{81}$$

If the deviations in density and volume are small, this expression can be linearized to yield the approximation

$$F_1 \approx \bar{\rho} \frac{dV}{dt} + \bar{V} \frac{d\rho}{dt} \tag{82}$$

where $\bar{\rho}$ = normal density
$\bar{V}$ = normal volume

Consider the second term on the right. For an ideal gas at constant temperature, the density is proportional to the pressure:

$$\rho = \frac{\text{SCF}}{\text{ft}^3 \text{ of tank volume}} = \frac{P, \text{ psi}}{14.7, \text{ psi (ft}^3/\text{SCF)}} \tag{83}$$

Therefore

$$\frac{d\rho}{dt} = \frac{1}{14.7}\frac{dP}{dt} \tag{84}$$

Having related the second term on the right-hand side to the pressure, we now turn to the first term. As shown in Figure 25, one end of the bellows is in a fixed position and the other moves; the position of the movable end is given by the distance, x, relative to the origin indicated at the fixed end. If the spring constant of the bellows is k_b lb_f/ft and the cross-sectional area normal to the direction of movement is A_b ft^2, then a force balance yields

$$PA_b = k_b x + m \tag{85}$$

where m = constant, lb_f

At the same time, the volume of the bellows is given by

$$V = V_0 + A_b x \tag{86}$$

where V_0 = volume when $x = 0$

Using the chain rule and Equations 83, 85, and 86, we have

$$\bar{\rho}\frac{dV}{dt} = \bar{\rho}\frac{dV}{dx}\frac{dx}{dP}\frac{dP}{dt} = \frac{\bar{P}}{14.7}\frac{A_b}{1}\frac{A_b}{k_b}\frac{dP}{dt}$$

Hence, Equation 81 reduces to

$$F_1 = \frac{\bar{V}}{14.7}\left(\frac{\bar{P}A_b^2}{\bar{V}k_b} + 1\right)\frac{dP}{dt} \tag{87}$$

The capacitance, C_e, is the ratio of F_1 to dP/dt:

$$C_e = \frac{\bar{V}}{14.7}\left(\frac{\bar{P}A_b^2}{\bar{V}k_b} + 1\right) \tag{88}$$

When the tank is not expandable, k_b is infinite and the result is the same as Equation 67 with $n = 1$ for isothermal behavior.

From Equations 87 and 88, we see that

$$F_1 = C_e\frac{dP}{dt} \tag{89}$$

Transformation yields the following typical transfer function for a capacitance:

$$\frac{P}{F_1} = \frac{1}{C_e s} \tag{90}$$

In many applications, however, the transfer function of interest is P/P_1. The pressure drop is usually small so that the flow is subsonic. Therefore, the flow rate can be expressed in terms of a pressure drop and a single resistance:

$$F_1 = \frac{P_1 - P}{R} \tag{91}$$

Substituting into Equation 89 gives

$$\frac{P_1 - P}{R} = C_e \frac{dP}{dt} \tag{92}$$

Hence, the desired transfer function is:

$$\frac{P}{P_1} = \frac{1}{\tau s + 1} \tag{93}$$

where $\tau = RC_e$

A pneumatic valve operates in much the same way as a bellows. A typical valve is shown in Figure 7 of Chapter 10. The upper chamber of the valve contains a diaphragm that moves when the pressure, P, inside the valve motor changes. The pressure from the controller, P_V, is analogous to P_1 in the bellows example just discussed. As P changes, the stem moves and the area for flow, A_v, changes. If the mass of the moving parts is insignificant and there is no friction, then changes in A_v will be approximately proportional to changes in P over small ranges of pressure. Thus, an approximate transfer function for the behavior of this valve is

$$\frac{A_v}{P_V} = \frac{K_V}{\tau_V s + 1} \tag{94}$$

where $\tau_V = R_L C_e$

R_L = resistance in the pneumatic transmission line

D. Continuous-Flow, Stirred-Tank Reactors

The previous discussion of level and pressure control in tanks is relevant to the behavior of continuous-flow, stirred-tank reactors, since both of these variables will probably be important to its operation. However, they are of secondary significance in comparison with the concentrations of the reactants and products. These concentration variables are closely related to the thermal behavior of the reactor, which is discussed in Chapter 9. This type of reactor is one of the simplest to analyze because it falls in the category of a lumped parameter process.

1. *Behavior of the Process without Reaction*

Consider the reactor shown in Figure 26. A particular component, Component A, enters the reactor with concentration, c_F, and leaves the reactor with a concentration, c. To assess the effect of the reaction rate on the concentration behavior, we shall begin the analysis by first examining what will happen if there is no reaction at all. Then the tank will act merely as a surge tank, smoothing out the variations in c_F to produce a more nearly uniform leaving concentration, c.

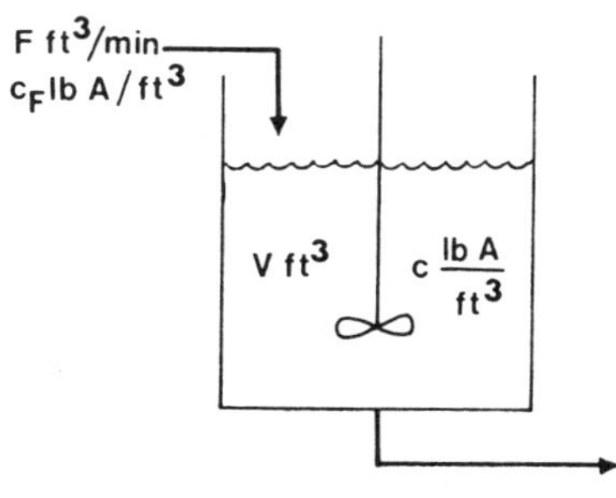

Figure 26. Continuous-flow, stirred-tank reactor.

In order to simplify the analysis, the following assumptions will be made: (1) As the feed enters, it is instantaneously dispersed throughout the tank by the impeller. Therefore the contents of the tank are uniform at all times and the leaving stream has the same composition as the tank contents; (2) The tank holdup, V, is constant. Physically, this could be accomplished by a level control system, or the liquid could leave through an overflow line inside the tank.

We wish to determine how the concentration in the tank varies with feed rate and feed concentration. Hence, a mass balance in terms of the concentration of Component A is made. Since F, c, and c_F are variables, each of these is written as the sum of its normal value plus a deviation; the unsteady-state mass balance is:

$$\text{input} \quad - \quad \text{output} \quad = \text{Accumulation}$$

$$(\bar{F} + \Delta F)(\bar{c}_F + \Delta c_F) - (\bar{F} + \Delta F)(\bar{c} + \Delta c) = V\,\frac{d\,\Delta c}{dt} \tag{95}$$

The initial steady-state equation is:

$$\bar{F}\bar{c}_F - \bar{F}\bar{c} = 0 \tag{96}$$

Expanding Equation 95, neglecting the second-order terms, subtracting Equation 96, and rearranging yields

$$\frac{V}{\bar{F}} \frac{d\,\Delta c}{dt} + \Delta c = \frac{\bar{c}_F - \bar{c}}{\bar{F}} \Delta F + \Delta c_F \tag{97}$$

Hence, the time constant and gain factors are:

$$\tau = \frac{V}{\bar{F}} \tag{98}$$

$$K_F = \frac{\bar{c}_F - \bar{c}}{\bar{F}} = 0 \tag{99}$$

$$K_{c_F} = 1 \tag{100}$$

Thus, the system is first order for small changes in feed rate and inlet concentration. The time constant is simply the holdup time of the tank. Specifically, the transfer functions are:

$$\frac{c}{F} = \frac{K_F}{\tau s + 1} = 0 \tag{101}$$

$$\frac{c}{c_F} = \frac{1}{\tau s + 1} \tag{102}$$

The electrical analogies for this system can be deduced from Equation 95 since each term represents a flow rate—in particular, the flow rate of Component A with units of lb of A/min. The analogy of voltage or potential is the concentration, c, with units of lb of A/ft^3. Having determined the analogies of current and voltage, it follows from Table 2 that the analog of capacitance is the tank holdup, V ft^3, and the analog of resistance is the reciprocal of the normal flow rate, $1/\bar{F}$. In retrospect, a glance at Equation 98 would have led to the same conclusions since the time constant is the product of a capacitance and a resistance.

2. Behavior with First-Order Reaction

We turn now to a consideration of a tank-flow reactor in which a first-order reaction is occurring:

$$A \rightarrow B \tag{103}$$

As indicated, the reaction is assumed to be irreversible. The picture of the process is the same as that of the previous case. The reaction rate is given by the expression:

$$r = \text{reaction rate} = k\,Vc \qquad \text{lb } A/\text{hr} \tag{104}$$

where k = rate constant, 1/hr

The rate constant is a function of temperature and the Arrhenius equation is usually used to express this relationship:

$$k = ae^{-E/R'T} \tag{105}$$

where a = constant
E = activation energy
R' = gas constant
T = absolute temperature

In this analysis, we shall assume the temperature to be constant. The mass balance is given by Equation 95 with the addition of the reaction term:

$$(\bar{F} + \Delta F)(\bar{c}_F + \Delta c_F) - (\bar{F} + \Delta F)(\bar{c} + \Delta c) - kV(\bar{c} + \Delta c) = V\frac{d\,\Delta c}{dt} \tag{106}$$

The steady-state equation is:

$$\bar{F}\bar{c}_F - \bar{F}\bar{c} - kV\bar{c} = 0 \tag{107}$$

Expanding Equation 106, neglecting the second-order terms, subtracting Equation 107, and rearranging yields

$$\frac{V}{\bar{F} + kV}\frac{d\,\Delta c}{dt} + \Delta c = \frac{\bar{c}_F - \bar{c}}{\bar{F} + kV}\Delta F + \frac{\bar{F}}{\bar{F} + kV}\Delta c_F \tag{108}$$

Comparing this result with Equation 97, we see that the presence of the reaction has reduced the time constant; the capacitance analog—namely, the tank volume—is the same, of course, but the resistance analog, $1/(\bar{F} + kV)$, is smaller. Thus the flow rate and reaction rate are completely parallel in their effects.

SUMMARY

In this chapter we have reviewed a pure time delay, looked at an example of a pure capacitance, and developed transfer functions for the first-order systems, which are based on mass balances. In Chapter 9, first-order systems derived from energy balances will be studied. Many analogies will be evident between the transfer functions of these two chapters.

REFERENCES

1. M. Tyner and F. P. May, *Process Engineering Control*, The Ronald Press Co., New York, 1968, Ch. 3.
2. F. G. Shinskey, *Process-Control Systems*, McGraw-Hill Book Co., Inc., New York, 1967, p. 6.

3. L. A. Gould, *Chemical Process Control: Theory and Applications*, Addison-Wesley Publishing Co., Reading, Mass., 1969, pp. 137–140, 344–349.

4. B. F. Dodge, *Chemical Engineering Thermodynamics*, McGraw-Hill Book Co., Inc., New York, 1944, Ch. VIII.

5. R. B. Bird, W. E. Stewart, and E. N. Lightfoot, *Transport Phenomena*, John Wiley and Sons, Inc., New York, 1960, pp. 480–482.

6. G. D. Shilling, *Process Dynamics and Control*, Holt, Rinehart, and Winston, New York, 1963, pp. 72–74, 90–94.

7. P. S. Buckley, *Techniques of Process Control*, John Wiley and Sons, Inc., New York, 1964, pp. 116–118.

8. P. Harriott, *Process Control*, McGraw-Hill Book Co., Inc., New York, 1964, p. 38.

9. D. P. Eckman, *Automatic Process Control*, John Wiley and Sons, Inc., New York, 1958, p. 23.

PROBLEMS

PROBLEMS CONCERNING LIQUID STORAGE TANKS

Problem 1.   A tank is being used for surge capacity for a very thick liquid. The normal level is 6 ft above the bottom of the tank and the liquid leaves through a 4-ft vertical line at the bottom. The exit end of this line is open to the atmosphere. The liquid passes through the line in *laminar* flow. If the normal flow rate is 1 ft^3/min and the cross-sectional area of the tank is 30 ft^2, what is the time constant?

Answer: 300 min

Problem 2.   A storage tank for water is 8 ft deep and open to the atmosphere. The cross-sectional area of the tank is 100 ft^2. The water enters the tank through a valved line located at a level 4 ft above the bottom of the tank. The pressure in the supply line upstream of the valve is 50 psig. The water leaves the tank at the bottom through a horizontal, valved line which vents to the atmosphere.

(a)   When the flow rate to the tank is 100 ft^3/min, the normal level is 6 ft. The flow through the supply and exit lines is turbulent. Calculate the time constant.

(b)   Suppose the valve in the exit line is opened sufficiently so that at the same feed rate of 100 ft^3/min, the level drops to 3 ft. Now what will the time constant be, assuming the flow through the valves is still turbulent?

Answers: (a) 11.4 min
(b) 6.00 min

Problem 3.   Consider the liquid storage tank shown in Figure 1 of this chapter. The transfer functions relating the level, h, to changes in P_V and F_i are given by Equations 35 and 36. Find the transfer functions relating F_o

to changes in F_i and P_V. If the leaving flow rate is the controlled variable, draw the block diagram containing these transfer functions.

$$Answers: \quad \frac{F_o}{F_i} = \frac{1}{\tau s + 1}$$

$$\frac{F_o}{P_V} = \frac{K_V \tau s}{\tau s + 1}$$

Problem 4. A surge tank for an aqueous stream operates under a partial vacuum at an absolute pressure of 8.2 psia. The normal feed rate to the tank is 100 ft^3/min and the normal level is 17 ft. The liquid leaves through a valved line at the bottom and vents to the atmosphere at the same elevation as the bottom of the tank. If the cross-sectional area of the tank is 1400 ft^2, calculate the time constant of the process.

Answer: 56.0 min

Problem 5. The transfer function, h/F_i, for a surge tank was presented in Equation 36. Using Equations 39 and 41, this can be expressed as:

$$\frac{h}{F_i} = \frac{R}{CRs + 1}$$

In the discussion of this tank, it was mentioned that the phenomena of self-regulation is a manifestation of negative feedback.

(a) Deduce a feedback arrangement of these capacitive and resistive elements that yields this transfer function.
(b) Deduce a second feedback arrangement that yields the same result.
(c) Which of the two diagrams actually coincides with the physical situation?

Problem 6. Suppose that the flow through the valve in Figure 17 is sonic. How does this alter the electrical analogy shown in Figure 21?

Problem 7. Water enters the top of a 1000-gal tank at a rate of 100 gpm. A constant-displacement pump removes 60 gpm and the rest passes through a valved line. The normal level is 8 ft and the cross-sectional area of the tank is 15 ft^2. Find the transfer function relating the liquid level to the feed rate.

Answers: Gain = 0.40 ft/gpm
Time constant = 44.9 min

Problem 8. Consider the tank shown in Figure 27.

(a) Find an expression for the transfer function, h_2/F_2.
(b) Find an expression for the transfer function, h_2/h_1.
(c) Find an expression for the transfer function, h_2/P_1.
(d) What kind of "sources" are F_2 and P_1?

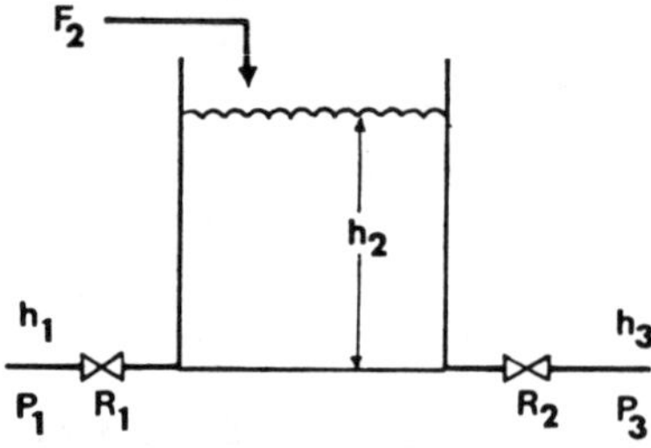

Figure 27. Physical situation in Problems 8 and 9.

Problem 9. Suppose that F_2 in Figure 27 is zero and the pressure source, P_1, is replaced by a constant-displacement pump feeding through the valve signified by R_1. Denoting this new feed, F_1, find the transfer function, h_2/F_1.

Problem 10. A surge tank of cross-sectional area, A, supplies a constant displacement pump with liquid at flow rate, F_o. The level in the tank, h, is controlled by manipulating the pressure, P_V, to a control valve in the inlet line to the tank. The valve is described by the transfer function $F_i/P_V = K_V$, which is the valve gain. The feed enters at the top of the tank as in Figure 3.

(a) If the system is under proportional control, find the transfer function relating the level, h, to the level set point, h_s.

(b) If a unit change in set point is made, sketch the response of the level in the tank. What is the offset, if any? Are you surprised by your answer to this question?

Problem 11. A sump pump is used to remove water that accumulates in a reservoir below the level of a basement floor. This pump operates under on-off control in which the level in the reservoir is measured by a float; the float in turn is attached to an electrical switch in the motor circuit of the pump. The switch has a differential gap.

(a) Draw a block diagram for this system and identify the nature of all transfer functions.

(b) Sketch the time behavior of the level if the drain tiles supply water at a constant rate to the reservoir. Assume that it takes 2 min to fill the reservoir and 15 sec to empty it.

Problem 12. Suppose a liquid storage tank is used for surge capacity between two processing units. The level is controlled by manipulating the leaving flow rate. This is sometimes referred to as "averaging control." Since the idea is to provide for smooth and gradual changes in flow rates in order to avoid upsetting processing equipment, the level control action is deliberately made slow. Therefore, the dynamics of the valve and the level sensor and transmitter are not important. Under these assumptions, investigate the following statements made by P. S. Buckley in his book, *Techniques of Process Control*, John Wiley and Sons, Inc., New York, 1964, p. 168:

(a) "In response to a sustained change in inflow, the proportional controller will not allow the change in outflow to exceed the change in inflow at any time."

Hint: The tank is described by a first-order transfer function. All other elements are approximately described by pure gains. What is the closed-loop response for such a system?

(b) "Any controller with reset, on the other hand, will, in response to sustained load changes, *always* cause the outflow to temporarily exceed the inflow change *regardless of controller proportional band and reset adjustment*. In other words, if a controller with reset is used, process equipment downstream of the tank must have a larger overdesign factor . . . for peak transient loads than it would have if a proportional controller were used."

Hint: The answer should be evident if you consider: (1) what the final level must be in relation to the initial level, and (2) if there is a step increase in feed rate, the feed rate will exceed the leaving flow rate, at least for a while at the start.

PROBLEMS CONCERNING GAS STORAGE TANKS AND PRESSURIZED VESSELS

Problem 13. A gas tank is supplied by a compressor at a rate of 4 SCFM. The normal pressure in the tank is 30 psig and the volume of the tank is 10 ft^3. The gas is then throttled through a valve to a process downstream.

(a) The normal pressure of the downstream process is 25 psig. What is the time constant of the tank?

(b) The normal pressure of the downstream process is 5 psig. What is the time constant?

Answers: (a) Time constant = 1.70 min

(b) Time constant = 7.6 min

Problem 14.   A gas is supplied at a constant rate of 20 SCFM (standard conditions are 1 atmosphere and 72°F) to a surge tank having a volume of 50 ft^3. The normal pressure in the tank is 35 psig. The gas then passes through a control valve to the following processing unit which operates at 25 psig. The valve is normally 50% open. If the opening is suddenly increased to 60%, how long will it take for the pressure in the tank to reach 80% of its ultimate final value?

Answer: 5.47 min

Problem 15.   A gas surge tank having a volume of 50 ft^3 operates at a normal pressure of 25 psig. It is supplied from a process at 85 psig and the normal flow rate is 100 SCFM. The tank in turn supplies gas to a process at a normal pressure of 1.0 atmosphere.

(a)   Calculate the time constant.
(b)   If atmospheric pressure changes from 14.69 to 14.50 psia, what will be the ultimate change in the pressure in the tank?

Answers: (a) Time constant = 1.35 min
(b) No change

Problem 16.   Consider the gas storage tank of Figure 16.

(a)   Find an expression for the time constant assuming subsonic flow through the valves.
(b)   Draw the electrical analogy of this process. Discuss the major differences between your circuit and that shown in Figure 4 of this chapter for the liquid storage tank.

Problem 17.   Draw the electrical analogy for Equation 78 of this chapter.

PROBLEMS CONCERNING STIRRED-TANK REACTORS

Problem 18.   A stirred tank is being used to mix two brine solutions at rates F_1 and F_2 lb/hr. The respective concentrations of the feeds are c_1 and c_2 lb salt/lb of feed. Assuming that the tank holdup is constant, find the transfer function, c/F_1, where c is the concentration of the product stream and M is the total holdup in lb.

Answer: $\text{Gain} = \dfrac{\bar{c}_1 - \bar{c}}{\bar{F}_1 + \bar{F}_2}$

$\text{Time constant} = \dfrac{M}{\bar{F}_1 + \bar{F}_2}$

Problem 19.   A first-order reaction is carried out in a continuous-flow, stirred-tank reactor. The main load variable is the feed concentration, c_F, and the controlled variable is the product concentration, c. It is proposed to control the concentration of the reactant, c, in the reactor through manipulation of the feed rate, F, in a simultaneous feedback-feedforward arrangement.

(a) Draw the block diagram of the system. Find the expressions for the transfer functions, c/F and c/c_F.

(b) The normal operating conditions are as follows:

$$\begin{aligned}
\bar{F} &= 5000 \text{ ft}^3/\text{hr} \\
\bar{V} &= 5000 \text{ ft}^3 \\
k &= 1 \text{ hr}^{-1} \\
\bar{c}_F &= 20 \text{ lb/ft}^3
\end{aligned}$$

The valve has linear trim and the normal pressure to the valve is 9 psig. What must be the gain of the computer-transmitter combination in the feedforward loop if the control is to be perfect?

(c) Under the operating conditions given above, calculate the reactor time constant.

(d) The composition transmitter in the feedback loop has a range of 50 lb/ft^3 and the controller gain is 15 psi/psi. Calculate the time constant for the closed loop.

(e) Suppose a level control system were added to the reactor with the leaving flow rate as the manipulated variable. What is the main load variable? Are there any interactions between the level control loop and the concentration loop? If so, describe them.

(f) Calculate the gain for the transfer function, c/h, where h is the liquid level in the reactor. The cross-sectional area of the tank is 333 ft^2.

$$\begin{aligned}
\textit{Answers:} \quad &\text{(b) } K_{CT} = -0.6 \text{ psi/lb/ft}^3 \\
&\text{(c) Time constant} = 0.5 \text{ hr} \\
&\text{(d) Closed-loop time constant} = 7.5 \text{ min} \\
&\text{(f) Gain} = -0.333 \text{ lb/ft}^3/\text{ft}
\end{aligned}$$

CHAPTER IX

Capacitance and First-Order Processes Based on Energy and Momentum Balances

In Chapter 8 we found that first-order behavior resulted when a process could be modeled by a single lumped capacitance in combination with a single lumped resistance. In the case of processes involving a mass balance, the capacitance usually takes the physical form of a tank, and the resistance is associated with valves and lines. In a similar way, we shall discover in this chapter that first-order thermal behavior results from a single lumped thermal capacitance in combination with a single lumped thermal resistance.

First-order behavior is not confined to resistance-capacitance combinations. It can also arise from the combination of a single lumped inertance and a single lumped resistance. The corresponding physical situation involves a momentum balance and will be illustrated at the end of this chapter.

I. THERMAL ANALOGIES TO RESISTANCE AND CAPACITANCE

There are only two passive elements in heat transfer—resistance and capacitance. These two properties already bear the names of their electrical analogs and require little in the way of interpretation.

A. Resistance

We found in Chapter 8 that the resistance in fluid flow cannot, in general, be defined as the simple ratio of the driving force to the flow rate since this ratio may not be constant. While it is approximately constant in laminar flow, it is far from constant in turbulent flow. In heat transfer, the ratio is essentially constant in simple situations such as heat transfer through a slab, but in convective heat transfer, the ratio is variable. Hence, for generality, we define:

$$R = \frac{\text{change in the driving force for heat transfer, °F}}{\text{change in the rate of heat transfer, Btu/hr}} \tag{1}$$

The units in this definition are given only as examples.

For conduction, the preceding definition leads to the expression

$$R = \frac{\Delta x}{kA} \tag{2}$$

where Δx = thickness of material, ft

k = thermal conductivity, Btu/hr, ft^2, °F/ft

A = area normal to the direction of heat transfer, ft^2

For convection, the resistance is defined in terms of the coefficient of heat transfer:

$$R = \frac{1}{hA} \tag{3}$$

where h = coefficient of heat transfer, Btu/hr, ft^2, °F

In the case of radiation, the radiation equation may be linearized for small changes in the driving force and then Equation 3 can be used for obtaining the resistance.

B. Capacitance

For generality again, the capacitance is also defined in terms of the ratio of changes:

$$C = \frac{\text{change in the flow rate of thermal energy, Btu/hr}}{\text{rate of change of the change in driving force, }^\circ\text{F/hr}} \tag{4}$$

As in the cases of fluid flow and electricity, this property can be looked upon as the rate of change of a stored quantity with respect to a driving force:

$$C = \frac{\text{change in thermal energy stored, Btu}}{\text{change in temperature, }^\circ\text{F}} \tag{5}$$

This equation is analogous to Equation 14 of Chapter 8. For most lumped parameter systems, this capacitance is simply the product of a mass and a specific heat.

Note that in the case of pure conduction, as, for example, through a slab of metal, unsteady-state heat transfer cannot be purely resistive. The resistance and capacitance are uniformly distributed throughout the material in which the conduction is taking place; hence, this is a "distributed parameter" situation rather than a lumped parameter one. However, in many practical situations, the capacitive aspect is relatively insignificant compared to the resistive, and the capacitance can be neglected. In some other situations, it may be permissible to lump all of the resistance and all of the capacitance; this possibility is discussed in Chapter 11.

For convective heat transfer, the resistance is assumed to reside in a thin film that has negligible capacitance; therefore this resistance can be treated as a lumped property. In radiation heat transfer, the space between the radiating surfaces is assumed to have negligible capacitance also and a lumped resistance treatment is satisfactory.

II. PURELY CAPACITIVE PROCESSES

Just as there are very few purely capacitive processes for mass, there are very few in thermal systems. This is because there is usually some thermal resistance associated with thermal capacitance in processes. However, it is possible to conceive of some purely capacitive thermal processes.

Consider a solid mass of very high thermal conductivity that is well insulated, but heated either by an imbedded electrical coil or by induction. If the rate of heat production is constant, the temperature will rise nearly linearly with time—it will be linear if the specific heat is constant and there are no heat losses. Thus, the solid acts as an "integrator" for the energy and is therefore a pure capacitor. When the source of energy is turned off, the temperature of the solid will theoretically remain constant. As a practical matter, there are no perfect insulators, so heat losses will occur and these represent heat transfer through a resistance. So in reality, this example is a combination of resistance and capacitance, although over short periods of time, the mass does behave as nearly pure capacitance.

As a second example, consider an insulated stirred tank containing a fixed quantity of water which is heated by sparging live steam into it. No water is added to or removed from the tank. This process is purely capacitive. If the steam rate is constant, the water temperature will rise linearly with time until it boils.

III. FIRST-ORDER PROCESSES BASED ON ENERGY BALANCES

As noted earlier, first-order thermal processes arise from a combination of a single lumped resistance with a single lumped capacitance. Various degrees of complexity occur, mostly because of the forms that the resistance can take. The capacitance is normally quite easy to calculate because the mass of the capacitance is usually constant. However, one example of a changing-mass capacitor will be examined later in this chapter. The simplest first-order example is a thermometer, to which we turn our attention now.

A. Temperature Bulb Behavior

1. Physical Situation

Consider the temperature bulb shown in Figure 1. The bulb is made of glass and contains a liquid that expands as its temperature rises. Alternatively,

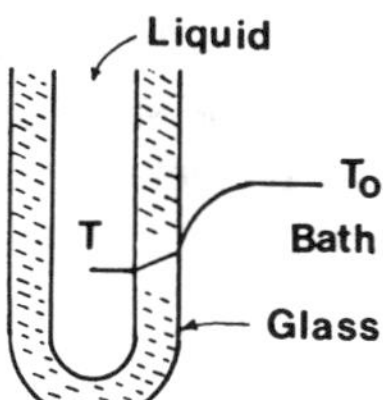

Figure 1.   Temperature bulb in a constant-temperature bath.

the bulb could be part of a temperature transmitter, in which case the bulb would be metallic and would contain a gas.

Suppose that the bulb has been in the bath for some time so that the glass and liquid are at the same temperature as the bath. The problem is to find the response of the *indicated* temperature after the bulb is suddenly moved to another bath at a higher temperature. This will be achieved by determining the transfer function, T/T_o, which relates the indicated temperature to the bath temperature.

When the bulb is placed in the second bath, heat is immediately transferred from the bath to the bulb, and the liquid begins to rise in the stem. There is resistance to heat transfer across the fictitious thin film surrounding the bulb and across the thermal resistances of the glass and liquid. For this simple analysis, these three resistances can be lumped together and expressed in terms of an overall coefficient, U, based on the outside area of the bulb, A. The capacitance is that of the glass shell and the liquid in the bulb. As the liquid rises, the mass of it within the bulb changes. However, as a first approximation, the amount of liquid in the bulb can be considered constant. Also, the capacitances of the liquid and glass can be lumped together. Thus, in effect, we have one lumped resistance and one lumped capacitance.

As indicated in Figure 1, most of the resistance to heat transfer is in the outside film. There is only a slight drop in temperature across the glass and the temperature of the liquid is essentially uniform. Thus, the outside film is the controlling resistance.

2. Transfer Function

The analysis begins with the unsteady-state energy balance:

$$UA(T_o - T) = mc_p \frac{dT}{dt} \tag{6}$$

where U = overall coefficient of heat transfer, Btu/hr, ft^2, °F
A = outside area of bulb, ft^2
m = mass of bulb (glass plus liquid), lb
c_p = average heat capacity of bulb, Btu/lb, °F
T = indicated temperature, °F
T_o = temperature of second bath, °F

The temperatures in Equation 6 are the actual temperatures. Prior to time zero when the bulb is moved, the indicated temperature is the same as the actual temperature of the first bath. If the temperatures are defined as the sum of the temperature of the first bath plus a deviation, it is easily shown that Equation 6 still applies in terms of the deviations:

$$UA(\Delta T_o - \Delta T) = mc_p \frac{d\,\Delta T}{dt} \tag{6a}$$

where ΔT_o = difference between the two bath temperatures, °F

ΔT = deviation in indicated temperature from the first bath temperature, °F

Thus, Equation 6 could be interpreted as a deviation equation, and this is a consequence of the linearity of the equation with respect to the variables, T and T_o.

Rearrangement to the standard form and transformation yields the transfer function

$$\frac{T}{T_o} = \frac{1}{\tau_b s + 1} \tag{7}$$

where $\tau_b = mc_p/UA$ = time constant of bulb, hr

The gain factor is unity, as it should be, since a 1°F rise in bath temperature will result in a 1°F rise in indicated temperature.

3. Electrical Analogy

The electrical analog can be deduced from Equation 6 in which the term on the left is the flow rate of thermal energy. Referring back to Equation 3, we see that the factor, UA, can be interpreted as the reciprocal of a resistance, and Equation 5 indicates that the product on the right-hand side, mc_p, is a capacitance. Thus, in terms of analogies, Equation 6 may be written

$$\frac{T_o - T}{R} = C\frac{dT}{dt} \tag{8}$$

where T_o and T are now deviations from the temperature of the first bath, and

$$R = \frac{1}{UA} \quad \text{and} \quad C = mc_p$$

The driving force, $T_o - T$, appears across the resistor as indicated in the equivalent circuit shown in Figure 2.

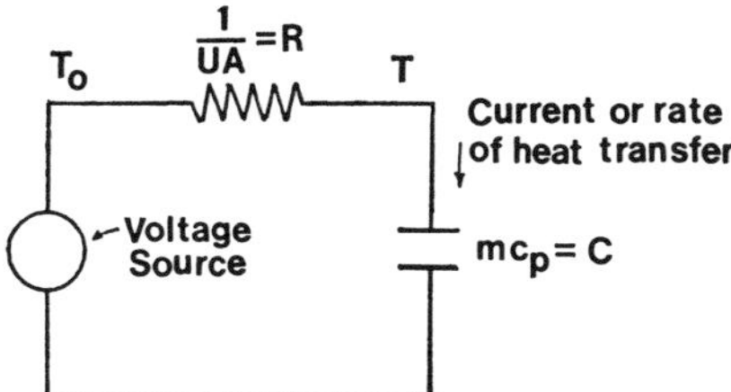

Figure 2.   Electrical analogy of a temperature bulb.

In this example, the difference between the two bath temperatures, T_o, is analogous to a voltage source, which is the active element of this circuit. As we saw in Chapter 8, current and voltage sources are roughly equally common in systems involving mass balances. In thermal systems, voltage-source situations, which arise from temperature driving forces, are probably more common.

Reflecting back to Chapter 8, the mass-balance analogy to this thermal system would be a constant-volume, dead-ended tank in a pressure system as discussed in Section IV.C.4 of that chapter. The pressure inside the tank, P, is analogous to the indicated temperature, T, and the pressure outside the tank, P_1, is analogous to the bath temperature, T_o.

B. Continuous-Flow, Stirred-Tank Heater

Consider the continuous-flow, stirred-tank heater shown in Figure 3. This heater was the focal point for Chapters 4 and 5. The liquid feed enters at a rate of F lb/hr and at a temperature, T_F, and leaves at temperature, T. The

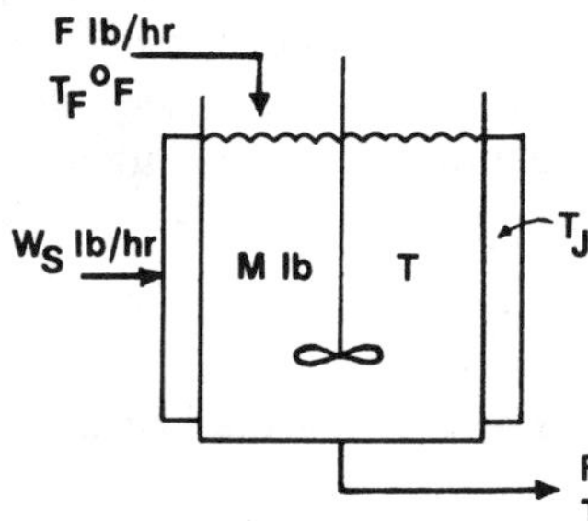

Figure 3.   Continuous-flow, stirred-tank heater.

tank holdup is M lb. The thermal energy is provided by steam, which is throttled through a control valve. The rate of heat transfer between the steam in the jacket and fluid in the tank is given by the equation:

$$Q = W_S\lambda = UA(T_J - T) \tag{9}$$

where Q = rate of heat transfer, Btu/hr
$\quad\quad W_S$ = steam rate, lb/hr
$\quad\quad \lambda$ = heat of condensation, Btu/lb
$\quad\quad U$ = overall coefficient of heat transfer, Btu/hr, ft^2, °F
$\quad\quad A$ = jacket area for heat transfer, ft^2
$\quad\quad T_J$ = effective jacket temperature, °F
$\quad\quad T$ = temperature of tank contents, °F

Since $1/UA$ is analogous to a thermal resistance, this equation can be written

$$Q = W_S \lambda = \frac{T_J - T}{R_1} \tag{9a}$$

where

$$R_1 = \frac{1}{UA} \tag{10}$$

The temperature of the heater will be controlled by manipulating the rate of heat transfer, Q. As we shall see momentarily, this can be done by manipulating either W_S or T_J. Since the equality of Equation 9a must be satisfied, it might seem at first that it would make no difference which of these was manipulated. However, this is not the case.

To understand why, it is helpful to also examine the fluid flow analogies of the heat transfer cases. Consider the current-source situation of Case (a) in Figure 4. For fluid flow, a constant-displacement pump is a current source.

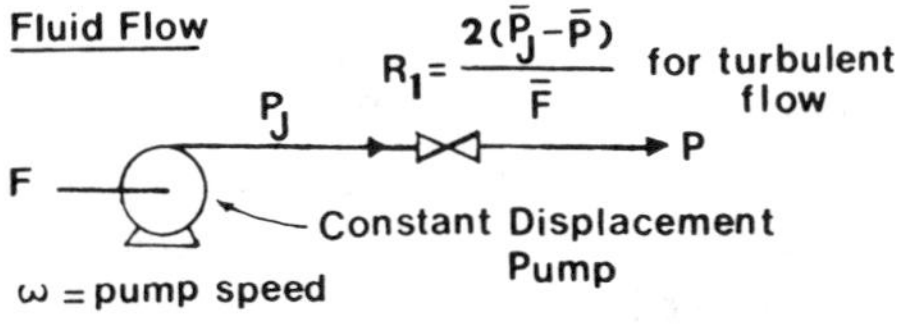

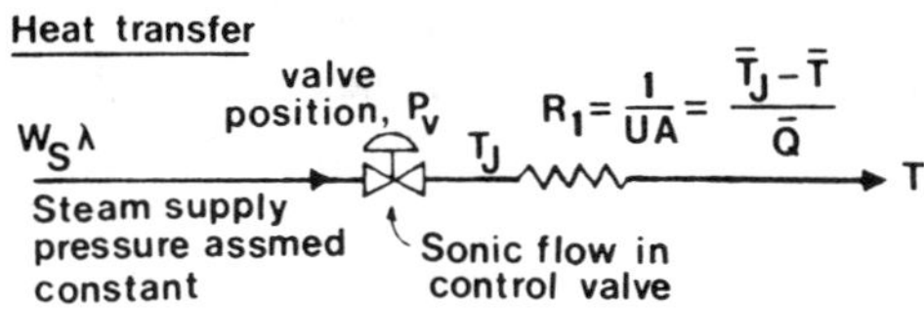

Case (a)   Current-source Situation

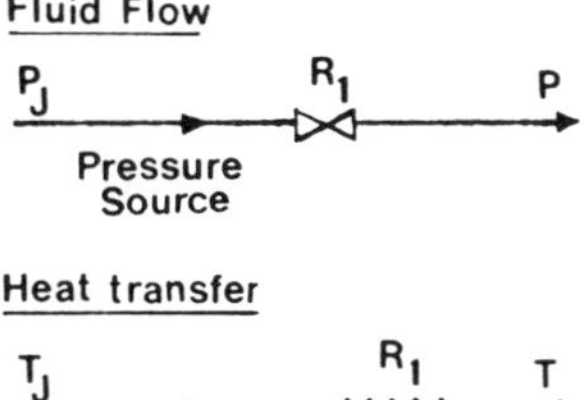

Case (b)   Voltage-source Situation

Figure 4.   Fluid-flow analogies of Equation 9a.

The flow rate through the line and valve is determined by the speed of the pump. As long as the speed remains fixed, the valve and line resistance, R_1, or the exit pressure, P, can be changed, but there will still be no change in the flow rate. If R_1 is changed, for example, then either or both P_J and P must change to satisfy the fluid-flow analog of Equation 9a. Thus, the flow rate is under direct control of the pump speed.

Now consider the parallel case for heat transfer. If the velocity through the valve is sonic, then the flow rate, W_S, and hence, the rate of heat transfer, is only dependent upon the valve position, and independent of the conditions downstream of the valve. Suppose that the valve position is fixed, but there is a change in the tank temperature, T. This means that T_J will have to adjust to keep the rate of heat transfer constant. Thus, the heat transfer rate is directly controlled by the valve position and not by the driving force across the thermal resistance. We see that the valve position here is analogous to the pump speed in the case of fluid flow.

The voltage-source situations are shown in Figure 4b. For fluid flow, the active element is the pressure source, P_J. For example, in a system in which water was the fluid, this source might be the water mains. The flow rate here is partially dependent upon the source pressure, P_J, and is proportional to the pressure drop, $P_J - P$. In a similar way, for heat transfer, the rate is partially dependent upon the jacket temperature, T_J, and proportional to the temperature drop, $T_J - T$. Thus, the rate of heat transfer is only indirectly controlled by T_J.

There are several situations in which the jacket temperature is the manipulated variable. As a first example, suppose that the steam is throttled through a hand-operated valve and the pressure in the jacket is manipulated by a vent valve. If the steam in the jacket is saturated, then this pressure manipulation amounts to a manipulation of the jacket temperature. A second example arises when the heating fluid is a hot liquid passing through the jacket. The *effective* jacket temperature then is varied by manipulating either the entering temperature of the heating fluid or its flow rate. The word "effective" is used here since the temperature of the liquid varies as it passes through the jacket.

With this background, we turn now to the development of the transfer functions and electrical analogies for these current-source and voltage-source cases.

1. Current-Source Case

a. TRANSFER FUNCTIONS

This was the case used in the illustrations of Chapters 4 and 5. Since the heat transfer rate is under direct control, it is expressed in terms of the steam rate, W_S. Letting W_S, T, and T_F be deviation variables, the unsteady-state energy balance is

$$\lambda W_S + F c_p (T_F - T) = M c_p \frac{dT}{dt} \tag{11}$$

After rearrangement and transformation, the pertinent transfer functions are:

$$\frac{T}{W_S} = \frac{K_{P1}}{\tau_1 s + 1} \tag{12}$$

$$\frac{T}{T_F} = \frac{K_{L1}}{\tau_1 s + 1} \tag{13}$$

where

$$\tau_1 = \frac{M}{F} = \text{holdup time} \tag{14}$$

$$K_{P1} = \frac{\lambda}{F c_p} \tag{15}$$

$$K_{L1} = 1 \tag{16}$$

b. ELECTRICAL ANALOGY

The electrical analogy is obtained by referring to Equation 11, since the terms have the units of flow rate of thermal energy, Btu/hr. The product, $F c_p$, is in the nature of an inverse resistance and the thermal capacitance of the tank contents, $M c_p$, is analogous to a capacitance. Therefore, Equation 11 can be written

$$\lambda W_S + \frac{T_F - T}{R_2} = C \frac{dT}{dt} \tag{17}$$

where

$$R_2 = \frac{1}{F c_p} \tag{18}$$

$$C = M c_p \tag{19}$$

Accordingly, the diagram for the process is shown in Figure 5a.

The load variable, T_F, appears in the circuit as a voltage source. If there are no deviations in this variable, the middle term on the left-hand side of Equation 11 drops out. This has the effect of bypassing the voltage source in Figure 5a with the result that the diagram reduces to that of Figure 5b.

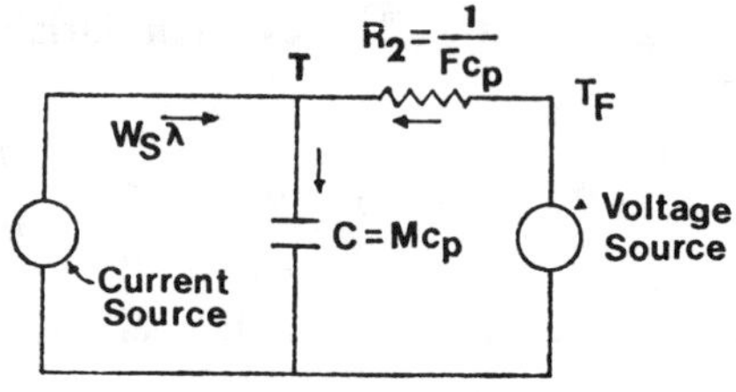

(a) Variable Feed Temperature

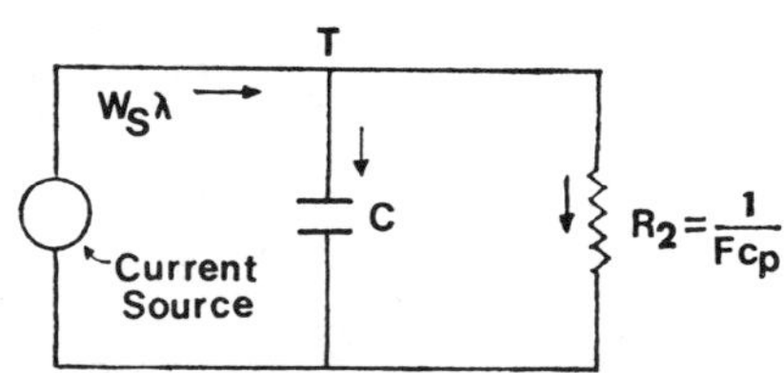

(b) Constant Feed Temperature

Figure 5.   Electrical analogies for current-source case.

2. *Voltage-Source Case*

a. TRANSFER FUNCTIONS

In this case, the rate of heat transfer is indirectly controlled by manipulating T_J and the unsteady-state energy balance is expressed as follows:

$$UA(T_J - T) + Fc_p(T_F - T) = Mc_p \frac{dT}{dt} \tag{20}$$

where T_J, T, and T_F are deviation variables. After transformation and rearrangement, the transfer functions are:

$$\frac{T}{T_J} = \frac{K_{P2}}{\tau_2 s + 1} \tag{21}$$

$$\frac{T}{T_F} = \frac{K_{L2}}{\tau_2 s + 1} \tag{22}$$

where

$$\tau_2 = \frac{Mc_p}{UA + Fc_p} \tag{23}$$

$$K_{P2} = \frac{UA}{UA + Fc_p} \tag{24}$$

$$K_{L2} = \frac{Fc_p}{UA + Fc_p} \tag{25}$$

It is important to note that the time constant and gain factors differ from those of the current-source case.

b. ELECTRICAL ANALOGY

The electrical analogy is obtained by reference to Equation 20. This time there are two resistance analogies:

$$R_1 = \frac{1}{UA} \tag{10}$$

$$R_2 = \frac{1}{Fc_p} \tag{18}$$

and

$$C = Mc_p \tag{19}$$

Hence, Equation 20 may be written as

$$\frac{T_J - T}{R_1} + \frac{T_F - T}{R_2} = C\frac{dT}{dt} \tag{26}$$

Accordingly, the electrical circuit is that shown in Figure 6a. If there is no variation in feed temperature, the diagram reduces to that of Figure 6b.

Comparing Figure 6b with that of Figure 2 for the temperature bulb, we see that they are almost the same except for the shunting resistance, R_2. This resistance approaches infinity as the flow rate, F, approaches zero, in which case this stirred-tank heater becomes completely analogous to the temperature bulb.

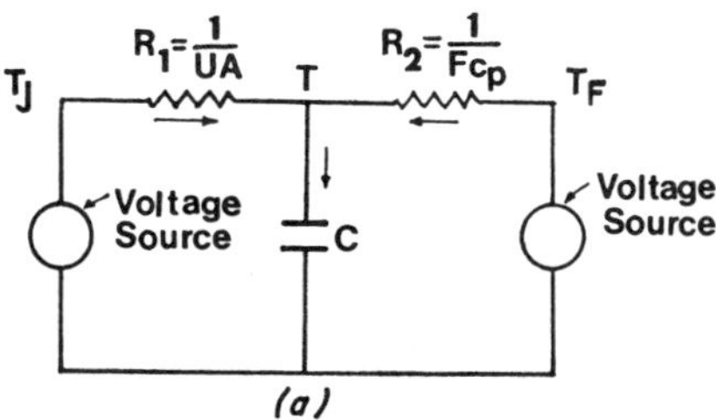

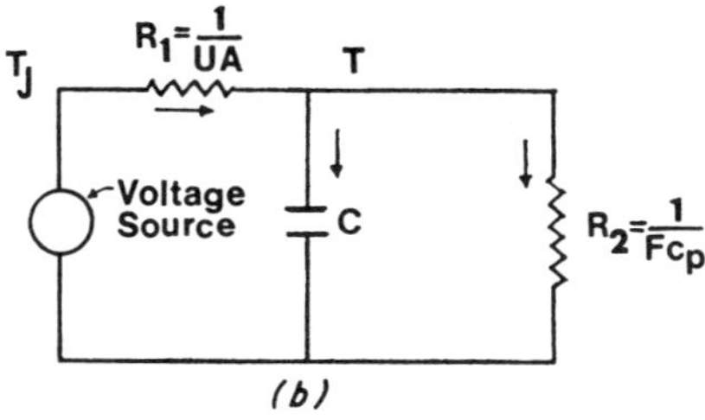

Figure 6. Electrical analogies for voltage-source case. (a) Variable feed temperature. (b) Constant feed temperature.

C. Continuous-Flow, Stirred-Tank Reactor

The continuous stirred-tank reactor, sometimes abbreviated "C.S.T.R.," has received a great deal of attention in the research field of reactor stability and control. One of the reasons for this will become apparent shortly as we look at the case where the reaction is exothermic.

Consider the reactor shown in Figure 7. The feed and product stream

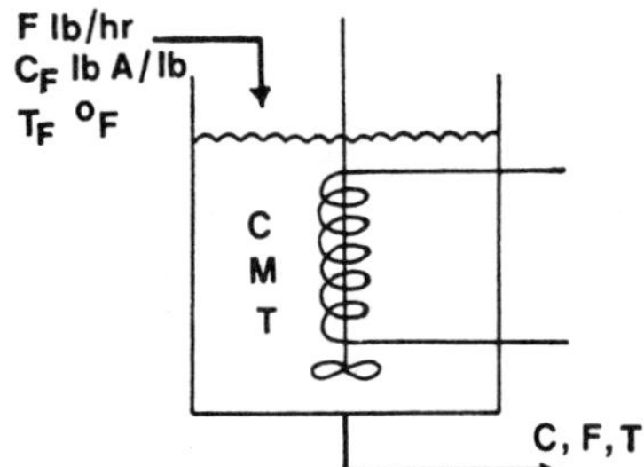

Figure 7.   Continuous-flow, stirred-tank reactor.

rates are constant and equal to F lb/hr. The feed contains c_F lb of A/lb of feed and the concentration in the reactor itself is c. The reaction is assumed to be sufficiently exothermic that a cooling coil or jacket is required to carry away some of the heat in addition to that removed as sensible heat in the product. We wish to determine the transfer functions, T/T_F and T/T_C, where T_C is the effective coolant temperature.

The unsteady-state energy balance for the reactor is

$$UA(T_C - T) + Fc_p(T_F - T) + Q_G = Mc_p \frac{dT}{dt} \tag{27}$$

where Q_G is the rate of heat generation from the reaction and is a positive number since the reaction is exothermic. For this analysis, the independent variables are T_F, T, and T_C. Equation 27 can be interpreted as a deviation equation since it is linear in the variables, T_F, T, Q_G, and T_C.

To obtain the desired transfer functions, the change in Q_G with temperature must be determined. For purposes of discussion, suppose that the reaction is first order. The reaction rate is as follows:

$$\text{reaction rate} = k_1 Mc \quad \text{lb } A/\text{hr} \tag{28}$$

where k_1 = reaction rate constant, 1/hr
This constant is related to the reactor temperature by the Arrhenius equation, which was given in Equation 105 of Chapter 8. The rate of heat generation is

just the reaction rate multiplied by the heat of reaction per unit of reactant consumed:

$$Q_G = k_1 Mc\gamma = ae^{-E/R'T} Mc\gamma \tag{29}$$

where γ = heat of reaction, Btu/lb A reacted

The heat of reaction varies only slightly with temperature and therefore can be assumed constant.

Since the transfer function relating T to T_C assumes that all other variables are held constant, this means that the deviation in Q_G is given by

$$Q_G = \frac{\partial Q_G}{\partial T} T \tag{30}$$

The partial derivative is evaluated at the normal concentration, $\bar{c}$, and normal temperature, $\bar{T}$. The symbol, T, is the deviation in tank temperature. For a first-order reaction, the partial derivative turns out to be

$$\frac{\partial Q_G}{\partial T} = \left(\frac{E}{R'\bar{T}^2}\right) \bar{Q}_G \tag{31}$$

After substituting Equation 30 into Equation 27, the resulting equation can be rearranged into the form

$$Mc_p \frac{dT}{dt} + \left(Fc_p + UA - \frac{\partial Q_G}{\partial T}\right) T = Fc_p T_F + UAT_C \tag{32}$$

Transformation and rearrangement gives the desired transfer functions:

$$\frac{T}{T_C} = \frac{K_{T_C}}{\tau_r s + 1} \tag{33}$$

$$\frac{T}{T_F} = \frac{K_{T_F}}{\tau_r s + 1} \tag{34}$$

where

$$\tau_r = \frac{Mc_p}{UA + Fc_p - \dfrac{\partial Q_G}{\partial T}} \tag{35}$$

$$K_{T_C} = \frac{UA}{UA + Fc_p - \dfrac{\partial Q_G}{\partial T}} \tag{36}$$

$$K_{T_F} = \frac{Fc_p}{UA + Fc_p - \dfrac{\partial Q_G}{\partial T}} \tag{37}$$

Comparing Equations 33 to 37 with Equations 21 to 25, we see that they are identical except for the presence of the partial derivative.

The denominators of Equations 35 to 37 are particularly interesting. The two terms, UA and Fc_p, are both positive and are related to the removal of heat from the reactor. In fact, their sum is the rate of change of heat removal per degree rise in reactor temperature. The partial derivative, $\partial Q_G/\partial T$, is also a positive quantity and is the rate of change of heat generation per degree rise in reactor temperature. Since this derivative is preceded by a minus sign, there is the possibility of the denominator being negative, in which case the time constant is also negative.

If the time constant is negative, this implies that the reactor will be inherently unstable. This can be seen as follows. Suppose there is a unit step increase in feed temperature. The corresponding transform of T_F is $1/s$. When this is used with Equations 34 and 37 and the resulting transform of T is inverted, one finds that the tank temperature will rise exponentially. On the other hand, a similar analysis for a positive time constant shows that the temperature will also increase, but the increase is bounded. Thus, a negative time constant means that the reactor is inherently unstable.

This analysis of stability is somewhat simplified and it leads to a condition for stability that is more severe than necessary. It is "simplified" in the sense that it does not take into account the effect of temperature on the concentration in the reactor. Suppose, for example, that the temperature rises slightly. The rate of reaction will increase as indicated by the Arrhenius equation; however, this increase will cause the concentration of reactant to fall and this tends to decrease the reaction rate. Hence, these two effects counteract each other to some extent. The preceding simplified analysis neglects the effect of changing concentration and assumes that the concentration remains fixed at the normal value. A complete analysis must take into account the dynamics of the reactor with respect to both temperature and concentration. This interaction leads to two stability criteria, both of which must be met. See Reference 1 for further details.

D. An Example of Variable Thermal Capacitance

It was noted in Section III.A of Chapter 8 that variable capacitance for mass occurs in the case of a storage tank for liquids in which the cross-sectional area of the tank is variable. A tank having the shape of a truncated cone would be an example. An analogous thermal situation can arise if the thermal capacitance changes as the temperature changes.

Looking back over the examples of this chapter, in all cases the thermal capacitance was constant because the mass of material was constant. Of course, it was assumed that the heat capacity was constant, but this assumption is reasonable since this property changes only very slightly with temperature. It is apparent, then, that a case of variable thermal capacitance can only arise if the mass of the process changes with temperature. Hence, we should expect to find variable capacitance if the process involves a gas, since its density changes significantly with temperature.

Consider the heat exchanger shown in Figure 8. The shell contains saturated steam and the tubes contain a slurry of ice and water. The problem is to find the differential equation relating the steam temperature, T_S, to the valve pressure, P_V.

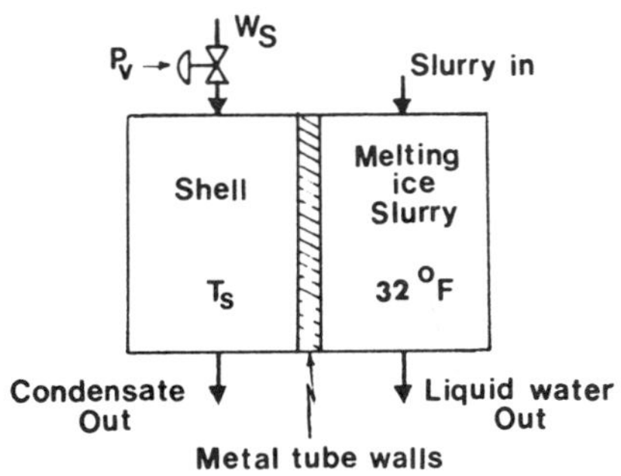

Figure 8.   Heat exchanger illustrating variable thermal capacitance.

For a complete analysis, two unsteady-state energy balances would be required, one for the steam and one for the condensate holdup. If we assume that the condensate holdup is negligible, then only the following one for the steam is required:

$$W_S\lambda - UA(T_S - T_I) = \frac{d(V\lambda\rho)}{dt} \tag{38}$$

where W_S = steam rate, lb/hr
λ = heat of condensation, Btu/lb
U = overall coefficient of heat transfer, Btu/hr, ft², °F
A = area for heat transfer, ft²
T_S = temperature of steam, °R
T_I = temperature of water-ice slurry, 32°F = 492°R
V = volume of shell occupied by steam, ft³
ρ = density of steam, lb/ft³
t = time, hr

The variables in this equation are W_S, T_S, and ρ. The heat of condensation can be assumed constant for small changes in the variables, and the temperature of the slurry is, of course, constant. Therefore, the equation is linear in the variables and can be interpreted as a deviation equation if the term containing T_I is dropped. Thus

$$\lambda W_S - UAT_S = V\lambda \frac{d\rho}{dt} \tag{39}$$

where W_S, T_S, and ρ are now deviations.

If we assume sonic flow through the steam valve and a constant steam supply pressure, then the flow rate is only a function of the pressure to the valve, P_V. The change in steam rate is related to the change in valve pressure by the valve gain:

$$W_S = K_V P_V$$

where K_V = valve gain, lb steam/hr, psi
   P_V = deviation in valve pressure, psi

Hence, Equation 39 may be written

$$K_V\lambda P_V - UAT_S = \lambda V \frac{d\rho}{dt} \tag{40}$$

We must now relate the density of the steam to its temperature, T_S. Using the chain rule

$$\frac{d\rho}{dt} = \frac{\partial\rho}{\partial P}\left(\frac{dP}{dT}\right)_S \frac{dT_S}{dt}$$

The derivative, $(dP/dT)_S$, is the slope of the saturated vapor line on a steam chart and it can be accurately calculated from a set of steam tables. The partial derivative, $\partial\rho/\partial P$, is evaluated at the normal steam temperature. If ideal gas behavior is assumed, this derivative is equal to the normal density divided by the normal absolute pressure. Hence, Equation 40 can be written

$$K_V\lambda P_V - UAT_S = \lambda V \frac{\rho}{P}\left(\frac{dP}{dT}\right)_S \frac{dT_S}{dt} = \frac{\lambda m}{P}\left(\frac{dP}{dT}\right)_S \frac{dT_S}{dt} \tag{41}$$

where m = mass of steam in the shell at the normal pressure and temperature, lb

The left-hand side of this equation is a flow rate of thermal energy and the derivative, dT_S/dt, is the rate of change of the driving force. Hence, in accordance with its definition, the capacitance is

$$C_s = \frac{\lambda m}{P}\left(\frac{dP}{dT}\right)_S \tag{42a}$$

Again, if it is assumed that the steam behaves as an ideal gas, the perfect gas law can be used to obtain the following alternate expression for the capacitance:

$$C_s = \lambda V \,\frac{\dot{M}}{R'\overline{T}_S}\left(\frac{dP}{dT}\right)_S \tag{42b}$$

where M = molecular weight of water
R' = gas constant
$\overline{T}_S$ = normal temperature of steam, $°R$

This expression shows that the capacitance is inversely proportional to the *absolute* temperature, $\overline{T}_S$, and directly proportional to the slope of the vapor pressure curve, and therefore the capacitance will change significantly as the temperature changes.

Using Equation 42a, Equation 41 can be rearranged to the standard form:

$$\frac{C_s}{UA}\frac{dT_S}{dt} + T_S = \frac{K_V\lambda}{UA}\,P_V \tag{43}$$

The time constant and gain factor can be calculated from the exchanger dimensions and design parameters.

This concludes the discussion of first-order thermal systems. We now turn to an example of first-order behavior that arises from the inertance property of a fluid.

IV. FIRST-ORDER PROCESSES BASED ON MOMENTUM BALANCES

Inertance rarely plays a significant role in chemical processes. However, one should be aware of those situations where it might be important, and several examples will be discussed here and in Chapters 10 and 11. We shall begin with one of the simplest examples.

A. Unsteady Flow Behavior of a Liquid in a Pipe

1. Physical Situation

The inertance property of a process only becomes important when unbalanced forces and accelerations occur. Consider the case in which a liquid is flowing through a line containing a control valve, as shown in Figure 9. The line

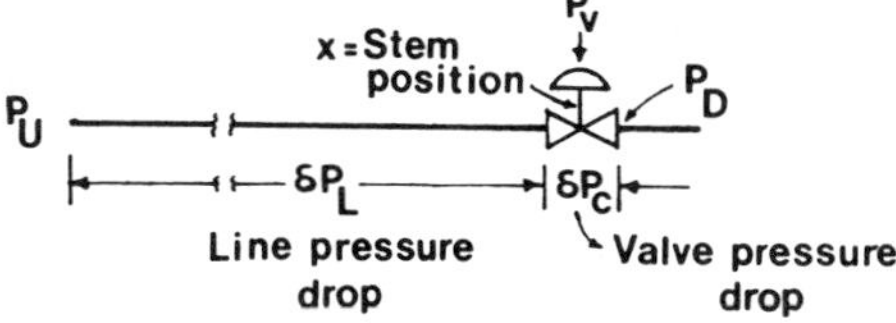

Figure 9. Physical situation for unsteady flow in a pipe.

pressure drop, δP_L, accounts for the drops across all lengths of straight pipe in the process as well as the drops across pipe fittings such as elbows and T's. The pressure drop, δP_C, is only that across the control valve. If the flow is steady, then the sum of the pressure drops across the line and control valve is equal to the driving force, $P_U - P_D$:

$$P_U - P_D = \delta P_L + \delta P_C \qquad \text{(steady state)} \qquad (44)$$

For constant supply and outlet pressures, suppose that the valve opening is suddenly increased, thus decreasing the valve resistance. This causes a corresponding sudden decrease in the pressure drop across the valve; hence, the sum of the pressure drops across the line and the valve is no longer equal to the driving force for flow, $P_U - P_D$. The unbalanced force causes the liquid to accelerate in accordance with Newton's law:

$$S[(P_U - P_D) - (\delta P_L + \delta P_C)] = \frac{M}{g_c} \frac{dv}{dt} \qquad (45)$$

where M = mass of fluid in the line, lb_m
    S = cross-sectional area of line, ft^2
    g_c = dimensional conversion constant, 32.17 lb_m, ft/lb_f, sec^2
    v = average fluid velocity, ft/sec

In writing Equation 45, it has been assumed that the velocity profile of the fluid is flat—that is, the fluid is moving in "plug flow." This assumption is valid for turbulent flow, which is the usual situation.

2. Transfer Functions

In the most general case, not only is the valve position a possible variable, but also the supply and outlet pressures for the line. The objective is to obtain transfer functions relating the velocity to each of these three variables. Equation 45 is linear in the variables, P_U, P_D, δP_L, δP_C, and v, and therefore could be interpreted as a deviation equation. However, it may be a bit clearer if we carry through the analysis with all variables initially expressed as the sum of the normal value plus a deviation:

$$P_U = \bar{P}_U + \Delta P_U \qquad (46)$$

$$P_D = \bar{P}_D + \Delta P_D \qquad (47)$$

$$\delta P_L = \overline{\delta P_L} + \Delta \delta P_L \qquad (48)$$

$$\delta P_C = \overline{\delta P_C} + \Delta \delta P_C \qquad (49)$$

$$v = \bar{v} + \Delta v \qquad (50)$$

At steady state:

$$\bar{P}_U - \bar{P}_D = \overline{\delta P_L} + \overline{\delta P_C} \qquad (51)$$

If Equations 46 to 50 are substituted into Equation 45, and Equation 51 is subtracted from that result, the following deviation equation results:

$$S[(\Delta P_U - \Delta P_D) - (\Delta \delta P_L + \Delta \delta P_C)] = \frac{M}{g_c} \frac{d \Delta v}{dt} \tag{52}$$

The line drop is only a function of velocity but the valve drop is a function of both velocity and stem position, x. Linearization of the deviations in terms of their respective variables yields:

$$\Delta \delta P_L = \frac{d \delta P_L}{dv} \Delta v \tag{53}$$

$$\Delta \delta P_C = \frac{\partial \delta P_C}{\partial v} \Delta v + \frac{\partial \delta P_C}{\partial x} \Delta x \tag{54}$$

Substitution of Equations 53 and 54 into Equation 52, followed by rearrangement, gives

$$\frac{M}{g_c} \frac{d \Delta v}{dt} + S \left(\frac{\partial \delta P_C}{\partial v} + \frac{d \delta P_L}{dv} \right) \Delta v = S \left(- \frac{\partial \delta P_C}{\partial x} \Delta x + \Delta P_U - \Delta P_D \right) \tag{55}$$

The change in velocity is first order with respect to changes in stem position, and in the supply and outlet pressures. After transformation of this equation, the following transfer functions result:

$$\frac{v}{x} = \frac{K_x}{\tau_x s + 1} \tag{56}$$

$$\frac{v}{P_U} = \frac{K_{P_U}}{\tau_x s + 1} \tag{57}$$

$$\frac{v}{P_D} = \frac{K_{P_D}}{\tau_x s + 1} \tag{58}$$

where, as usual, the Δ notation has been omitted in the expressions for the transfer functions. The time constant and gains are:

$$\tau_x = \frac{M/g_c}{S \left(\dfrac{\partial \delta P_C}{\partial v} + \dfrac{d \delta P_L}{dv} \right)} \tag{59}$$

$$K_x = \frac{- \dfrac{\partial \delta P_C}{\partial x}}{\dfrac{\partial \delta P_C}{\partial v} + \dfrac{d \delta P_L}{dv}} \tag{60}$$

$$K_{P_U} = \cfrac{1}{\dfrac{\partial \delta P_C}{\partial v} + \dfrac{d \delta P_L}{dv}} \qquad (61)$$

$$K_{P_D} = \cfrac{-1}{\dfrac{\partial \delta P_C}{\partial v} + \dfrac{d \delta P_L}{dv}} \qquad (62)$$

3. Electrical Analogy

Equation 55 is in terms of the velocity of the fluid. In accordance with the current analogy given in Table 1 of Chapter 8, the flow rate must be converted from a velocity to a volumetric rate. The conversion is simple here since plug flow has been assumed:

$$\Delta F = S \, \Delta v = \text{volumetric flow rate, ft}^3/\text{sec} \qquad (63)$$

Thus, Equation 55 can be written as follows:

$$\frac{M}{S^2 g_c} \frac{d \, \Delta F}{dt} + \frac{1}{S} \left(\frac{\partial \, \delta P_C}{\partial v} + \frac{d \, \delta P_L}{dv} \right) \Delta F = - \frac{\partial \, \delta P_C}{\partial x} \Delta x + \Delta P_U - \Delta P_D \quad (64)$$

The first term on the left arises from the inertance property of the fluid. The coefficient, $M/(S^2 g_c)$, has the units of a driving force in lb_f/ft^2 divided by the rate of change of the flow rate in ft^3/sec^2, and therefore this coefficient is the inertance:

$$I = \text{inertance} = \frac{M}{S^2 g_c} \qquad (65)$$

The second term on the left results from the resistance to fluid flow through the control valve and lines. The coefficient of ΔF has the units of a driving force in lb_f/ft^2 divided by the flow rate in ft^3/sec, so this coefficient is the resistance:

$$R = \text{resistance} = \frac{1}{S} \left(\frac{\partial \, \delta P_C}{\partial v} + \frac{d \, \delta P_L}{dv} \right) \qquad (66)$$

When written in terms of these analogies, Equation 64 takes the following form:

$$I \frac{d \, \Delta F}{dt} + R \, \Delta F = - \frac{\partial \, \delta P_C}{\partial x} \Delta x + \Delta P_U - \Delta P_D \qquad (67)$$

Each term has the units of pressure, which is analogous to voltage.

In drawing electrically equivalent diagrams before, we began with a differential equation in which each term has the units of flow rate or current. Even though Equation 67 is in a "voltage form," it can be used to draw the electrical diagram. It will be recalled that Kirchhoff's voltage law states that the sum of the voltage drops around a loop must be zero. The three terms on

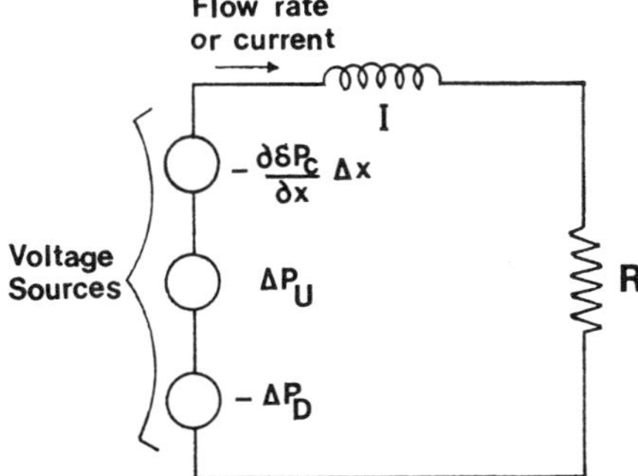

Figure 10. Electrical analogy for unsteady flow in a pipe.

the right-hand side are voltage sources and the inertance and resistance terms on the left are passive. Thus, the diagram takes the form shown in Figure 10.

4. Effect of System Parameters on the Time Constant and Valve Gain Factor

The time constant and gain factors given in Equations 59 to 62 are not in particularly convenient forms for calculations. Turning first to the time constant, the factor, M/S, is related to the line length and liquid density by the equation

$$\frac{M}{S} = L\rho \tag{68}$$

where L = length of line, ft

ρ = density of liquid, lb_m/ft^3

Looking at Equations 59 to 62, we see that their denominators contain the sum of two partial derivatives. This sum is related to the normal pressure drops across the valve and line. This can be seen in the following way. For straight pipe, the pressure drop is given by the Fanning equation:

$$\delta P_L = \frac{2f\rho v^2 L}{g_c D} \qquad \text{for straight pipe} \tag{69}$$

where f = friction factor

D = diameter of line, ft

Pressure drops in valves and fittings are usually expressed in terms of the equivalent length of straight pipe, so Equation 69 applies to the line and valve if L is replaced by the sum of the line length plus the equivalent length. Furthermore, for small changes in velocity, the friction factor is nearly constant in turbulent flow. Hence, the pressure drop across the valve and line will be proportional to the square of the velocity:

$$\delta P_C + \delta P_L = \alpha v^2 \tag{70}$$

where α = proportionality constant

Therefore

$$\frac{\partial(\delta P_C + \delta P_L)}{\partial v} = 2\alpha\bar{v} \qquad (71)$$

The partial derivative is evaluated at the normal velocity, $\bar{v}$. From Equation 70, it can be seen that

$$\alpha = \frac{\overline{\delta P_C} + \overline{\delta P_L}}{\bar{v}^2} \qquad (72)$$

Therefore

$$\frac{\partial(\delta P_C + \delta P_L)}{\partial v} = \frac{2(\overline{\delta P_C} + \overline{\delta P_L})}{\bar{v}} \qquad (73)$$

Using Equations 68 and 73, the time constant can be expressed as

$$\tau_x = \frac{\rho L\bar{v}/g_c}{2(\overline{\delta P_C} + \overline{\delta P_L})} \qquad (74)$$

Equation 73 can similarly be used to simplify the gains given in Equations 60 to 62 to yield the following expressions:

$$K_x = \frac{-(\partial\, \delta P_C/\partial x)\bar{v}}{2(\overline{\delta P_C} + \overline{\delta P_L})} \qquad (75)$$

$$K_{P_U} = \frac{\bar{v}}{2(\overline{\delta P_C} + \overline{\delta P_L})} \qquad (76)$$

$$K_{P_D} = \frac{-\bar{v}}{2(\overline{\delta P_C} + \overline{\delta P_L})} \qquad (77)$$

Equation 75 indicates the effect of the pressure drop in the line upon the valve gain. If there were negligible drop in the line, the gain would be

$$K_x = \frac{-(\partial\, \delta P_C/\partial x)\bar{v}}{2\, \overline{\delta P_C}} \qquad (78)$$

But from Equation 73 it can be seen that

$$\frac{\partial\, \delta P_C}{\partial v} = \frac{2\, \overline{\delta P_C}}{\bar{v}} \qquad (79)$$

Therefore Equation 78 can be written

$$K_x = \frac{-\partial\, \delta P_C/\partial x}{\partial\, \delta P_C/\partial v} = \frac{\partial v}{\partial x} \qquad \text{for negligible pressure drop in the line} \quad (80)$$

This is the inherent gain of the valve since it is the rate of change of velocity with respect to stem position for constant pressure drop *across the valve*.

Equation 75, on the other hand, represents the effective gain since it is the rate of change of velocity with respect to stem position for constant pressure drop *across the entire flow process*. As the pressure drop across the line decreases relative to that across the valve, the effective gain increases; in other words, as the resistance across the valve becomes more "controlling" relative to the resistance of the line, the valve has more effect on the flow rate.

Equation 74 indicates that the time constant is increased as the pressure drop across the control valve is decreased. If this drop is negligible and the drops through fittings are small compared with those through the lengths of straight pipe, then the time constant becomes independent of the length of the line. This can be shown by decomposing the pressure drop, $\overline{\delta P}_L$, as follows:

$$\overline{\delta P}_L = \overline{\delta P}_F + L\,\overline{\delta P}_{sp} \tag{81}$$

where $\overline{\delta P}_F$ = pressure drop across fittings

$\overline{\delta P}_{sp}$ = pressure drop *per foot* of straight pipe

Thus, when both $\overline{\delta P}_C$ and $\overline{\delta P}_F$ are negligible, Equation 74 reduces to

$$\tau_{sp} = \frac{\rho\bar{v}}{2g_c\,\overline{\delta P}_{sp}} \quad \begin{cases} \text{for negligible pressure drop across} \\ \text{the control valve and fittings} \end{cases} \tag{82}$$

This is the maximum value that the time constant can have.

5. Example

A brine having a specific gravity of 1.18 and a viscosity of 2.5 cp is carried through 100 ft of smooth tubing at a rate of 25 gpm. The cross-sectional area of the tube is 0.777 in.^{2} and its inside diameter is 0.995 in. What is the time constant of the line?

Solution:

$$\bar{v} = \frac{25\text{ gal/min} \times 231\text{ in.}^3/\text{gal}}{60\text{ sec/min} \times 0.777\text{ in.}^2 \times 12\text{ in./ft}} = 10.3\text{ ft/sec}$$

$$N_{Re} = \frac{(0.995/12)\text{ft} \times 10.3\text{ ft/sec} \times 1.18 \times 62.4\text{ lb}_m/\text{ft}^3}{6.72 \times 10^{-4}\text{ lb}_m/\text{ft, sec/cp} \times 2.5\text{ cp}} = 3.75 \times 10^4$$

$$f = 0.0056 \qquad \text{[from friction factor plot in McCabe and Smith (2)]}$$

$$\overline{\delta P}_{sp} = \frac{2f\rho\bar{v}^2}{g_c D} = \frac{2 \times 0.0056 \times 1.18 \times 62.4 \times 10.3^2}{32.17 \times 0.995/12} = 32.9\text{ lb}_f/\text{ft}^2$$

$$\tau_{sp} = \frac{\rho\bar{v}}{2g_c\,\overline{\delta P}_{sp}} = \frac{1.18 \times 62.4 \times 10.3}{2 \times 32.17 \times 32.9} = 0.358\text{ sec}$$

SUMMARY

This chapter concludes our study of first-order processes. In all of the cases that were covered, the elements of resistance, capacitance, and inertance were identified so as to stress the points of commonality between the various processes. First-order processes were found to arise when there is a combination of a single lumped resistance with a single lumped capacitance, or a combination of a single lumped resistance with a single lumped inertance. We shall learn in Chapter 10 that a combination of a single lumped capacitance with a single lumped inertance would not give rise to first-order behavior, but rather would result in second-order behavior.

The property of inertance assumes importance only rarely in process control. As we shall see in Chapter 11, it plays a role in the analytical treatment of pneumatic transmission lines and it might be important in very unusual flow control systems. However, ordinarily, capacitance and resistance are the main elements of concern in process dynamics.

REFERENCES

1. P. Harriott, *Process Control*, McGraw-Hill Book Co., Inc., New York, 1964, pp. 309–312.
2. W. L. McCabe and J. C. Smith, *Unit Operations of Chemical Engineering*, McGraw-Hill Book Co., Inc., New York, 1956, p. 68.

PROBLEMS

PROBLEMS CONCERNING TEMPERATURE BULBS

Problem 1.   A thermometer having a time constant of 0.1 min is at a steady-state temperature of 90°F. At time $t = 0$, the thermometer is placed in a temperature bath maintained at 100°F. Determine the time needed for the thermometer to read 98°F.

Answer: 0.161 min

Problem 2.   A mercury-filled thermometer indicates that the temperature of a room is 68°F. The thermometer is placed in a pan of boiling water. After 6 sec, it indicates a temperature of 180°F and after an additional 6 sec, 205°F. Finally—in roughly half a minute—the indicated temperature is 212°F. The thermometer is then removed from the pan and hung on a hook on the wall. After 6 sec, the indicated temperature is 181°F and after an additional 6 sec, 157°F.

(a)   Does the thermometer appear to be exhibiting first-order behavior in the boiling water? If so, what is the time constant?

(b)   Does it appear to be exhibiting first-order behavior in the room? If so, what is the time constant?

(c)   Explain any differences in behavior between parts *a* and *b*.

Answers: (a) Time constant $= 3.97$ sec
(b) Time constant $= 24.8$ sec

Problem 3.   A thermometer bulb of a temperature transmitter is used to measure the temperature of a liquid stream passing through a pipe 1 ft in diameter. The pipe is well insulated and the temperature is uniform throughout the liquid. The flow is in the laminar regime.

The bulb, $\frac{1}{2}$ in. long and $\frac{1}{8}$ in. in diameter, is fairly close to the wall— about $\frac{1}{2}$ in. from it. It is suspected that the response of the bulb is slower than it should be and the following three suggestions have been made for improving the response:

(a)   Use the next larger size bulb, which is 1 in. long and $\frac{1}{8}$ in. in diameter.

(b)   Use a slightly fatter bulb, $\frac{1}{2}$ in. long and $\frac{3}{16}$ in. in diameter.

(c)   Continue to use the same bulb but move it to the center of the pipe.

Comment on these suggestions.

Problem 4.

(a)   The transfer function of a temperature bulb was presented in Equation 7 as follows:

$$\frac{T}{T_o} = \frac{1}{\tau_b s + 1} \tag{A}$$

Here, T_o and T are deviations from original values. For a unit step in outside temperature, the change of T with time is given by the equation

$$T = 1 - e^{-t/\tau_b} \tag{B}$$

Sketch this response.

(b)   Now consider the case of unsteady flow in a pipe. Suppose the only variable is the upstream supply pressure, P_U.

   (1)   Find an expression for F/P_U in terms of the electrical analogies of inertance and resistance.

   (2)   For a unit step change in P_U, find the response of F and sketch this.

   (3)   Multiply R by F and call the result P. Sketch the response of P.

   (4)   Based on the expression for T in Equation B and your result for Part *b*3, is it correct to say that the processes behave analogously?

   (5)   What is the physical interpretation of P?

(c) (1) For a temperature bulb, find the expression for the rate of heat transfer, Q, as a function of time and sketch this.
  (2) Compare your result from Part $c1$ with the appropriate analogous equation for flow in a pipe. Based on this comparison is it correct to say that the processes behave analogously?

(d) Based on your results for parts $b4$ and $c2$, are the two processes completely analogous?

PROBLEMS CONCERNING CONTINUOUS-FLOW, STIRRED-TANK HEATERS AND REACTORS

Problem 5. Review the continuous-flow steam heater illustration in Section II.A of Chapter 6.

(a) Draw the electrical analogies that apply during the on and off portions of the cycle. Include the feed temperature as a load variable. Indicate the directions of the deviations of current flow that are analogous to the directions of the deviations in the rates of heat transfer.
(b) If the feed temperature is constant, how does this alter the analog diagrams of part a? Again, draw the directions of heat flow.

Problem 6. A chemical reaction is carried out in a stirred, jacketed tank. The reaction is essentially zero order and the conversion is normally 30%. Other reaction conditions are:

Feed: 4000 lb/hr containing 50% of component A
Heat capacity: 1 Btu/lb, °F for feed and products
Tank capacity: 8000 lb
Heat of reaction: 1200 Btu/lb A reacting
Normal temperatures: feed 80°F, reactor 130°F, jacket 110°F

The rate of heat generation is proportional to the fourth power of the absolute temperature over the range of 110 to 150°F. The jacket fluid is maintained at a uniform temperature by a high circulation rate.

(a) What is the time constant of the tank for slight changes in feed temperature?
(b) What is the approximate tank temperature 10 min after the feed temperature suddenly drops to 75°F?

Answers: (a) 0.319 hr
(b) 129.675°F

Problem 7. An exothermic reaction is carried out in a continuous-flow, stirred-tank reactor. The normal operating conditions are:

$$F = 6000 \text{ lb/hr}$$
$$U = 50 \text{ Btu/hr, ft}^2, \text{ }^\circ\text{F}$$
$$A = 100 \text{ ft}^2$$
$$c_p = 1 \text{ Btu/lb, }^\circ\text{F}$$
$$T_F = 184^\circ\text{F}$$
$$T_C = 150^\circ\text{F}$$

The heat of reaction is given by the expression:

$$Q = 3 \times 10^{15} e^{-E/R'T} \quad \text{Btu/hr}$$

where E = activation energy, 30,000 Btu/lb mol
R' = gas constant = 1.987 Btu/lb mol, $^\circ$R
T = absolute temperature, $^\circ$R = 459.7 + $^\circ$F

Find the three possible steady states and determine the stability of each.

Problem 8.   Draw the electrical analogy for a continuous-flow, stirred-tank reactor.

Problem 9.   The property of self-regulation was pointed out in conjunction with the liquid surge tank in Chapter 8. If appropriate, discuss the meaning of this term with reference to:

(a)   A temperature bulb.
(b)   A continuous-flow, stirred-tank heater.
(c)   A continuous-flow, stirred-tank reactor.

MISCELLANEOUS THERMAL ENERGY PROBLEMS

Problem 10.   A corner room in a house has two walls exposed to the outside. A top view is shown in Figure 11. The walls are 8 ft high. Two walls are essentially insulated as indicated by the cross-hatched lines; heat transfer through the ceiling and floor is also negligible. A hot water heating system is used and the normal temperature of the water in the finned tubes is about

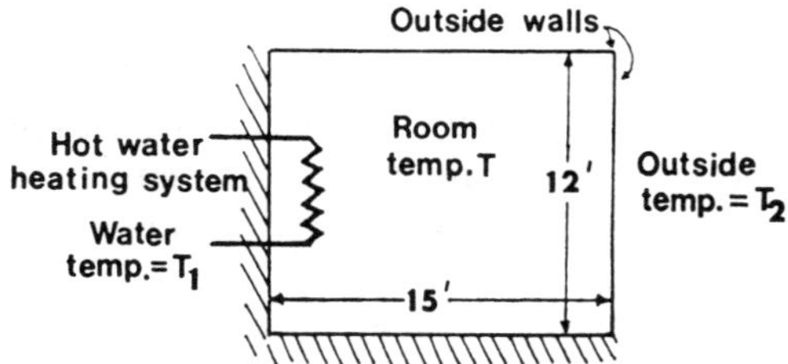

Figure 11.   Physical situation for Problem 10.

180°F. The effective area for heat transfer is 20 ft^2 and the overall coefficient of heat transfer is 10 Btu/hr, ft^2, °F. The only heat transfer is through the two outside walls and the overall coefficient of heat transfer is 0.1 Btu/hr, ft^2, °F. The heat capacity and density of air are:

$$c_p = 0.24 \text{ Btu/lb, }°F$$
$$\rho = 0.075 \text{ lb/ft}^3$$

(a) Estimate the time constant for changes in hot water temperature.
(b) Estimate the time constant for changes in outside air temperature.
(c) What assumption is probably most serious in making these calculations?

Answer: (a) 7.0 min

Problem 11. Referring to the example of variable thermal capacitance that was discussed in Section III.D of this chapter, suppose that the volume of the shell is 30 ft^3 and contains saturated steam at a pressure of 40 psia.

(a) In the derivation of the capacitance, it was assumed that the heat of condensation is constant. The resulting expression for the capacitance was:

$$C_s = \frac{\lambda m}{P}\left(\frac{dP}{dT}\right)_s \qquad \text{(Equation 42a)}$$

Using this expression, find the capacitance of the shell.
(b) If it is not assumed that the heat of condensation is constant, find an expression for the capacitance and calculate its value.

Answers: (a) 44.2 Btu/°F
(b) 39.7 Btu/°F

PROBLEMS CONCERNING UNSTEADY LIQUID FLOW THROUGH PIPES

Problem 12.   Water is flowing at a velocity of 5 ft/sec in a steel pipe having a nominal diameter of 2 in. The pipeline is 250 ft long and in addition it has fittings and valves whose equivalent length is 60 ft. Estimate the time constant for this flow process.

Answer: 1.01 sec

Problem 13.   The time constant for liquid flow in a line, given by Equation 59, was derived on the basis of plug flow, which is a reasonable assumption for the turbulent regime. When the flow is laminar, the velocity profile is parabolic if the flow is steady, but in unsteady flow, the profile is neither parabolic nor flat. Although some analytical solutions have been developed for this case (see, for example, Reference 5 of Chapter 11), an approximate value of the time constant can be calculated using Equation 59. The derivative, $d\,\delta P_L/dv$, is calculated from the Hagen-Poiseuille equation for laminar flow.

(a) Assuming that all of the resistance to flow is in the line and that resistances from fittings are negligible, find an expression for the time constant of the line.

(b) Based on your result for part *a*, is a time constant as large as 1 min reasonably possible?

$$\textit{Answer:} \ \text{(a)} \ \tau_x = \frac{\rho D^2}{32\mu}$$

CHAPTER   X

Second-Order Processes

A second-order process is one that is described by a second-order differential equation. As was true for first-order processes, second-order processes arise from particular combinations of passive elements. It is useful to classify these processes in three categories: (1) inherently second order (2) interacting second order; and (3) noninteracting second order. Before examining these, we shall first study some general properties of second-order behavior.

240

I. DEFINITION AND BEHAVIOR OF A SECOND-ORDER PROCESS

A. Standard Form

The differential equation for a second-order process can be arranged into the following standard form, similar to that given in Chapter 8 for a first-order process:

$$a \frac{d^2y}{dt^2} + b \frac{dy}{dt} + y = Kx \tag{1}$$

where y = output or response variable
  t = time
  x = input or forcing function
  K = gain factor

In the situations to be studied in this book, the coefficients, a and b, are constants. With zero initial conditions for y and its first derivative, Equation 1 may be Laplace transformed to yield the transfer function

$$\frac{Y}{X} = \frac{K}{as^2 + bs + 1} \tag{2}$$

Thus, we see that the process is characterized by three parameters, a, b, and K, and the characteristic polynomial is second order.

B. Response to a Unit Step Input

The response of a second-order process to a unit step input will be discussed here, and the response to a sinusoidal input will be taken up in Chapter 14. For a unit step, $X = 1/s$ so the transform, Y, is

$$Y = \frac{K}{s(as^2 + bs + 1)} = \frac{K/a}{s\left(s^2 + \frac{b}{a}s + \frac{1}{a}\right)} \tag{3}$$

The roots of the quadratic are obtained using the quadratic formula:

$$s_1 = -\frac{b}{2a} + \sqrt{\frac{1}{a}\left(\frac{b^2}{4a} - 1\right)} \tag{4a}$$

$$s_2 = -\frac{b}{2a} - \sqrt{\frac{1}{a}\left(\frac{b^2}{4a} - 1\right)} \tag{4b}$$

and

$$Y = \frac{K/a}{s(s - s_1)(s - s_2)} \tag{5}$$

Equation 5 can be expanded by partial fractions:

$$Y = \frac{A}{s} + \frac{B}{s - s_1} + \frac{C}{s - s_2} \tag{6}$$

where A, B, and C are constants. The roots depend only on the two constants, a and b; the form of the roots as exhibited by Equations 4a and 4b suggests defining these two constants in terms of two others:
Let

$$\frac{1}{a} = \omega_n{}^2 \tag{7a}$$

and

$$\frac{b^2}{4a} = \zeta^2 \tag{7b}$$

Then

$$s_1 = -\zeta\omega_n + \omega_n\sqrt{\zeta^2 - 1} \tag{8a}$$

and

$$s_2 = -\zeta\omega_n - \omega_n\sqrt{\zeta^2 - 1} \tag{8b}$$

In terms of ω_n and ζ, Equation 1 becomes

$$\frac{1}{\omega_n{}^2} \frac{d^2y}{dt^2} + \frac{2\zeta}{\omega_n} \frac{dy}{dt} + y = Kx \tag{9}$$

and the transfer function is

$$\frac{Y}{X} = \frac{K}{\dfrac{1}{\omega_n{}^2} s^2 + \dfrac{2\zeta}{\omega_n} s + 1} \tag{10}$$

Equations 6, 8a, and 8b indicate that the form of the solution depends on the magnitude of ζ relative to unity. Thus, three cases are possible.

1. Underdamped $\zeta < 1$—Case 1

In this case, the roots of the quadratic are complex and this leads to damped-oscillatory behavior, which is described by the following equation:

$$y = K\left[1 + \frac{1}{\sqrt{1 - \zeta^2}} e^{-\zeta\omega_n t} \sin\left(\omega_n\sqrt{1 - \zeta^2}\, t - \psi\right)\right] \tag{11a}$$

where

$$\psi = \tan^{-1} \frac{\sqrt{1 - \zeta^2}}{-\zeta}$$

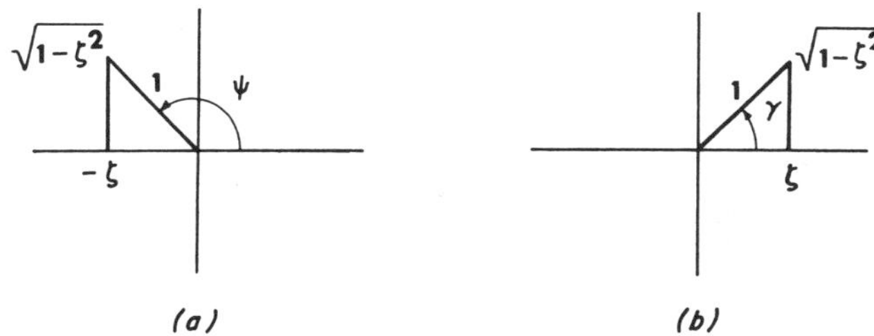

Figure 1.   Phase angle locations for second-order underdamped solutions. (*a*) Phase angle for Equation 11a. (*b*) Phase angle for Equation 11b.

The phase angle, ψ, is in the second quadrant, as illustrated by Figure 1*a*. An equivalent expression for y is sometimes used:

$$y = K\left[1 - \frac{1}{\sqrt{1 - \zeta^2}}\, e^{-\zeta\omega_n t} \sin\left(\omega_n\sqrt{1 - \zeta^2}\,t + \gamma\right)\right] \qquad (11b)$$

where

$$\gamma = \tan^{-1}\frac{\sqrt{1 - \zeta^2}}{\zeta}$$

and γ is in the first quadrant, as shown in Figure 1*b*.

If the parameter, ζ, is zero, then the exponential decay factor, $e^{-\zeta\omega_n t}$, is unity and the response is oscillatory with constant amplitude; in other words, no decay or damping occurs. Therefore, this parameter is called the "damping factor" or "damping coefficient." Also, when there is no damping, the frequency of oscillation is ω_n, and this is called the "natural frequency." With damping present, the frequency is $\omega_n\sqrt{1 - \zeta^2}$, and the effect of damping is to reduce the frequency below the natural value.

Although y could be plotted against t, a generalized plot is obtained by plotting y/K versus $\omega_n t$. A set of graphs is shown in Figure 2 for several values of the damping factor. Underdamped behavior is fairly common in some types of second-order processes and systems, and the important features of it are illustrated in Figure 3.

a. DERIVATIVE AT TIME ZERO

It can be shown from either Equation 11a or 11b that the derivative of y at time zero is zero. This is in marked contrast to the behavior of a first-order process that has its maximum slope at time zero when subjected to a unit step input.

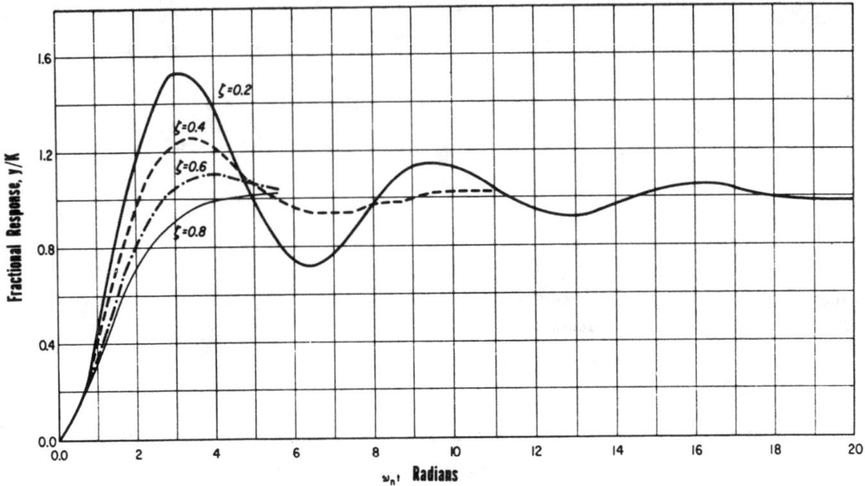

Figure 2.   Step response for underdamped second-order process.

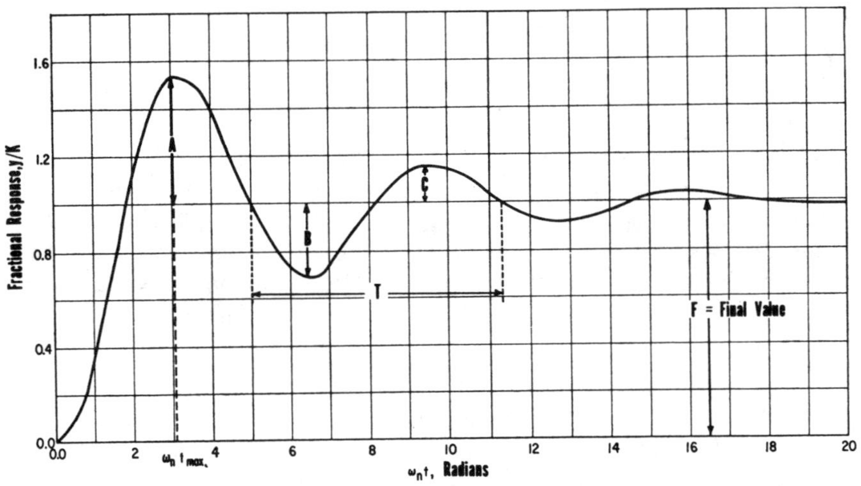

Figure 3.   Typical step response for underdamped behavior.

244

b. OVERSHOOT

The overshoot is the ratio, A/F, where A is the amount by which the first maximum differs from the final value or deviation. It can be shown that the overshoot is given by the expression:

$$\text{overshoot} = \exp\left(\frac{-\pi\zeta}{\sqrt{1-\zeta^2}}\right) \tag{12}$$

Thus, when there is no damping, the overshoot is unity; as the damping factor approaches unity, the overshoot approaches zero.

c. TIME OF FIRST MAXIMUM

The time of the first maximum is the time at which the overshoot is reached and is given by the following equation:

$$\omega_n t_{\max} = \frac{\pi}{\sqrt{1-\zeta^2}} \tag{13}$$

d. DECAY RATIO

The decay ratio is C/A, the ratio of successive maxima; in terms of the damping factor, it is:

$$\text{decay ratio} = \exp\left(\frac{-2\pi\zeta}{\sqrt{1-\zeta^2}}\right) = (\text{overshoot})^2 \tag{14}$$

e. PERIOD OF OSCILLATION

As mentioned before, the frequency of oscillation is $\omega_n\sqrt{1-\zeta^2}$ radians per unit time. Hence, the period is:

$$\text{period} = \frac{2\pi}{\omega_n\sqrt{1-\zeta^2}} \tag{15}$$

As the damping factor approaches zero, the period approaches $2\pi/\omega_n$. It is interesting to note that even when the damping factor is as high as 0.5, the frequency is still only about 13% less than the natural, undamped frequency.

2. Critically Damped $\zeta = 1$—Case 2

When the damping factor is unity, Equations 8a and 8b indicate that the two roots are identical and have the value $-\zeta\omega_n$. This double root results in the following response:

$$y = K[1 - (1 + \omega_n t)e^{-\omega_n t}] \tag{16}$$

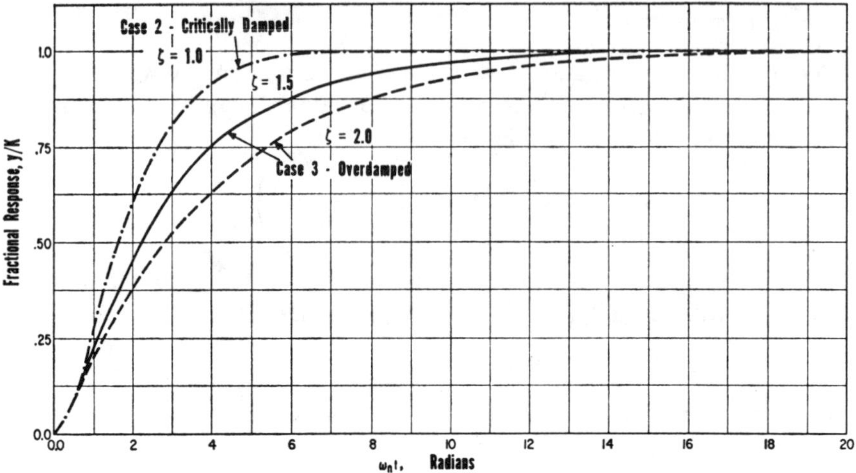

Figure 4.   Step response for critically damped and overdamped second-order process.

As shown in Figure 4, there is no overshoot or oscillation. The response approaches the final value asymptotically and in this respect, the behavior is similar to that of a first-order system. However, as for all second-order processes, the derivative at time zero is zero.

3. Overdamped $\zeta > 1$—Case 3

Equations 8a and 8b show that when the damping factor exceeds unity, both roots are real and negative. When this is the case, it is possible to break down the original second-order form of Equation 3 into the product of two first-order forms. This can be shown in the following way. We begin with Equation 3, which is repeated here for convenience:

$$Y = \frac{\dfrac{K}{a}}{s\left(s^2 + \dfrac{b}{a}s + \dfrac{1}{a}\right)} \tag{3}$$

Calling the two roots s_1 and s_2, Equation 3 can be expressed as

$$Y = \frac{\dfrac{K}{a}}{s(s - s_1)(s - s_2)} = \frac{\dfrac{K}{a}}{s[s^2 + (-s_1 - s_2)s + (-s_1)(-s_2)]} \tag{17}$$

Comparing this result with Equation 3, it can be seen that

$$a = \frac{1}{(-s_1)(-s_2)} \tag{18}$$

Thus, Equation 17 can be written

$$Y = \frac{K}{s\left(-\dfrac{s}{s_1} + 1\right)\left(-\dfrac{s}{s_2} + 1\right)} \tag{19}$$

Since the roots are real and negative, we can define two positive time constants:

$$\tau_1 = -\frac{1}{s_1} \tag{20a}$$

$$\tau_2 = -\frac{1}{s_2} \tag{20b}$$

These time constants are sometimes referred to as the "effective" time constants. Hence, the final transform has the form

$$Y = \frac{K}{s(\tau_1 s + 1)(\tau_2 s + 1)} \tag{21}$$

This is the product of two first-order forms.

Using transform pair 19 in Table 2 of Chapter 7, the response is

$$y = K\left[1 + \frac{1}{\tau_2 - \tau_1}(\tau_1 e^{-t/\tau_1} - \tau_2 e^{-t/\tau_2})\right] \tag{22a}$$

An alternative expression for y is

$$y = K\left\{1 - e^{-\zeta \omega_n t}\left[\cosh\left(\omega_n\sqrt{1 - \zeta^2}\,t\right) + \frac{\zeta}{\sqrt{1 - \zeta^2}}\sinh\left(\omega_n\sqrt{1 - \zeta^2}\,t\right)\right]\right\} \tag{22b}$$

It can be shown that the damping factor for this case is related to the effective time constants by the equation

$$\zeta = \frac{\tau_1 + \tau_2}{2\sqrt{\tau_1 \tau_2}} \tag{23}$$

Responses for two overdamped examples are shown in Figure 4. As the damping factor increases, the behavior approaches that of a single first-order process, but the derivative at time zero always remains zero.

II. INHERENTLY SECOND-ORDER PROCESSES

We turn now to an examination of the first category of second-order processes, which is comprised of those which are inherently second order. A process of this kind is defined as one whose analog can be reduced to a combination of a single lumped capacitance with a single lumped resistance and a single lumped inertance. Such combinations are frequently employed as "tuned circuits" in electrical networks but significant examples are comparatively rare in chemical processes. One example is a liquid-filled line that is terminated by an expandable bellows. The bellows provides the capacitance, and as we saw in Chapter 9, the mass of the liquid in the line has inertance and the fluid friction in the line accounts for the resistance. This combination would rarely occur in a process unless it was used for damping pulsations in the line.

Inherently second-order processes frequently arise in situations involving mechanical translation. The vibration of a mass that is suspended by a spring and damped by a dash pot is familiar to engineers from their training in physics and dynamics. The mass has inertance, the spring provides the capacitance, and the dash pot resists the motion. The mechanical parts of a pneumatic controller and those of a pneumatic valve are examples of this kind of process and the latter will be discussed momentarily.

As mentioned in Chapter 8, there is no analog of inertance in heat transfer, so there are no inherently second-order processes in that area. Two examples of inherently second-order processes follow.

A. Manometer

A sketch of a vertical, U-tube manometer is presented in Figure 5. Initially, the pressures in the two legs are equal so the two liquid levels are at the datum

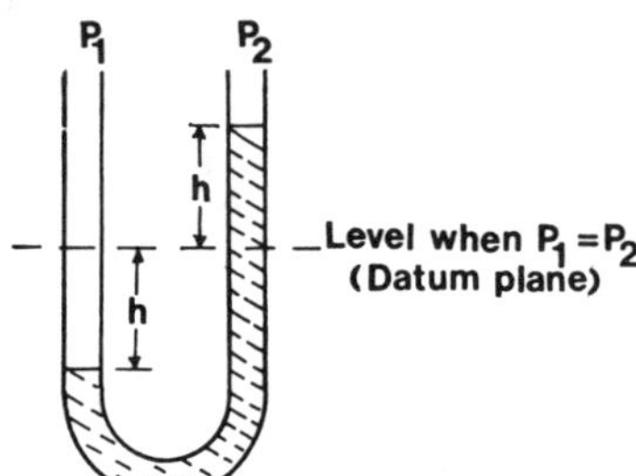

Figure 5.   U-tube manometer.

plane. We shall suppose that there is a sudden step increase in the pressure difference, $P_1 - P_2$, and the object of the analysis is to determine the resulting response of the level.

The step increase in pressure difference immediately results in an unbalanced force on the liquid, causing it to accelerate. This is impeded by a frictional resistance to fluid flow. To simplify the analysis, it will be assumed that the liquid accelerates with a flat velocity profile and that the resistance is independent of the velocity.

In accordance with Newton's Law, the following equation can be written:

$$\begin{pmatrix}\text{input force} \\ \text{difference}\end{pmatrix} - \begin{pmatrix}\text{force from} \\ \text{head} \\ \text{difference}\end{pmatrix} - \begin{pmatrix}\text{force from} \\ \text{fluid} \\ \text{friction}\end{pmatrix} = \text{mass} \times \text{acceleration}$$

or

$$P_i S \quad - \quad \frac{2\rho g S}{g_c} h \quad - \quad SRF \quad = \quad \frac{M}{g_c}\frac{dv}{dt} \tag{24}$$

where P_i = input pressure differential, $P_1 - P_2$, $\mathrm{lb_f/ft^2}$
  S = cross-sectional area of tube, $\mathrm{ft^2}$
  ρ = liquid density, $\mathrm{lb_m/ft^3}$
  g = acceleration due to gravity, $\mathrm{ft/sec^2}$
  g_c = conversion constant, 32.17 $\mathrm{lb_m}$, $\mathrm{ft/lb_f}$, $\mathrm{sec^2}$
  h = deviation in liquid level from datum plane, ft
  R = resistance, $(\mathrm{lb_f/ft^2})/(\mathrm{ft^3/sec}) = \mathrm{lb_f}$, $\mathrm{sec/ft^5}$
  F = flow rate, $\mathrm{ft^3/sec}$
  M = ρSL = mass of liquid in tube, $\mathrm{lb_m}$
  v = average velocity of liquid, ft/sec
  t = time, sec
  L = length of liquid column, ft

The level, h, is related to F by:

$$F = S\frac{dh}{dt} \tag{25}$$

Transforming Equation 25, we obtain

$$F = Ssh \tag{26}$$

or

$$h = \frac{1}{Ss}F \tag{27}$$

Also

$$S\frac{dv}{dt} = \frac{dF}{dt} \tag{28}$$

Using Equations 27 and 28, the transformation of Equation 24 is

$$\left(\frac{M}{S^2 g_c}s + \frac{2\rho g}{Sg_c}\frac{1}{s} + R\right)F = P_i \tag{29}$$

The first factor on the left, $M/(S^2 g_c)$, is the same as that found in the fluid flow example of Section IV.A.3 of Chapter 9 and is the inertance, I:

$$I = \frac{M}{S^2 g_c} \tag{30}$$

The middle factor, $2\rho g/(S g_c)$, has units of lb_f/ft^5, which are those of reciprocal capacitance for a fluid flow system. Hence

$$C = \frac{S g_c}{2\rho g} \tag{31}$$

A physical interpretation of this capacitance can be developed as follows. Consider a stationary column of liquid of height, $2h$. The pressure difference, ΔP, between the top and the bottom of this column is related to the height of the column by the expression

$$\Delta P = \frac{2h\rho g}{g_c} \qquad lb_f/ft^2$$

This can be rewritten as

$$\frac{hS}{\Delta P} = \frac{S g_c}{2\rho g} \tag{32}$$

Referring to the sketch of the manometer, we see that hS is the volume of fluid stored above the datum plane and this storage is caused by the pressure drop, ΔP. Since the ratio of these quantities conforms to the definition of capacitance, the physical interpretation of this property for this process is now clear.

Using Equations 30 and 31, Equation 29 can be expressed as

$$\frac{F}{P_i} = \frac{1}{Is + \dfrac{1}{Cs} + R} = \frac{Cs}{ICs^2 + RCs + 1} \tag{33}$$

This equation can be rewritten as follows:

$$IsF + \frac{F}{Cs} + RF = P_i \tag{33a}$$

Each term then has the units of lb_f/ft^2, which are analogous to those of a voltage drop. The corresponding electrical analogy appears in Figure 6. We see that the three passive elements are in series.

The unbalanced input pressure, P_i, can be expressed in terms of an unbalanced head, h_i:

$$h_i = \frac{P_i g_c}{2\rho g} \tag{34}$$

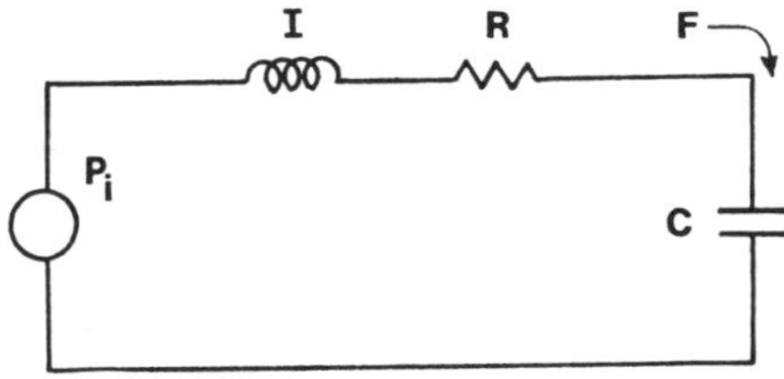

Figure 6. Electrical analogy of a manometer.

and

$$M = \rho SL \tag{35}$$

Using Equations 26, 34, and 35, Equation 29 can be rearranged and expressed as

$$\left(\frac{L}{2g} s^2 + \frac{RSg_c}{2\rho g} s + 1 \right) h = h_i \tag{36}$$

Thus, the transfer function is

$$\frac{h}{h_i} = \frac{1}{\dfrac{L}{2g} s^2 + \dfrac{RSg_c}{2\rho g} s + 1} \tag{37}$$

Using Equations 30 and 31, this equation can be written

$$\frac{h}{h_i} = \frac{1}{ICs^2 + RCs + 1} \tag{38}$$

Note that if the resistance were zero, only the middle term would vanish. Hence, second-order behavior would still result from the combination of a single lumped capacitance with a single lumped inertance.

The resistance can be estimated using the Hagen-Poiseuille equation. This results in the following expression for R:

$$R = \frac{32\mu L}{SD^2 g_c} \tag{39}$$

The natural frequency and damping factor are given by the following expressions:

$$\omega_n = \sqrt{\frac{2g}{L}} \tag{40}$$

$$\zeta = \frac{8L\mu}{\rho g D^2} \sqrt{\frac{2g}{L}} \tag{41}$$

Harriott (1i) gives several typical examples of manometers filled with water or mercury and in all cases they are very underdamped with damping coefficients between 0.024 and 0.15. Some commercial manometers in flow control systems have one leg of relatively large cross section and another of much smaller cross section. These are connected by a small valved line. The valve introduces additional resistance into the manometer and if the valve is closed sufficiently, overdamped behavior can be obtained.

B. Pneumatic Valve

Consider the typical single-seat control valve shown in Figure 7. This is an air-to-close type and the position of the stem is shown by the scale attached to the side of the body. Since the mechanical parts of the valve involve a

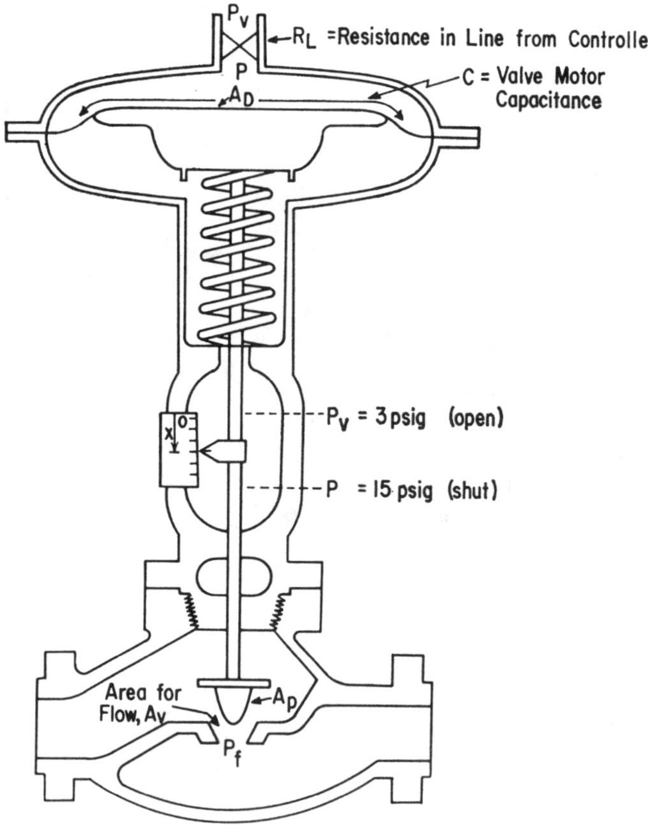

Figure 7.   Pneumatic control valve.

mass attached to a spring, second-order behavior is to be expected. As indicated in the drawing, the actual pressure in the valve motor is P and this will differ some from the output pressure at the controller, P_V, partially because of the resistance in the pneumatic transmission line and the capacitance of the valve motor.

The analysis of the dynamic behavior of the valve centers around the application of Newton's Law to the stem position of the valve. There are five forces to consider. Referring to the figure, the first is the force exerted by the diaphragm on the stem. The air supplied by the controller output acts against this diaphragm. The standard range of pressures is 3 to 15 psig, so if the area of the diaphragm were 50 in.^{2}, then the downward force on the stem could vary between 150 and 750 lb$_f$. If the valve is mounted vertically, then a second downward acting force is that of gravity.

A third force is that of the spring that acts upward and is proportional to the displacement, x. The maximum displacement is governed by the diaphragm construction and may be as small as $\frac{1}{4}$ in. in small valves, or perhaps as large as 3 in. in large ones. A fourth force, acting upward, is the so-called "thrust force" provided by the pressure of the fluid acting against the area of the plug, A_P. The fifth force is frictional and arises mainly from the close contact between the stem and the valve packing. The frictional force is roughly proportional to the velocity of the stem, dx/dt.

With this background, we can now write the equation for the stem in accordance with Newton's Law. The downward direction is taken as positive.

$$PA_D + M\frac{g}{g_c} - Kx - P_f A_P - R\frac{dx}{dt} = \frac{M}{g_c}\frac{d^2x}{dt^2} \qquad (42)$$

where P = pressure in the valve motor, lb$_f$/ft^2
  A_D = area of diaphragm, ft^2
  g = acceleration of gravity, ft/sec^2
  g_c = conversion constant, 32.17 lb$_m$, ft/lb$_f$, sec^2
  K = spring constant, lb$_f$/ft
  x = stem position, ft
  P_f = fluid pressure at valve seat, lb$_f$/ft^2
  A_P = area normal to valve plug, ft^2
  R = coefficient of friction between stem and packing, lb$_f$, sec/ft

Since this is a case of mechanical translation, the resistance, R, is defined as the ratio of the force to the velocity, as indicated in Table 3 of Chapter 8.

Possible input variables in Equation 42 are the pressure in the valve motor, P, and the pressure in the fluid at the valve seat, P_f. The equation is linear in these variables as well as in the dependent variable, x. Thus, Equation 42 can be considered to be a deviation equation in these variables if the gravitational force term is dropped since it is a constant. Our interest is mainly in the transfer function, x/P, rather than in x/P_f. Therefore, taking the deviation in P_f to be zero, the desired transfer function is readily obtained by transformation and rearrangement to yield

$$\frac{x}{P} = \frac{A_D/K}{\dfrac{M}{Kg_c}\, s^2 + \dfrac{R}{K}\, s + 1} \tag{43}$$

This transfer function indicates that the mechanical behavior of the valve is second order with the following natural frequency:

$$\omega_n = \sqrt{\frac{Kg_c}{M}} \qquad \text{rad/sec} \tag{44}$$

or

$$f_n = \frac{1}{2\pi}\sqrt{\frac{Kg_c}{M}} \qquad \text{cps} \tag{45}$$

The natural frequency is of some concern since oscillations can occur if there is little frictional resistance. Eckman (2i) points out that the natural frequency of a valve should be less than 25 cps.

The analysis here has neglected static frictional forces, which cause hysteresis effects. If these are serious and if the response of the valve is too slow, a valve positioner is used. This is a small feedback device that greatly reduces these problems. A description of its operation can be found in References 1ii and 2ii.

For small changes in stem position, the change in area, A_v, for flow is linearly related to the change in stem position:

$$A_v = K_s x \tag{46}$$

Combining this with Equation 43, the transfer function relating the change in area to the change in pressure in the valve motor is

$$\frac{A_v}{P} = \frac{K_s A_D/K}{\dfrac{M}{Kg_c}\, s^2 + \dfrac{R}{K}\, s + 1} \tag{47}$$

Using Table 3 of Chapter 8, it can be seen that the inertance analog is M/g_c, the capacitance analog is $1/K$, and the resistance analog is R. Hence, the analog form of the denominator of Equation 47 is exactly the same as that of Equation 38.

III. INTERACTING AND NONINTERACTING SECOND-ORDER PROCESSES

A. Two Liquid Tanks in Series

Although inherently second-order processes were discussed under a separate heading, the second and third categories—interacting and noninteracting second-order processes—are best understood if they are studied in parallel rather than sequentially. The classic example to illustrate the difference is that of two tanks in series, as shown in Figure 8. In the noninteracting case,

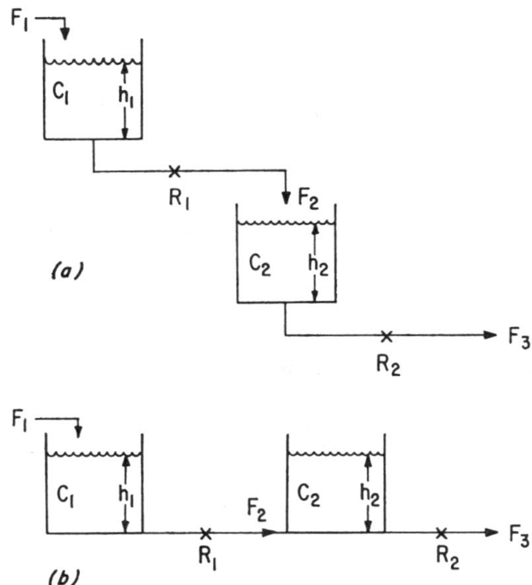

Figure 8.   (a) Noninteracting tanks. (b) Interacting tanks.

the behavior of the first tank is completely independent of the behavior of the second. On the other hand, in the interacting case, the flow rate, F_2, depends on the difference in level between the two tanks, so the behavior of the first tank does depend on the behavior of the second. The cases are physically quite different, and as will be shown shortly, the derivation of the transfer function, h_2/F_1, is simply obtained for the noninteracting case, but requires some algebra for the interacting one.

1. Noninteracting Tanks

Referring to Figure 8*a*, the first tank behaves independently of the second, so the treatment of Chapter 8 applies. From Equations 36 and 41 of Chapter 8, the transfer function relating h_1 to F_1 is

$$\frac{h_1}{F_1} = \frac{R_1}{\tau_1 s + 1} \tag{48}$$

where

$$\tau_1 = R_1 C_1 \tag{49}$$

and

$$R_1 = \frac{\partial h_1}{\partial F_2} \tag{50}$$

$$C_1 = \text{cross-sectional area of Tank 1} \tag{51}$$

The resistance, R_1, is the ratio of the change in level to the change in flow. Hence, the transfer function, F_2/h_1, is just the inverse of this resistance:

$$\frac{F_2}{h_1} = \frac{1}{R_1} \tag{52}$$

Multiplying the transfer functions of Equations 48 and 52, we obtain

$$\frac{F_2}{F_1} = \frac{h_1}{F_1} \times \frac{F_2}{h_1} = \frac{1}{\tau_1 s + 1} \tag{53}$$

Looking now at the second tank, the flow rate, F_2, is completely independent of the level, h_2. Therefore, Equation 48 also applies to this tank with appropriate changes in symbols:

$$\frac{h_2}{F_2} = \frac{R_2}{\tau_2 s + 1} \tag{54}$$

where

$$\tau_2 = R_2 C_2 \tag{55}$$

and

$$R_2 = \frac{\partial h_2}{\partial F_3} \tag{56}$$

$$C_2 = \text{cross-sectional area of Tank 2} \tag{57}$$

Combining Equations 53 and 54, we obtain the desired transfer function:

$$\frac{h_2}{F_1} = \frac{F_2}{F_1} \times \frac{h_2}{F_2} = \frac{R_2}{(\tau_1 s + 1)(\tau_2 s + 1)}$$

$$= \frac{R_2}{\tau_1 \tau_2 s^2 + (\tau_1 + \tau_2)s + 1} \tag{58}$$

The second-order parameters are:

$$\omega_n = \frac{1}{\sqrt{\tau_1 \tau_2}} \tag{59}$$

and

$$\zeta = \frac{\tau_1 + \tau_2}{2\sqrt{\tau_1 \tau_2}} \tag{60}$$

The damping factor is the ratio of the arithmetic mean of the time constants to their geometric mean. Since this ratio is always greater than or equal to unity, the damping factor is greater than unity; hence, the response is always overdamped. This conclusion can also be reached by observing that the roots of the characteristic equation are $-1/\tau_1$ and $-1/\tau_2$, both real and negative, and therefore they result in decaying exponential solutions without oscillations.

2. Interacting Tanks

The derivation of this case will be sketched out building upon our background from Chapter 8. The deviation equation for Tank 1 is

$$F_1 - F_2 = C_1 \frac{dh_1}{dt} \tag{61}$$

In terms of the driving force and resistance, F_2 is:

$$F_2 = \frac{h_1 - h_2}{R_1} \tag{62}$$

Hence, Equation 61 can be written

$$F_1 - \frac{h_1 - h_2}{R_1} = C_1 \frac{dh_1}{dt} \tag{63}$$

Transforming and rearranging, we obtain

$$h_1 = \frac{R_1 F_1 + h_2}{\tau_1 s + 1} \tag{64}$$

where $\tau_1 = R_1 C_1$, as in the noninteracting case

Turning now to Tank 2, we have

$$F_2 - F_3 = C_2 \frac{dh_2}{dt} \tag{65}$$

But

$$F_3 = \frac{h_2}{R_2} \tag{66}$$

Substitution of Equations 62 and 66 in Equation 65 yields

$$\frac{h_1 - h_2}{R_1} - \frac{h_2}{R_2} = C_2 \frac{dh_2}{dt} \tag{67}$$

Transforming and rearranging this equation, we obtain

$$h_1 = \left(C_2 R_1 s + 1 + \frac{R_1}{R_2} \right) h_2 \tag{68}$$

Equation 64 can now be used to eliminate h_1. After some algebra, the final transfer function is

$$\frac{h_2}{F_1} = \frac{R_2}{\tau_1 \tau_2 s^2 + (\tau_1 + \tau_2 + R_2 C_1)s + 1} \tag{69}$$

where $\tau_2 = R_2 C_2$

Comparing Equations 69 and 58, we see that they differ only by the term, $R_2 C_1$, in the middle coefficient of the denominator. This term might be called the "interaction factor" and a comparison of its magnitude with those of τ_1 and τ_2 will indicate the degree of interaction between the two tanks.

For a second-order process, the damping factor is proportional to the middle coefficient in the denominator. It was pointed out that the damping factor for the noninteracting case is always greater than one. Since the middle coefficient for the interacting case exceeds that of the noninteracting case by $R_2 C_1$, the interacting case will also always be overdamped. Hence, the behavior can be characterized by two effective time constants:

$$\frac{h_2}{F_1} = \frac{R_2}{(\tau_a s + 1)(\tau_b s + 1)} \tag{70}$$

where τ_a and τ_b are the effective time constants.

3. Electrical Analogies

a. NONINTERACTING TANKS

The analogy for this case can be drawn using the development for a single tank in Chapter 8. For convenience, the physical picture and the corresponding analogy for a single tank are presented in Figure 9. For two tanks in series, as shown in Figure 10a, there is a temptation to draw the electrical analogy as indicated in Figure 10b. However, it will be noticed that the circuit indicates that there is a driving force, $h_1 - h_2$, across the resistance, R_1. This is not the physical situation since the tanks are completely isolated from each other. The key point is that the flow from the first tank acts as a current source for the second tank, and hence, the correct diagram is that shown in Figure 10c.

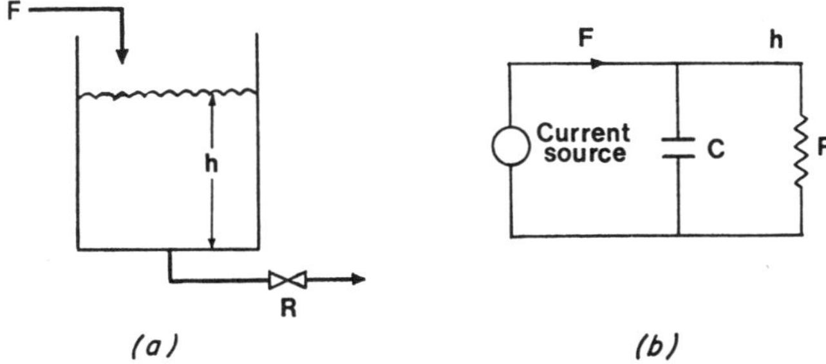

Figure 9.   Single tank. (*a*) Physical picture. (*b*) Electrical analogy.

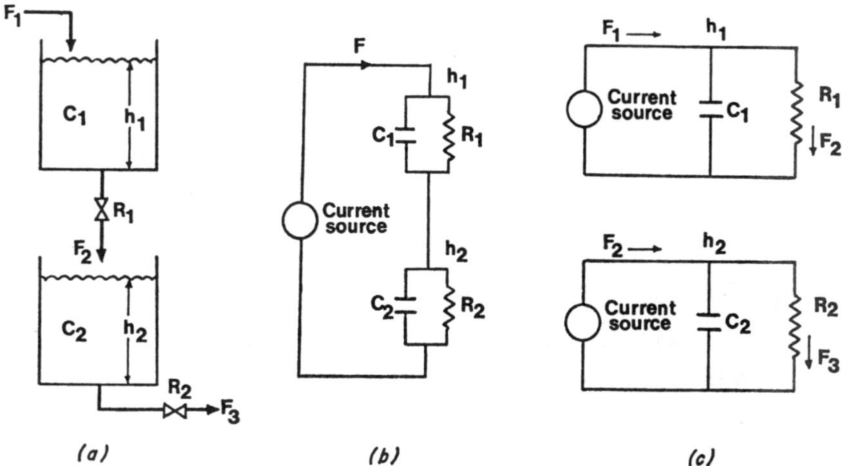

Figure 10.   Two noninteracting tanks. (*a*) Physical picture. (*b*) Incorrect electrical analogy. (*c*) Correct electrical analogy.

b. INTERACTING TANKS

The electrical analogy is best deduced from Equations 63 and 67. Each term has the units of flow rate that is analogous to current. The physical situation and electrical analogy are shown in Figure 11 and the two equations are repeated in the figure as an aid in understanding the analogy. Here, the potential, $h_1 - h_2$, is across the resistance, R_1, in accordance with the physical situation.

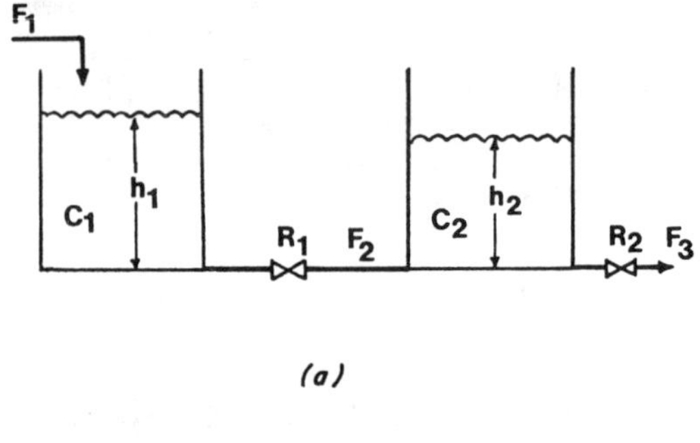

(a)

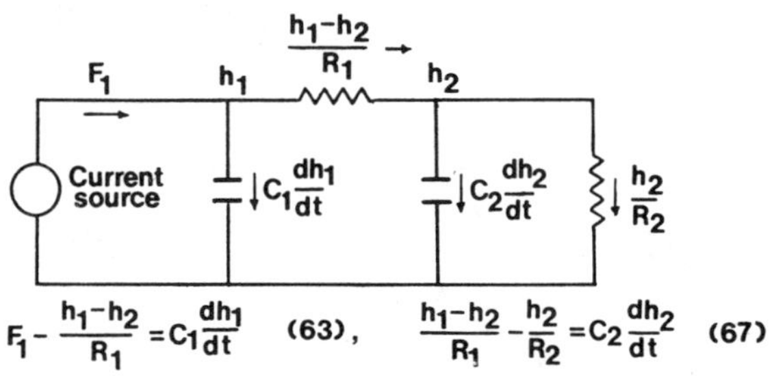

$$F_1 - \frac{h_1 - h_2}{R_1} = C_1 \frac{dh_1}{dt} \quad (63), \qquad \frac{h_1 - h_2}{R_1} - \frac{h_2}{R_2} = C_2 \frac{dh_2}{dt} \quad (67)$$

(b)

Figure 11.   Two interacting tanks. (*a*) Physical Picture. (*b*) Electrical analogy.

B. Temperature Bulb in a Well

As a second example of an interacting process, we shall examine the behavior of a temperature bulb when placed in a well, as illustrated in Figure 12. A well is frequently used to protect the bulb from corrosion and erosion. We learned in Chapter 9 that a bare bulb behaves as a first-order process. When the bulb is placed in the well, there is a small gap between the two that results in a resistance to heat transfer. As we shall see momentarily, this combination of two thermal masses, separated by a resistance, results in second-order behavior.

When the well and bulb are suddenly exposed to fluid at a higher temperature, a gradient occurs, as shown in the figure. The gradients in the bulb and well are small compared to those in the outside film and gap. Therefore, to simplify the analysis, average temperatures, T_w and T_b, can be used for them. Turning first to the well, the energy balance is

$$h_oA_o(T_o - T_w) - h_iA_i(T_w - T_b) = C_w \frac{dT_w}{dt} \qquad (71)$$

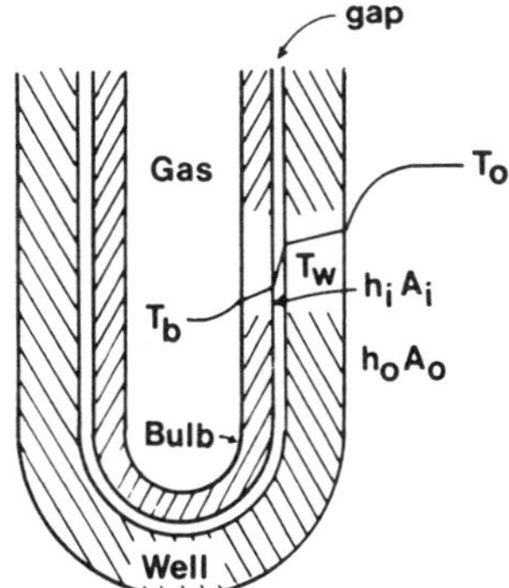

Figure 12.   Temperature bulb in a well.

This is a deviation equation for the several temperatures and the symbols have the following meanings:

h_o = outside film coefficient of heat transfer, Btu/hr, ft², °F
A_o = outside area, ft²
T_o = deviation in outside temperature from original steady-state value, °F
T_w = deviation in well temperature from original steady-state value, °F
h_i = inside film coefficient of heat transfer, Btu/hr, ft², °F
A_i = inside area, ft²
T_b = deviation in bulb temperature from original steady-state value, °F
C_w = heat capacitance of well, Btu/°F
t = time, hr

Anticipating some possible analogies, we define:

$$R_1 = \frac{1}{h_o A_o} \qquad \text{outside film} \tag{72}$$

$$R_2 = \frac{1}{h_i A_i} \qquad \text{gap film} \tag{73}$$

$$C_1 = C_w \qquad \text{well} \tag{74}$$

Thus Equation 71 can be written as

$$\frac{T_o - T_w}{R_1} - \frac{T_w - T_b}{R_2} = C_1 \frac{dT_w}{dt} \tag{75}$$

Transforming and rearranging Equation 75, we obtain

$$\left(C_1 s + \frac{1}{R_1} + \frac{1}{R_2} \right) T_w = \frac{T_o}{R_1} + \frac{T_b}{R_2} \tag{76}$$

The heat balance on the bulb is

$$h_i A_i (T_w - T_b) = C_b \frac{dT_b}{dt} \tag{77}$$

where C_b = heat capacitance of bulb, Btu/°F

Again, for purposes of analogy, we define:

$$C_2 = C_b \qquad \text{bulb} \tag{78}$$

Thus, Equation 77 can be written

$$\frac{T_w - T_b}{R_2} = C_2 \frac{dT_b}{dt} \tag{79}$$

Transforming and rearranging, we obtain

$$T_w = (C_2 R_2 s + 1) T_b \tag{80}$$

Since the transfer function T_b/T_o is desired, Equation 80 can be used to eliminate T_w in Equation 76; the final result is

$$\frac{T_b}{T_o} = \frac{1}{\tau_1 \tau_2 s^2 + (\tau_1 + \tau_2 + R_1 C_2)s + 1} \tag{81}$$

where $\tau_1 = R_1 C_1$
$\qquad \tau_2 = R_2 C_2$

This transfer function is quite similar in form to that for the interacting tanks, given by Equation 69. However, an examination of the electrical analogy reveals some striking differences. The analogy is obtained with the aid of Equations 75 and 79. These equations and the analogy are presented in Figure 13. In contrast to the tank example, this thermal analogy is driven by a voltage source, analogous to a temperature driving force; furthermore, the arrangement of resistances and capacitances is quite different. It is not

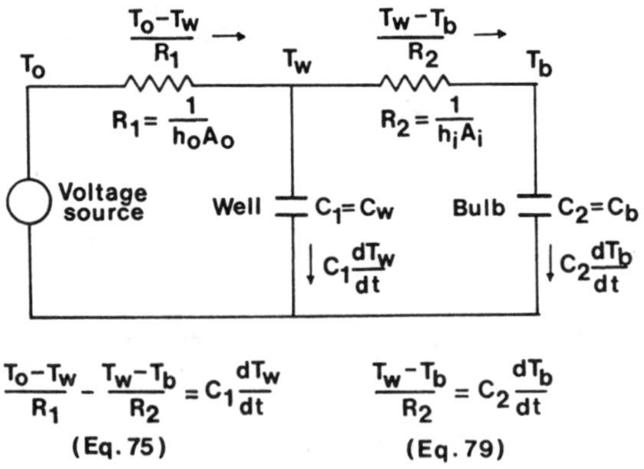

Figure 13.   Electrical analogy of a temperature bulb in a well.

surprising, therefore, to find that the "interaction factor" is R_1C_2 here, whereas it was R_2C_1 for the two tanks.

From purely physical considerations, it is obvious that the response will improve if the gap resistance, R_2, is decreased. Therefore, sometimes a tight-fitting sleeve is placed between the bulb and the well. Problem 10 at the end of the chapter is related to this.

C. Interacting Processes Having Weak Interactions

As mentioned earlier, the degree of interaction in second-order processes can be assessed by comparing the magnitude of the interaction factor with the two time constants, τ_1 and τ_2. While this comparison constitutes a quantitative test of the interaction, there are many interacting processes in which the interaction is obviously insignificant.

Consider, for example, the temperature bulb in the stirred tank shown in Figure 14. The size of the bulb relative to that of the tank is greatly exaggerated in the drawing. Fluid enters the tank at rate, F, and at temperature,

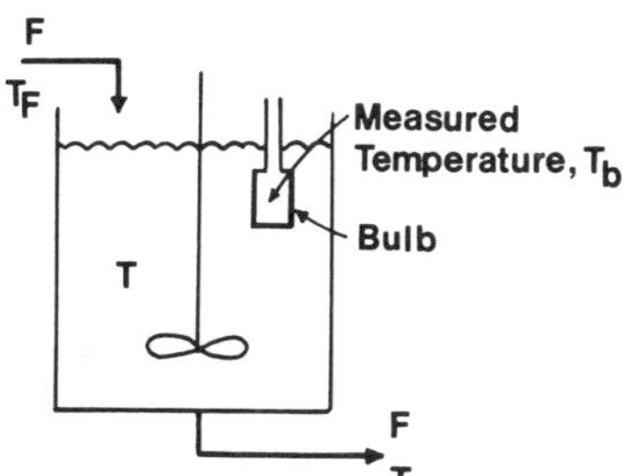

Figure 14.   An example of negligible interaction.

T_F. The fluid is thoroughly mixed in the tank and leaves at temperature, T. This temperature is sensed by the bulb and the indicated temperature is T_b. Suppose that the desired transfer function is T_b/T_F.

There is an interaction here between the fluid and the bulb, because the rate of heat transfer between them is proportional to their temperature difference, $T - T_b$. However, this flow rate of heat is clearly an insignificant term in the energy balance for the tank fluid, and therefore the interaction is negligible. Hence, it is legitimate to obtain the desired transfer function by multiplication of the transfer functions, T/T_F and T_b/T. Problem 11 at the end of this chapter concerns a quantitative investigation of this example.

SUMMARY

In this chapter, we have examined a number of second-order processes of various types. Inherently second-order processes assume importance only occasionally in chemical engineering. Interacting second-order cases occur frequently. The interaction is reflected in a resistance-capacitance product having no physical significance. Sometimes it is R_2C_1, as in the case of two interacting tanks in series, and sometimes it is R_1C_2, as in the case of the bulb in a well. At any rate, if this product is small compared with R_1C_1 and R_2C_2, then the process behaves as though it consists of two noninteracting first-order processes in series having time constants R_1C_1 and R_2C_2. On the other hand, if the interaction is significant, the transfer function can be factored into two first-order transfer functions having effective time constants which have no physical significance.

REFERENCES

1. P. Harriott, *Process Control*, McGraw-Hill Book Co., Inc., New York, 1964, (i) p. 47, (ii) p. 211.
2. D. P. Eckman, *Automatic Process Control*, John Wiley and Sons, Inc., New York, 1958, (i) p. 197, (ii) p. 198.

PROBLEMS

PROBLEMS CONCERNING GENERAL SECOND-ORDER FORMS

Problem 1. The response of an underdamped second-order system to a unit step input is given by Equation 11b. Show that the derivative of y at time zero is zero. This can also be demonstrated using Equation 10 since the transform of the derivative of y is sY. Thus, show that the initial derivative is zero using the initial-value theorem.

Problem 2. In each of the following, determine the natural frequency, damping factor, and gain factor. If the system is overdamped, find the effective time constants.

(a) $100\dfrac{d^2y}{dt^2} + 3.2\dfrac{dy}{dt} + 4y = 20x$

(b) $72\dfrac{d^2y}{dt^2} + 42\dfrac{dy}{dt} + 6y = 30x$

Answers: (a) $\omega_n = 0.2;\ \zeta = 0.08$
underdamped

(b) $\omega_n = 1/\sqrt{12};\ \zeta = 1.01$
overdamped
$\tau_1 = 3;\ \tau_2 = 4$

PROBLEMS CONCERNING INHERENTLY SECOND-ORDER PROCESSES

Problem 3.   A mercury-filled manometer is to be used to measure the pressure drop across an orifice meter. The meter will monitor the flow rate of air through a pipe. Specifications call for a decay ratio (not damping factor) of 0.3. The maximum pressure differential is such that a liquid column length of 30 in. should suffice.

(a)   Find the diameter (in cm) of suitable glass tubing for this meter. The specific gravity of mercury is 13.6 and the viscosity is 1.6 cp.
(b)   The calculation of part *a* having been made, you find that the only tubing available has a diameter of 0.2 cm. What should be done to meet the specification?

Answer: (a) 0.141 cm

Problem 4.   The treatment of the valve in this chapter culminated in the transfer function, x/P, given in Equation 43. In Chapter 8, the pressure in a bellows, P, was related to the input pressure, P_1, by Equation 93. It was pointed out that a valve motor is essentially a bellows.

(a)   The motor interacts with the mechanical parts of the valve. In the treatment of Chapter 8, the dynamics of these parts were neglected and the approximate transfer function of the valve was given by Equation 94. When the valve stem dynamics become important, the interaction cannot be neglected. Hence, it is not legitimate to obtain the transfer function, x/P_V, by multiplying the transfer functions, x/P and P/P_V (P_V is identically P_1 in the treatment of Chapter 8). In reviewing the treatment of Chapter 8, it turns out that the interaction is expressed in Equation 85, but this equation is incorrect if the mechanical parts of the valve have significant dynamic behavior. Using this hint, find the true transfer function, x/P_V.
(b)   If the interaction is neglected and the transfer function is obtained by multiplying the transfer functions, x/P and P/P_V, how does the result compare with that of part *a*?

$$\text{Answer:} \text{ (a) } \frac{x}{P_V} = \frac{\text{gain}}{a_0 s^3 + a_1 s^2 + a_2 s + 1}$$

$$\text{where gain} = A_D/K$$

$$a_0 = \frac{\overline{V} R_L}{14.7}\left(\frac{M}{Kg_c}\right)$$

$$a_1 = \frac{M}{Kg_c} + \frac{\overline{V} R_L R}{14.7K}$$

$$a_2 = \frac{R}{K} + \frac{P A_D^2 R_L}{14.7K} + \frac{\overline{V} R_L}{14.7}$$

PROBLEMS CONCERNING INTERACTING AND NONINTERACTING
SECOND-ORDER PROCESSES

Problem 5.   Liquid is flowing through a cascade arrangement of two non-interacting tanks, each open at the top. Tank dimensions and operating conditions are:

Tank 1
Cross-sectional area	20 ft^2
Tank height	9 ft
Normal level	6 ft
Normal flow rate	$10 \text{ ft}^3/\text{min}$

Tank 2
Cross-sectional area	20 ft^2
Tank height	9 ft
Normal level	8 ft

A workman carelessly dumps 50 ft^3 of liquid into the top of the first tank, thinking that there is no chance of any liquid overflowing the walls of either tank.

(a)   Does the first tank overflow? If not, go to part *b*.
(b)   Does the second tank overflow?

> *Answers:* (a) Tank 1 does not overflow
> (b) Tank 2 overflows about 19.5 min later

Problem 6.   Consider the derivation of the transfer function, h_2/F_1, for the interacting tanks, as given in Equations 61 to 69. Which equation in this derivation is basically responsible for the interaction and how would it be changed if there were no interaction?

Problem 7.   The parameters of an interacting tank system are as follows: $R_1 = 1 \text{ sec/ft}^2$; $R_2 = 2 \text{ sec/ft}^2$; $C_1 = 2 \text{ ft}^2$; and $C_2 = 3 \text{ ft}^2$.

(a)   Is the interaction significant?
(b)   Find the effective time constants.

Problem 8.   Two tanks are connected in an interacting fashion, as shown in Figure 8*b*. In addition, however, the second tank has a second exit line at the bottom, which vents to the atmosphere. This line contains a valve, R_3.

(a)   Find the transfer function, h_2/F_1.
(b)   Draw the electrical analogy for this process.
(c)   Suggest a heat-transfer analogy for the entire process including both tanks.

Problem 9.   Two tanks are connected in an interacting fashion as shown in Figure 8*b*. In addition, however, the first tank has a second exit line at the bottom, which vents to the atmosphere. This line contains a valve, R_3.

(a)   Find the transfer function, h_2/F_1. Partially check your result by showing that it reduces to Equation 69 when the valve, R_3, is shut.
(b)   Draw the electrical analogy for this process.

$$\text{Answer: (a)} \quad \frac{h_2}{F_1} = \frac{K}{as^2 + bs + 1}$$

$$\text{where } K = \frac{R_1 R_2 R_3}{R_1 R_2 + R_1^2 + R_1 R_3}$$

$$a = \frac{R_1^2 R_2 R_3 C_1 C_2}{R_1 R_2 + R_1^2 + R_1 R_3}$$

$$b = \frac{C_1(R_1 R_2 R_3 + R_1^2 R_3) + C_2(R_1^2 R_2 + R_1 R_2 R_3)}{R_1 R_2 + R_1^2 + R_1 R_3}$$

Problem 10.   Consider the transfer function for a bulb in a well given by Equation 81. As the resistance to heat transfer between the bulb and the well decreases, what limiting form does the transfer function approach? Does the result make sense?

Problem 11.

(a)   Consider the situation shown in Figure 14 of this chapter. Find the transfer function relating the bulb temperature, T_b, to the feed temperature, T_F. As a shortcut, investigate the possibility of using one of the transfer functions in this chapter after suitable modification.
(b)   Suppose the tank is the reflux drum of a distillation column. The holdup in the drum is 500 lb and the normal throughput is 5000 lb/hr. The heat capacity of the liquid is 1 Btu/lb, °F. The temperature of the liquid is sensed by a temperature bulb having the following properties:

Length	4 in.
Outside diameter	$\frac{1}{2}$ in.
Inside diameter	$\frac{3}{8}$ in.
Material	Steel
Density of steel	489 lb/ft^3
Heat capacity of steel	0.12 Btu/lb, °F

The bulb contains a gas of negligible mass and heat capacity. The overall coefficient of heat transfer between the bulb and the liquid is

200 Btu/hr, ft^2, °F. Estimate the importance of interaction between the bulb and the liquid in the drum and find the transfer function, T_b/T_F.

$$Answer: \quad \frac{T_b}{T_F} = \frac{1}{0.000130s^2 + 0.10130234s + 1}$$

Problem 12.   A thick-walled tank is used as the reflux drum of a distillation column. A sketch with appropriate symbols is shown in Figure 15. The heat transfer between the liquid and the wall is given in terms of the overall coefficient, U, and the area, A. The outside of the wall is well insulated from the surroundings.

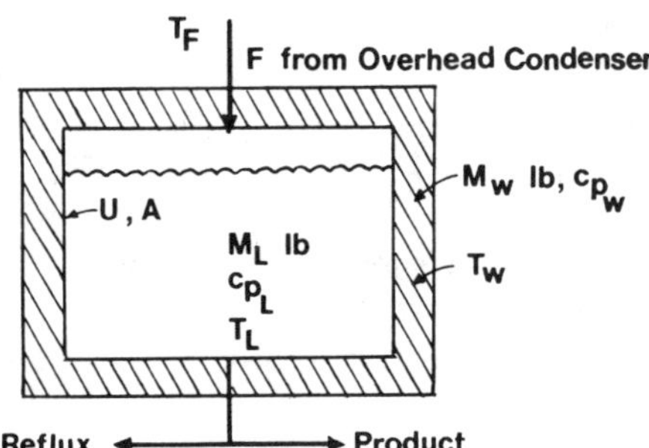

Figure   15. Physical situation for Problem 12.

(a)   Listing important assumptions, develop the transfer function, T_L/T_F.
(b)   Obtain the transfer function of part *a* in the following way.
   (1)   The transfer function, T_W/T_F, can be deduced from that for a bulb in a well if the lumped parameters for the latter case are suitably redefined in terms of this problem. What redefinitions are required?
   (2)   Using the result of part 1 and the transfer function for a bare bulb obtained in Chapter 9, the desired transfer function, T_L/T_F, is readily obtained. What redefinition is required for the parameters of the bare bulb? How does your resulting transfer function, T_L/T_F, compare with that found in part *a*?
(c)   Draw the equivalent electrical analogy.
(d)   If the tank were a reactor with a cooling coil having a coefficient, U_c, and area, A_c, how would this alter the transfer function found in part *a*? Assume that the rate of heat generation does not change with temperature.

Problem 13.

(a) Examine Equation 9 of Chapter 9 and its relationship to the example of the variable thermal capacitance in that chapter. What is implied by Equation 9?

(b) Develop a second-order model for a continuous-flow, stirred-tank heater relating the tank fluid temperature, T, to the control valve pressure, P_V. Assume critical flow through the valve and a constant supply pressure for the steam.

(c) Why is Equation 12 of Chapter 9 probably a reasonable approximation to your second-order model? Base your explanation on physical grounds.

(d) Draw an electrical analogy for your second-order model and compare it with that for the first-order model.

Problem 14. In Chapter 8, Equation 108 was an unsteady-state mass balance for Component A in the irreversible reaction of $A \to B$. Suppose that the feed rate, F, is constant.

(a) Find the transfer function, c_A/c_{FA}, where c_A is the concentration of Component A in the reactor and c_{FA} is the concentration of this component in the feed.

(b) Write a similar unsteady-state mass balance for Component B and find the transfer function, c_B/c_{FB}, where c_B is the concentration of Component B in the reactor and c_{FB} is the concentration of this component in the feed.

(c) Now find the transfer function, c_B/c_{FA}. What kind of second-order transfer function results, an interacting one or a noninteracting one? Why?

CHAPTER XI

Distributed Processes

Thus far, all of the processes we have examined have been treated in a lumped parameter way. In other words, it was assumed that all of the resistances, capacitances, and inertances of a process could be treated as though they existed at discrete points. Usually a lumped parameter analysis is to some extent an approximation, since there are usually some distributed elements.

This chapter has three objectives. The first is to examine some distributed processes so that the reader will gain an appreciation of how and where they arise. The illustrations will be confined to unsteady-state heat transfer in a slab, transmission of pneumatic signals in tubing, and the unsteady-state behavior of some simple heat exchangers.

The second objective is to show how distributed models are formulated and what the resulting transfer functions look like. Partial differential equations always arise in the development of these models, and therefore the mathematics is more sophisticated than that resulting from lumped parameter models. However, the emphasis will be more toward the results of the mathematical analyses than in the steps used in obtaining them. In a few simple cases, derivations will be outlined to illustrate the methods employed.

The third objective is to indicate some ways in which some distributed transfer functions can be approximated by lumped parameter forms. Distributed transfer functions frequently have forms that make them difficult to invert back into the time domain, and render them somewhat unwieldy in other applications, notably in frequency response analysis, covered in Chapter 14. Hence, a considerable amount of research has been directed at developing and verifying such approximations.

I. UNSTEADY-STATE HEAT TRANSFER IN A SLAB

A. Development of the Partial Differential Equation

In the example in Chapter 10 dealing with a temperature bulb in a well, the average, instantaneous temperatures of the well and bulb walls were used. This was justifiable because their temperature gradients were negligible compared to those in the gap and outside film. Heat conduction in metal walls cannot always be treated in such a simple way. Take the case of a thick-walled tank, for example. If the temperature of the fluid in the tank is changing, then unsteady-state heat transfer will occur in the wall, and the gradient in the wall may not be negligible. Furthermore, the thermal resistance and capacitance are uniformly distributed through the wall. Hence, a lumped parameter treatment for the wall may not be permissible.

To understand this more fully, consider the sketch of a slab of material shown in Figure 1. For simplicity, we shall assume that unsteady-state heat

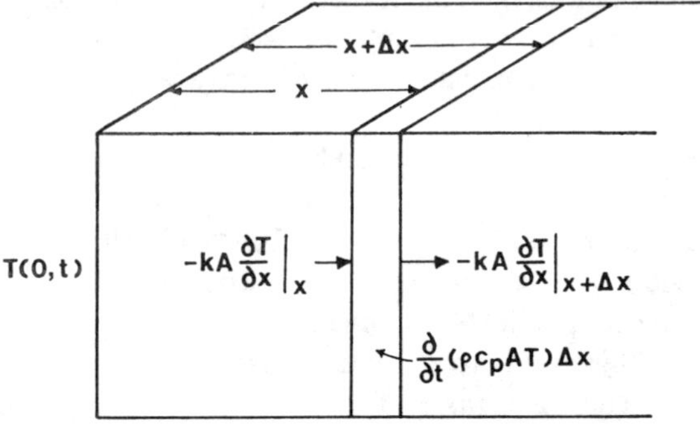

Figure 1.   Unsteady-state heat transfer in a slab.

transfer is occurring in the x-direction only. An energy balance over a slab of differential thickness, Δx, results in the following equation:

$$\text{in} \quad - \quad \text{out} \quad = \text{accumulation}$$

$$-kA\left.\frac{\partial T}{\partial x}\right|_x - \left(-kA\frac{\partial T}{\partial x}\right)\Bigg|_{x+\Delta x} = \frac{\partial}{\partial t}(\rho c_p A T)\,\Delta x \tag{1}$$

where k = thermal conductivity, Btu/hr, ft², °F/ft

A = cross-sectional area normal to the direction of heat transfer, ft²

T = temperature, °F

x = distance, ft

t = time, hr

ρ = density, lb$_m$/ft³

c_p = heat capacity, Btu/lb$_m$, °F

Equation 1 is divided by Δx and the limit is taken as Δx approaches zero. If the physical properties are constant, the result is

$$kA\frac{\partial^2 T}{\partial x^2} = \rho c_p A \frac{\partial T}{\partial t} \tag{2}$$

Since the resistance and capacitance are uniformly distributed in the slab, they are expressed on a per-foot-of-slab basis as follows:

$$R' = \frac{1}{kA} \quad \text{hr, °F/Btu, ft} \tag{3}$$

$$C' = \rho c_p A \quad \text{Btu/°F, ft} \tag{4}$$

Using these definitions, Equation 2 may be expressed as

$$\frac{\partial^2 T}{\partial x^2} = R'C' \frac{\partial T}{\partial t} \tag{5}$$

In problems of unsteady-state heat conduction, this equation usually appears in the following form:

$$\frac{\partial T}{\partial t} = \alpha \frac{\partial^2 T}{\partial x^2} \tag{6}$$

where $\alpha = 1/R'C'$ = thermal diffusivity, ft²/hr

B. Electrical Analogy

An electrical analogy for this case can be developed by expressing Equation 1 as follows:

$$-\frac{1}{R'} \frac{\partial T}{\partial x}\bigg|_x - \left(-\frac{1}{R'} \frac{\partial T}{\partial x}\right)\bigg|_{x+\Delta x} = (C' \Delta x) \frac{\partial T}{\partial t} \tag{7}$$

Each term has the units of flow rate of thermal energy, analogous to a current. Hence, one possible representation of the analog for a section of thickness, Δx, is that shown in Figure 2*a*. A second possible representation is shown in Figure 2*b*. Both of these are equally reasonable, but neither *exactly* represents

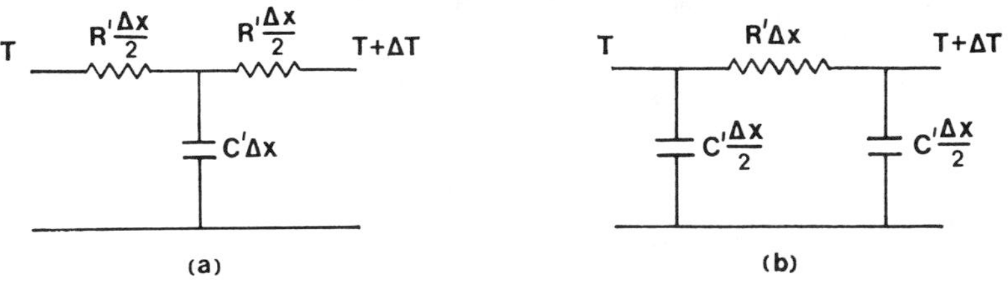

Figure 2.   Two possible electrical analogies for a slab of thickness, Δx.

the situation since each of these analogies is drawn as a lumped parameter representation of a truly distributed process. Since an exact picture of the actual case cannot be drawn, it is not surprising to learn that it is not possible to exactly match the properties of a distributed process with a lumped parameter model.

The entire slab consists of a series of lumped parameter approximations of the types shown in Figure 2, but of differential thickness. Combining a series of either of the two representations in Figure 2 leads to the representation of the slab drawn in Figure 3. This figure brings out the point

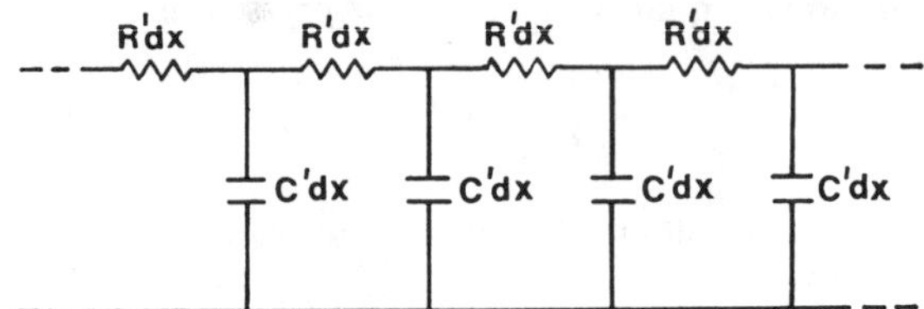

Figure 3. Possible analogy for the slab.

that the slab could be considered to be an infinite series of interacting first-order processes, each with a differentially small time constant, $R'C'(dx)^2$.

C. Development of Transfer Functions for Special Cases

As in the case of lumped parameter models, transfer functions are obtained by taking the Laplace transforms of the differential equations describing the process. Here, the process is described by the partial differential equation given in Equation 6. The transform of the left-hand side of the equation is that given in Equation 21 of Chapter 7. Although the derivative on the right-hand side of Equation 6 is second order with respect to x, it can be treated in exactly the same way as the first-order derivative was in Equation 22 of Chapter 7. Thus, the transform of Equation 6 is

$$s\overline{T}(x, s) - T(x, 0) = \alpha \frac{d^2\overline{T}(x, s)}{dx^2} \tag{8}$$

where $\overline{T}(x, s)$ = Laplace transformation of $T(x, t)$
$\quad\quad T(x, 0)$ = initial temperature distribution

This is as far as the solution can be carried without specifying the initial condition, $T(x, 0)$, and those at the boundaries of the slab. We shall now examine two special cases.

1. Semiinfinite Slab

Consider the sketch of a semiinfinite slab shown in Figure 4. The left face of the slab is at the origin and the slab extends infinitely in all directions of a vertical plane normal to this page. For simplicity, let us suppose that the initial temperature distribution is such that the temperature is completely uniform throughout the slab; this is shown as the solid line, $T(x, 0)$, in the figure. Now, since the original Equations 6 and 8 are linear in T, we can consider them to be deviation equations in which the deviation is measured

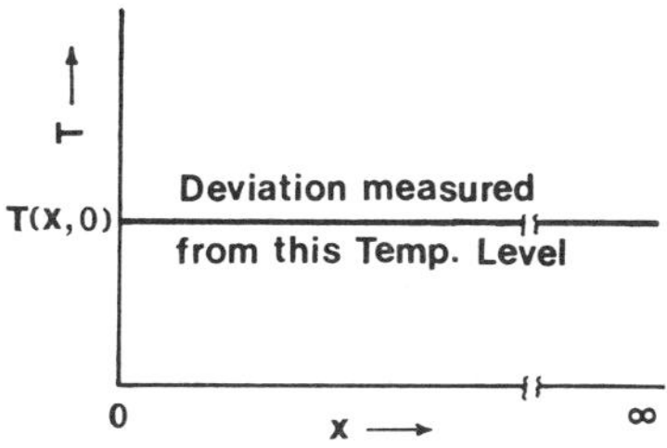

Figure 4.   Semiinfinite slab.

relative to the original condition, $T(x, 0)$. Hence, Equation 8 reduces to

$$\frac{d^2 \overline{T}}{dx^2} - \frac{s}{\alpha} \overline{T} = 0 \tag{9}$$

where $\overline{T}$ is the transform of the deviation in temperature.

Equation 9 is a linear, ordinary differential equation with s as a parameter. The general solution has the form

$$\overline{T} = A_1 e^{-\sqrt{s/\alpha}\, x} + A_2 e^{\sqrt{s/\alpha}\, x} \tag{10}$$

The constants, A_1 and A_2, are determined by the boundary conditions. One of these conditions is that the temperature deviation cannot become infinite as x approaches infinity; therefore, A_2 must be zero. The second boundary condition is selected to be the change in the temperature at the left face, $T(0, t)$. This variable has the transform, $\overline{T}(0, s)$. With this boundary condition, the solution of Equation 10 is

$$\overline{T}(x, s) = \overline{T}(0, s) e^{-\sqrt{s/\alpha}\, x} \tag{11}$$

Hence, the transfer function is

$$\frac{\overline{T}(x, s)}{\overline{T}(0, s)} = e^{-\sqrt{s/\alpha}\, x} \tag{12}$$

The presence of an exponential function of s is indicative of a distributed process.

2. Finite Slab, Perfectly Insulated on One Face

A finite slab of thickness L is depicted in Figure 5. The left face is at the origin and the right face is at a distance, L, in the positive x-direction. The slab extends infinitely in all directions of a vertical plane normal to this page. Again, for simplicity, we shall assume that initially the slab is of uniform temperature, $T(x, 0)$, as indicated by the solid line. If temperatures are

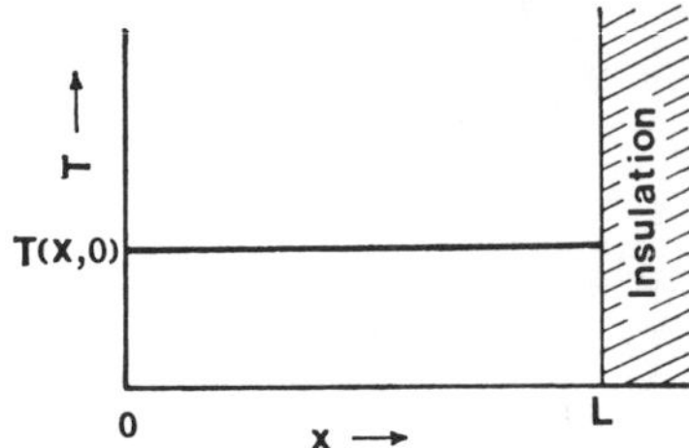

Figure 5.   Finite slab, perfectly insulated on one face.

measured as deviations from this initial state, then Equation 10 applies. The boundary condition at $x = L$ is that the derivative with respect to x is zero at all times since the insulation is perfect; hence

$$\frac{\partial T}{\partial x} = 0 \text{ at } x = L \tag{13}$$

This boundary condition must be transformed so that it can be used for establishing the constants in Equation 10. This transform is

$$\frac{d\overline{T}(L, s)}{dx} = 0 \tag{14}$$

As in the previous example, the second boundary condition is selected to be the change in the temperature at the left face, $T(0, t)$, whose transform is $\overline{T}(0, s)$. With these two boundary conditions, the constants have the following values:

$$A_1 = \overline{T}(0, s)/(1 + e^{-2\sqrt{s/\alpha}\,L}) \tag{15}$$

$$A_2 = \overline{T}(0, s)/(1 + e^{2\sqrt{s/\alpha}\,L}) \tag{16}$$

Substituting these constants into Equation 10 and carrying out some algebraic manipulations, the desired transfer function can be expressed in the following form:

$$\frac{\overline{T}(x, s)}{\overline{T}(0, s)} = \frac{\cosh\left[\sqrt{s/\alpha}\,(L - x)\right]}{\cosh\left(\sqrt{s/\alpha}\,L\right)} \tag{17}$$

It was mentioned earlier in this chapter that the wall of a thick-walled tank possibly could not be treated in a lumped parameter way. Problem 12 of Chapter 10 concerned the development of the thermal dynamics of such a tank based upon a lumped parameter model for the wall. Equation 17 can be used to model the same process in a relatively more exact way.

a. APPLICATION TO THE THERMAL DYNAMICS OF A THICK-WALLED TANK

Consider the sketch of a thick-walled, cylindrical tank shown in Figure 6. The liquid is assumed to be thoroughly mixed so that the temperature is uniform throughout. Although the wall is not exactly a flat slab, it can be

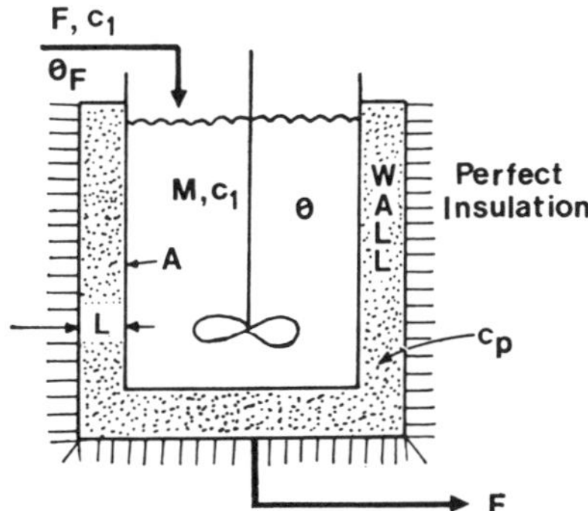

Figure 6. Thick-walled tank covered by perfect insulation.

treated as though it were with negligible error; we shall denote the cross-sectional area of this "slab" as A ft^2; this is the area for heat transfer between the fluid and the wall. In addition, the following symbols and units will be used in the analysis:

F = feed rate of liquid, lb$_m$/hr

c_1 = heat capacity of liquid, Btu/lb$_m$, °F

M = mass of liquid in tank, lb$_m$

θ_F = feed temperature, °F

θ = temperature of liquid in tank, °F

T = $T(x, t)$, temperature in wall as a function of position and time, °F

Q = rate of heat transfer from liquid to wall, Btu/hr

L = wall thickness, ft

c_p = heat capacity of wall, Btu/lb$_m$, °F

ρ = density of wall, lb$_m$/ft^3

t = time, hr

Our goal is to obtain the transfer function, θ/θ_F, which relates the temperature of the fluid in the tank to the feed temperature. We begin with an energy balance for the fluid in the tank:

$$Fc_1\theta_F - Fc_1\theta - Q = Mc_1 \frac{d\theta}{dt} \tag{18}$$

The equation is linear in the variables, θ_F, θ, and Q, so it can be interpreted as a deviation equation in these variables.

We must now express Q in terms of Equation 17. This transfer function can only be used here if it is assumed that the temperature of the wall at its inside surface is the same as that of the fluid. This assumption is reasonable if the fluid is highly agitated. The model could be further refined if the transfer function were modified to allow for a film coefficient of heat transfer between the fluid and the wall; this modification is the subject of Problem 9 at the end of this chapter.

The rate of heat transfer to the wall, Q, is equal to the time derivative of the thermal energy of the wall, E; that is:

$$Q = \frac{dE}{dt} \tag{19}$$

where E = deviation in thermal energy of the wall, Btu

The thermal energy, in turn, can be expressed in terms of the temperature distribution of the wall, as follows:

$$E = \int_0^L A\rho c_p T(x, t)\, dx \tag{20}$$

Therefore, the rate of heat transfer is

$$Q(t) = \frac{dE}{dt} = \frac{d}{dt} \int_0^L A\rho c_p T(x, t)\, dx \tag{21}$$

The initial value of the deviation in energy, $E(0)$, is assumed to be zero, so the Laplace transform of Equation 21 is

$$Q(s) = s \int_0^L A\rho c_p \overline{T}(x, s)\, dx \tag{22}$$

Noting that $\overline{T}(0, s) \equiv \theta(s)$ and substituting Equation 17 into Equation 22, we have:

$$Q(s) = \frac{A\rho c_p s\theta(s)}{\cosh(\sqrt{s/\alpha}\, L)} \int_0^L \cosh\left[\sqrt{s/\alpha}\,(L - x)\right] dx$$

$$= \frac{A\rho c_p s\theta(s)}{\sqrt{s/\alpha}} \tanh(\sqrt{s/\alpha}\, L) \tag{23}$$

Transformation and rearrangement of Equation 18 yields

$$\left(\frac{M}{F} s + 1\right)\theta + \frac{Q(s)}{Fc_1} = \theta_F \tag{24}$$

Substituting Equation 23 for $Q(s)$, and rearranging, we obtain the desired transfer function:

$$\frac{\theta}{\theta_F} = \frac{1}{\dfrac{M}{F} s + \dfrac{A\rho c_p s}{\sqrt{s/\alpha}\, Fc_1} \tanh\left(\sqrt{s/\alpha}\, L\right) + 1} \tag{25}$$

Heat transfer in a slab only involves two distributed parameters, resistance and capacitance. Three distributed parameters arise in the analysis of pneumatic transmission lines.

II. PNEUMATIC TRANSMISSION LINES

All of the discussions of control systems in this text have been in the context of pneumatic controllers and control valves. It has been implied in most cases that when there is a change in controller output pressure, there is an identical and immediate change in the pressure inside the valve motor. However, unless the controller is located right at the valve, there will be a pneumatic transmission line between the two that could be several hundred feet long.

The transmission of a pressure signal through such a line requires time since pressure waves travel at a rate closely related to the speed of sound. The speed of sound in free air is about 1100 ft/sec, but some studies (1) have shown that the speed in tubing is on the order of half the value in free air. Whether this transmission time is significant or not depends, of course, on the dynamic behavior of the process being controlled. If it is a relatively slow process, then the dynamic behavior of the transmission line would be an insignificant factor in the control system response.

A. Equations Describing the Behavior of the Air in the Lines

Consider a controller connected to a pneumatic valve through a transmission line. If there is a sudden increase in the controller pressure, a pressure wave will travel down the line. The increased pressure would at least momentarily cause a slight increase in temperature of the air and the thermal behavior of the air would fall somewhere between adiabatic and isothermal. Hence, a complete analysis would have to include not only the equations of continuity and motion, but also the equation of energy. A relatively detailed treatment of this type can be found in Reference 2. However, in most studies, isothermal behavior has been assumed, thus eliminating the need for the equation of energy.

1. Equation of Continuity

Let us first consider the equation of continuity for the line. Since transmission tubing is cylindrical, the equation in cylindrical coordinates is used (3i):

$$\frac{\partial \rho}{\partial t} + \frac{1}{r}\frac{\partial}{\partial r}(\rho r v_r) + \frac{1}{r}\frac{\partial}{\partial \theta}(\rho v_\theta) + \frac{\partial}{\partial x}(\rho v) = 0 \tag{26}$$

The radial and angular components of velocity, v_r and v_θ, will be negligible compared to that in the axial direction, v, with the result that the equation reduces to

$$\frac{\partial \rho}{\partial t} = -\frac{\partial(\rho v)}{\partial x} = -\rho\frac{\partial v}{\partial x} - v\frac{\partial \rho}{\partial x} \tag{27}$$

where ρ = gas density, lb_m/ft^3
$\quad t$ = time, sec
$\quad v$ = velocity in the x-direction, ft/sec
$\quad x$ = distance, ft

If the change in pressure at the inlet end of the line is small, then this equation can be linearized about the normal density and velocity to yield

$$\frac{\partial \rho}{\partial t} = -\bar{\rho}\frac{\partial v}{\partial x} - \bar{v}\frac{\partial \rho}{\partial x} \tag{28}$$

where $\bar{\rho}$ = normal density at a point in the line
$\quad \bar{v}$ = normal velocity at a point in the line

Furthermore, since the line is mainly used for the transmission of pressure signals rather than flows, the normal velocity at a point, $\bar{v}$, is zero and the normal density, $\bar{\rho}$, is essentially a constant along the entire line. Therefore Equation 28 reduces to

$$\frac{\partial \rho}{\partial t} = -\bar{\rho}\frac{\partial v}{\partial x} \tag{29}$$

Although the velocity profile is not flat, this equation still applies to the entire cross-sectional area of the tube if v is interpreted as the average velocity, v_{ave}. The actual volumetric flow rate, Q, is given by the equation:

$$Q = Sv_{ave} \tag{30}$$

where $\quad S$ = cross-sectional area of tube, ft^2
$\quad v_{ave}$ = volumetric average velocity, ft/sec

Furthermore, the density is proportional to the pressure. Hence, Equation 29 can be written in the form

$$\frac{S}{\bar{P}}\frac{\partial P}{\partial t} = -\frac{\partial Q}{\partial x} \tag{31}$$

where P = pressure, $\mathrm{lb_f/ft^2}$, and $\bar{P}$ is the normal pressure

In previous chapters, gas flows were expressed in terms of a standard cubic foot defined at 1 atmosphere and the normal temperature. Since ideal gas behavior is assumed, the actual volumetric flow rate can be converted to the standard basis by the equation

$$Q = \frac{2117}{\bar{P}} Q_s \qquad (2117 \ \mathrm{lb_f/ft^2} = 1 \ \text{atmosphere}) \tag{32}$$

where Q_s = volumetric flow rate at standard conditions, SCF/sec

Substituting Equation 32 into Equation 31, we obtain

$$\frac{S}{2117} \frac{\partial P}{\partial t} = - \frac{\partial Q_s}{\partial x} \tag{33}$$

Dividing through by $\partial P/\partial t$, it can be seen that $S/2117$ is the rate of change of the flow rate per unit length divided by the rate of change of the driving force with respect to time. Referring to Table 3 of Chapter 8, $S/2117$ then is the capacitance per unit length of tubing. Hence, the equation of continuity can be written

$$C' \frac{\partial P}{\partial t} = - \frac{\partial Q_s}{\partial x} \tag{34}$$

where

$$C' = \text{capacitance per foot of tubing} = \frac{S}{2117} \frac{\mathrm{ft^5}}{\mathrm{lb_f}} \ /\mathrm{ft} \tag{35}$$

2. Equation of Motion

It was pointed out earlier that since the analysis is confined to small changes in input pressure, the density of the air in the line is essentially constant. Furthermore, it is assumed that the flow is laminar, so the Navier-Stokes equation in cylindrical coordinates can be used (3ii):

$$\frac{\rho}{g_c}\left(\frac{\partial v}{\partial t} + v\,\frac{\partial v}{\partial x}\right) = -\frac{\partial P}{\partial x} + \mu\left[\frac{1}{r}\frac{\partial}{\partial r}\left(r\,\frac{\partial v}{\partial r}\right) + \frac{1}{r^2}\frac{\partial^2 v}{\partial \theta^2} + \frac{\partial^2 v}{\partial x^2}\right] + \rho g_x \tag{36}$$

As in the case of the equation of continuity, Equation 36 can be linearized about the average velocity, $\bar{v}$, which is zero; therefore the factor, $v(\partial v/\partial x)$, on the left-hand side drops out. The second partial derivative, $\partial^2 v/\partial x^2$, on the right-hand side is negligible (4) and the other second partial derivative, $\partial^2 v/\partial \theta^2$, drops out because there is axial symmetry. Finally, the gravitational force, ρg_x, is zero for a horizontal tube and negligible for a vertical one. With all of these assumptions, Equation 36 reduces to

$$\frac{\rho}{g_c}\frac{\partial v}{\partial t} = -\frac{\partial P}{\partial x} + \mu\,\frac{1}{r}\frac{\partial}{\partial r}\left(r\,\frac{\partial v}{\partial r}\right) \tag{37}$$

The term on the left-hand side can be interpreted as the pressure gradient arising from the inertia of the fluid and will be denoted $\partial P_i/\partial x$. The first term on the right-hand side, $\partial P/\partial x$, is the total pressure gradient, which is caused by the input pressure disturbance to the line. The last term is the pressure gradient arising from fluid friction and will be denoted $-\partial P_f/\partial x$. Thus, Equation 37 can be rearranged and expressed in the following way:

$$\underbrace{\frac{\partial P_i}{\partial x}}_{\substack{\text{Inertial}\\\text{Term}}} + \underbrace{\frac{\partial P_f}{\partial x}}_{\substack{\text{Fluid}\\\text{Friction}\\\text{Term}}} = \underbrace{-\frac{\partial P}{\partial x}}_{\substack{\text{Input}\\\text{Pressure}\\\text{Gradient}}} \tag{38}$$

Let us consider the frictional term. In an approximate analysis, it can be assumed that the flow is laminar so that the resistance can be estimated from the Hagen-Poiseuille equation. In terms of the volumetric average velocity, the frictional gradient then is

$$\frac{\partial P_f}{\partial x} = \frac{32\mu v_{\text{ave}}}{g_c D^2} \tag{39}$$

where μ = viscosity, $\text{lb}_\text{m}/\text{ft}$, sec
     D = tube diameter, ft

But

$$S v_{\text{ave}} = \frac{\pi D^2}{4} v_{\text{ave}} = Q \tag{40}$$

Therefore, using Equations 32 and 40, Equation 39 becomes

$$\frac{\partial P_f}{\partial x} = \frac{2117(128)\mu}{\pi D^4 \bar{P} g_c} Q_s \tag{41}$$

The factor multiplying Q_s is the frictional pressure drop per foot of tubing per unit flow rate. In accordance with Table 3 of Chapter 8, this is the resistance per unit length:

$$R' = \text{resistance per foot of tubing} = \frac{2117(128)\mu}{\pi D^4 \bar{P} g_c} \quad \frac{\text{lb}_\text{f}, \text{ sec}}{\text{ft}^5}\, /\text{ft} \tag{42}$$

Equation 41 may then be written

$$\frac{\partial P_f}{\partial x} = R' Q_s \tag{43}$$

Now consider the inertial term of Equation 38:

$$\frac{\partial P_i}{\partial x} = \frac{\rho}{g_c} \frac{\partial v}{\partial t} \tag{44}$$

Although the velocity profile is not flat, this gradient can be approximated in terms of the average velocity:

$$\frac{\partial P_i}{\partial x} = \frac{\rho}{g_c}\frac{\partial v_{\text{ave}}}{\partial t} \tag{45}$$

Using Equation 40 we obtain

$$\frac{\partial P_i}{\partial x} = \frac{\rho}{Sg_c}\frac{\partial Q}{\partial t} \tag{46}$$

The actual volumetric flow rate, Q, is now converted to the flow rate, Q_s, at standard conditions. Since the volumetric flow rate is inversely proportional to the density, it can be seen that

$$\rho Q = \rho_s Q_s \tag{47}$$

where ρ_s = density at standard conditions (1 atmosphere and normal temperature), $\text{lb}_{\text{m}}/\text{ft}^3$

The density, ρ, is assumed to be constant, so Equation 46 becomes

$$\frac{\partial P_i}{\partial x} = \frac{\rho_s}{Sg_c}\frac{\partial Q_s}{\partial t} \tag{48}$$

The factor, $\rho_s/(Sg_c)$, is the ratio of the inertial pressure gradient to the rate of change of flow rate. Referring to Table 3 of Chapter 8, it can be seen that it is the inertance per unit length:

$$I' = \text{inertance per foot of tubing} = \frac{\rho_s}{Sg_c}\frac{\text{lb}_{\text{f}}, \text{sec}^2}{\text{ft}^5}/\text{ft} \tag{49}$$

Equation 48 can then be expressed as follows:

$$\frac{\partial P_i}{\partial x} = I'\frac{\partial Q_s}{\partial t} \tag{50}$$

Combining Equations 43 and 50 with Equation 38, the analog form of the equation of motion is

$$R'Q_s + I'\frac{\partial Q_s}{\partial t} = -\frac{\partial P}{\partial x} \tag{51}$$

A number of approximations have been made in reaching this result. A more exact treatment of the equation of motion can be found in the paper of Rohmann and Grogan (5).

3. Electrical Analogy

The equation of continuity, Equation 34, and the equation of motion, Equation 51, are the bases for the electrical analogy. The flow rate, Q_s, is analogous to current, i, and the pressure, P, is analogous to voltage, e. Hence, in electrical terminology, these equations can be written as follows:

$$R'i + I'\frac{\partial i}{\partial t} = -\frac{\partial e}{\partial x} \tag{52}$$

$$C'\frac{\partial e}{\partial t} = -\frac{\partial i}{\partial x} \tag{53}$$

These two equations appear in the theory of electrical transmission lines, although there is usually an additional term, $G'e$, on the left-hand side of Equation 53 to account for the leakance between the two wires comprising the line. There is no analog for this factor in pneumatic transmission lines since there is only one tube, whereas an electrical transmission line consists of two parallel wires.

As in the case of heat transfer in a slab, an exact representation of the equations for this distributed process cannot be drawn; however, there are a number of possible approximate representations, one of which is shown in Figure 7. Comparing this with Figure 3 for the heat transfer case, we see that

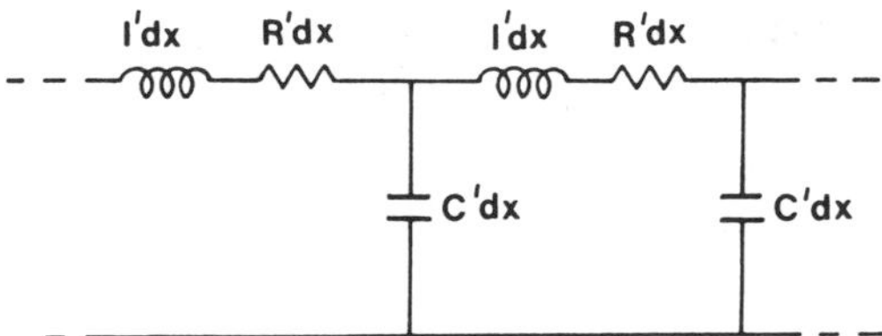

Figure 7.   A possible electrical analogy for a pneumatic transmission line.

the two differ only in the addition of the inertance in series with the resistance. Hence, we should expect the behavior of a transmission line to be closely related to that of heat transfer in a slab if the inertance is negligible.

B. Solution of the Equations for Special Cases

Equations 34 and 51 are linear in the variables, P and Q_s, and therefore they can be interpreted as deviation equations in these variables. These equations are transformed and the resulting ordinary differential equations are then solved simultaneously to obtain expressions for $P(x, s)$ and $Q(x, s)$; the resulting equations are:

$$P(x, s) = c_1' e^{-nx} + c_2' e^{nx} \tag{54}$$

$$Q_s(x, s) = c_3' e^{-nx} + c_4' e^{nx} \tag{55}$$

where

$$n = \text{``propagation function''} = \sqrt{(R' + sI')sC'} \tag{56}$$

The coefficients, c_1' through c_4', are constants determined by the terminal conditions of the line. Therefore, this is as far as the analysis can be carried without specifying these conditions.

The boundary conditions for transmission lines can take on a variety of forms of varying degrees of complexity. Consider the sketch of a transmission line in Figure 8. As indicated, at the sending end of the line there is

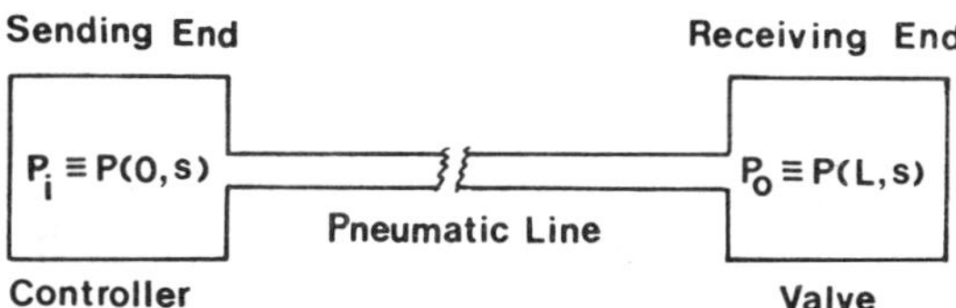

Figure 8.   Pneumatic transmission line with sending and receiving terminations.

usually a controller and the receiver is normally either a valve or a valve positioner, which was mentioned in Section II.B of Chapter 10. In electrical theory, these sending and receiving ends are characterized by their "impedances." In control, we are mainly interested in the behavior of the line with respect to the transmission of a pressure signal from the sending end to the receiving end. Thus, the discussion here will be confined to the properties of the transfer function, $P(L, s)/P(0, s)$.

To simplify the analysis, we shall first assume that the controller has a very large capacity to supply air to the receiving end of the line; in other words, when the controller output pressure increases, the controller can adequately supply the corresponding air requirements to the valve or other termination. We shall further assume that the controller has negligible flow resistances within it. This assumption amounts to saying that the "sending impedance" of the controller is zero. A more comprehensive discussion of these assumptions is given by Farrington (6i). Under the assumption of zero impedance, Equation 54 takes the form (6ii):

$$\frac{P(L, s)}{P(0, s)} = \frac{1 + r_R}{e^{nL} + r_R e^{-nL}} \tag{57}$$

where

$$L = \text{length of line, ft}$$

$$r_R = \frac{Z_R - Z_C}{Z_R + Z_C} \tag{58}$$

$$Z_R = \text{receiving end impedance}$$

$$Z_C = \text{characteristic impedance of line} = \sqrt{\frac{R' + sI'}{sC'}} \tag{59}$$

We shall now examine two common receiving end terminations.

1. Bellows Termination

The bellows used in control instruments are quite small, having volumes that vary between about 1 and 4 in.^{3}. A bellows of this type might be part of an indicator mechanism or a valve positioner. Referring to Equations 57 and 58, it can be seen that the final form of the transfer function depends upon how the impedance of the receiver compares with that of the line. When a line terminates in a small instrument bellows, it can usually be assumed that the impedance of the bellows is much greater than that of the line and can be considered infinite. Equation 57 then reduces to the following form:

$$\frac{P(L, s)}{P(0, s)} = \frac{1}{\cosh nL} \tag{60}$$

It is interesting to compare this result with that of Equation 17 for heat transfer in a finite slab, perfectly insulated at one face. When x is set equal to L in Equation 17, the transfer functions have the same form. The analogy is not quite complete since for heat transfer, the argument of the hyperbolic cosine is

$$\sqrt{\frac{s}{\alpha}} L = \sqrt{R'C's}\, L$$

whereas for the transmission line, the argument is

$$nL = \sqrt{(R' + sI')sC'}\, L$$

Thus, the problems are mathematically the same if the inertance, I', is zero. Since the inertance is not zero, it is a question of the significance of the term, sI', in comparison with R'. As demonstrated in Reference 6iii, the inertance is usually insignificant.

When the inertance is negligible, a lumped parameter approximation of the transfer function is fairly easily obtained. With the inertance omitted, Equation 60 can be written

$$\frac{P(L, s)}{P(0, s)} = \frac{1}{\cosh \sqrt{R'C's}\, L} \tag{61}$$

If the hyperbolic cosine is expanded in a series and only the first two terms are retained, the result is

$$\frac{P(L, s)}{P(0, s)} \cong \frac{1}{1 + \dfrac{R'C'L^2}{2}\, s} \tag{62}$$

This is a first-order representation with a time constant, $R'C'L^2/2$. Since the entire resistance of the line is $R'L$ and the entire capacitance is $C'L$, we see that the time constant is just half of the product of all of the resistance and capacitance.

In actual experimental studies with transmission lines terminated by bellows, it has been found that the response to a step input more closely resembles a pure time delay in series with a first-order element. The result of a typical test by Bradner (7) is shown in Figure 9. In this test, Bradner terminated a 1000-ft length of $\frac{3}{16}$-in. I.D. tubing with a bellows. The line pressure

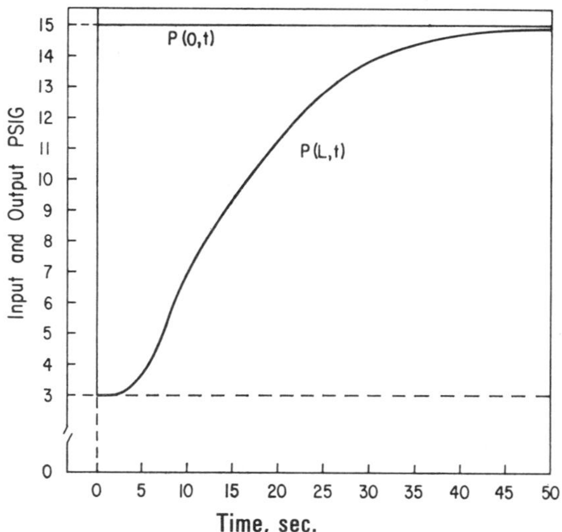

Figure 9.   Typical response of bellows pressure to step input at transmission line entrance. [From Bradner (7), courtesy of Instrument Society of America.]

was initially uniform at 3 psig and the input pressure was suddenly raised to 15 psig. By inspection, it can be seen that a reasonable approximation is of the following form:

$$\frac{P(L, s)}{P(0, s)} = \frac{e^{-\tau_D s}}{\tau s + 1} \tag{63}$$

where τ = effective time constant, sec

τ_D = effective pure delay, sec

This model was used in some studies by Moise (4) who correlated his results as shown in Figure 10. In these tests, the volume at the receiving end was negligible and the tubing was also $\frac{3}{16}$ in. in inside diameter.

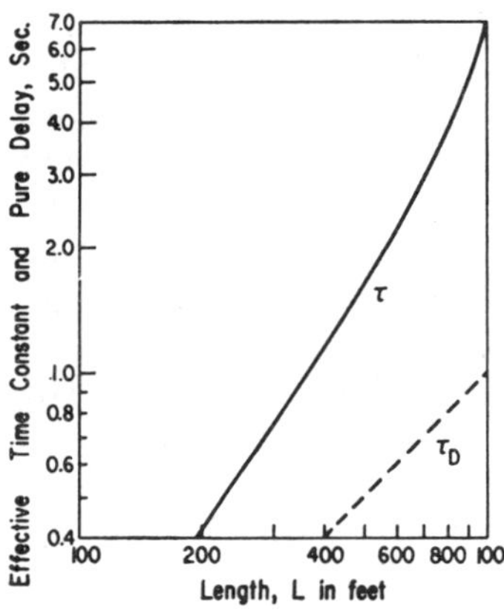

Figure 10. Parameters of Equation 63 as a function of line length. [From Moise (4), courtesy of Instrument Society of America.]

2. Valve Motor Termination

When the transmission line terminates with the motor of a valve, the capacitance of the motor can be important. Valve motors having volumes of several hundred cubic inches are not uncommon. While it would be possible to derive relatively exact transfer functions for valve motor terminations, lumped parameter approximations have commonly been employed.

One suggested approximation is that of Hougen et al. (8, 9), who took the theoretical results of Schuder and Binder (10) and developed a transfer function based on the analogy shown in Figure 11. This approximation is basically the same as that suggested by Harriott (11i). The resulting transfer function is

$$\frac{P(L, s)}{P(0, s)} = \frac{1}{I\left(\dfrac{C}{2} + C_T\right) s^2 + R\left(\dfrac{C}{2} + C_T\right) s + 1} \tag{64}$$

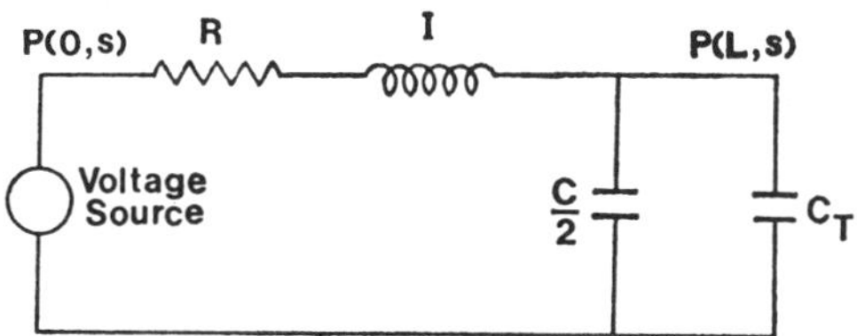

Figure 11. Lumped approximation for valve motor termination.

where $I = I'L =$ total line inertance
$C = C'L =$ total line capacitance
$R = R'L =$ total line resistance
$C_T =$ capacitance of transmission line receiver

When the terminal volume is a valve motor, then the capacitance must be treated like that of an expandable bellows. By definition, the capacitance of the motor is

$$C_m = \frac{d}{dP}\left(\frac{V_m P}{P_s}\right) = \frac{V_m}{P_s} + \frac{P}{P_s}\left(\frac{dV_m}{dP}\right) \tag{65}$$

where $C_m =$ valve capacitance, SCF/lb$_f$/ft^2
$P =$ pressure, lb$_f$/ft^2
$P_s =$ standard pressure $= 2117$ lb$_f$/ft^2
$V_m =$ motor volume, ft^3

A still simpler approximation has been to treat the valve and line as a first-order process. Thus, in the studies of Bradner (7) previously referred to, transmission lines of various lengths were tested with a number of different valves. The time required for 63.2% of the ultimate change in valve position was measured and this time was defined as the "time constant." The resulting correlation is shown in Figure 12.

As indicated in Figure 12, some valve motors are very large. Even for small increases in the controller output pressure, relatively large amounts of air would be required by the valve. If the controller were connected directly to the valve motor, then this air would have to pass through the transmission line and the response of the valve would be relatively sluggish. This disadvantage can be avoided by placing a valve positioner on the valve. This device senses the pressure signals from the controller but without demanding a large amount of air from it. The input to the positioner is a small bellows, connected to the transmission line. In effect, this bellows substitutes for the valve motor termination. The positioner in turn supplies the large air requirements called for by the controller. A positioner not only speeds up the response

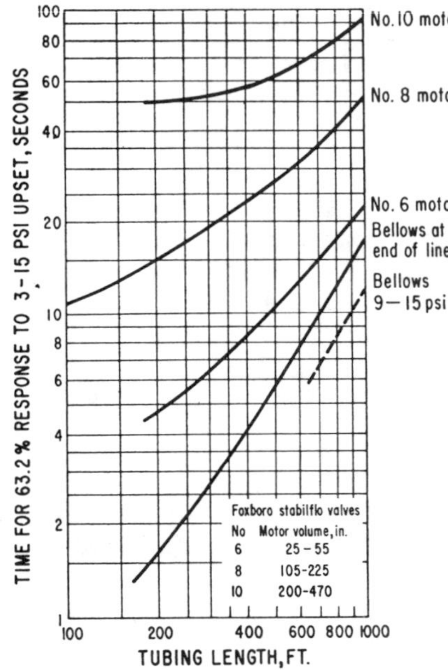

Figure 12. Transient response time for valves and bellows connected to $\frac{3}{16}$-in. I.D. tubing. [From Bradner (7), Courtesy of Instrument Society of America.]

of the valve, but as noted in Chapter 10, it also minimizes the effects of friction, which can cause hysteresis behavior.

III. HEAT EXCHANGERS

There are many types of heat exchangers, but the most common type is a shell-and-tube unit. From the standpoint of dynamic analysis, it is useful to classify such exchangers according to the way in which the temperatures of the fluids vary as they pass through. The simplest situation is that in which the temperature of each fluid is constant. This was the case in the example of variable thermal capacitance in Section III.D. of Chapter 9 in which steam in the shell was condensing by transferring heat to a slurry of ice and water passing through the tubes. A lumped parameter model was adequate because of the uniformity of temperatures in the shell and tubes.

The most complex situation arises when the temperature of each of the streams varies. As will be discussed later, the dynamic behavior must then be expressed in terms of at least two partial differential equations. Hence, a good starting point for a study of the dynamic behavior of exchangers is a condenser in which liquid passes through the tubes and the condensing vapor in the shell has a uniform temperature.

A. Condensers

For purposes of discussion, let us assume that steam is being condensed in the shell and water is passing through the tubes, as shown in Figure 13a. The temperature profiles at steady state are indicated in Figure 13b. The driving force for heat transfer at the entrance is δ_1 and that at the exit, δ_2. We wish to find the response of the temperature of the leaving water, T_2, to changes in its entrance temperature, T_1, and its velocity, v, and to changes in the temperature of the steam, T_s.

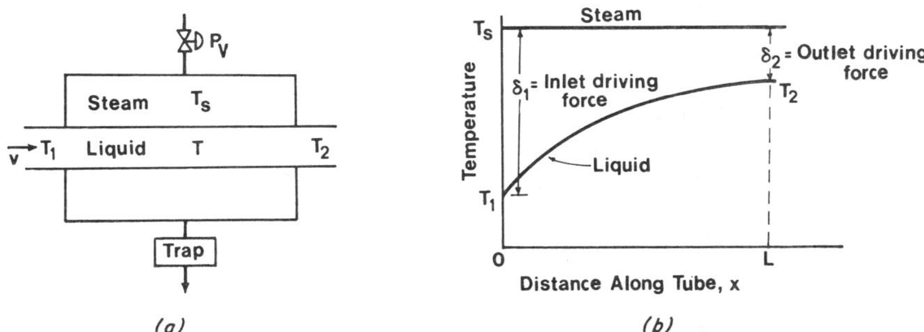

Figure 13.   Simple single-pass, shell-and-tube condenser. (a) Physical situation. (b) Steady-state temperature profiles.

The basic factors to consider in a study of a condenser are the dynamic behaviors of the tube and shell fluids. In addition to these, the effects of the tube and shell walls must also be taken into account. In the interests of simplicity, we shall focus our attention only on the fluids. Some discussion of the effects of the walls can be found in References 11ii and 12. To develop some understanding of the behavior of a condenser, first we shall disregard the interaction between the two fluids and examine only the dynamics of the tube fluid. Then we shall consider the effect of the interaction.

1. Thermal Dynamics of the Liquid in the Tube

Consider the small section of tube shown in Figure 14. If the flow is turbulent, then the velocity and temperature gradients in the water in the radial direction will be negligible. It is reasonable to assume that the velocity is sufficiently high that the convective flux of energy is much greater than that resulting from axial conduction, so that the latter can be neglected. With these assumptions, the energy balance over the section of length, Δx, is:

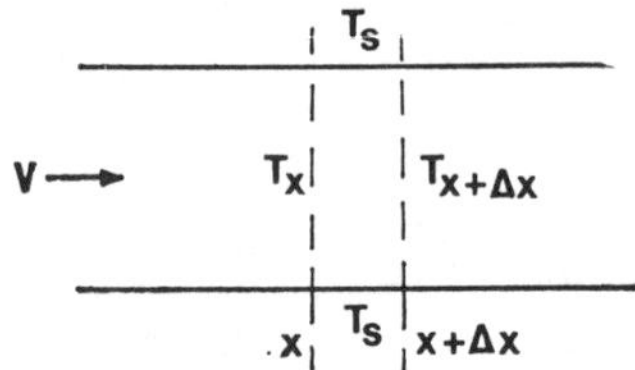

Figure 14.   Small section of condenser tube.

$$v\rho c_p S T_x + A'U\,\Delta x(T_s - T_{x+\Delta x}) - v\rho c_p S T_{x+\Delta x} = S\rho c_p\,\Delta x\,\frac{\partial T_{x+\Delta x}}{\partial t} \qquad (66)$$

sensible +   heat transfer   −   sensible   =   accumulation
heat in   heat out

where v = fluid velocity, ft/hr
ρ = fluid density, $\mathrm{lb_m/ft^3}$
c_p = heat capacity, $\mathrm{Btu/lb_m}$, °F
S = cross-sectional area of tube = $\pi D^2/4$, $\mathrm{ft^2}$
D = tube diameter, ft
T_x = temperature at x, °F
$T_{x+\Delta x}$ = temperature at $x + \Delta x$, °F
A' = heat transfer area per foot of pipe = πD, ft
U = heat transfer coefficient, Btu/hr, $\mathrm{ft^2}$, °F
x = distance, ft
T_s = steam temperature, °F
t = time, hr

It will be noticed that the temperature, $T_{x+\Delta x}$, has been used in the terms for the heat transfer and accumulation. Strictly speaking, it would be more correct to use some average temperature in the section. However, in the limit as Δx approaches zero, the error in this approximation approaches zero. Equation 66 can be rewritten as follows:

$$F c_p T_x - F c_p T_{x+\Delta x} + A'U\,\Delta x(T_s - T_{x+\Delta x}) = C'\,\Delta x\,\frac{\partial T_{x+\Delta x}}{\partial t} \qquad (67)$$

where F = flow rate, lb/hr = $v\rho S$
C' = capacitance per foot, Btu/°F/ft = $S\rho c_p$

a. DEVELOPMENT OF THE ELECTRICAL ANALOGY

The electrical analogy is developed by noting that this little section of the tube can be modeled as a small stirred-tank heater, as shown in Figure 15a. The corresponding electrical analogy is shown in Figure 15b, which is the same as given in Figure 6a of Chapter 9 but with appropriate changes in symbols and a slight rearrangement. Thus, the entire tube could be modeled

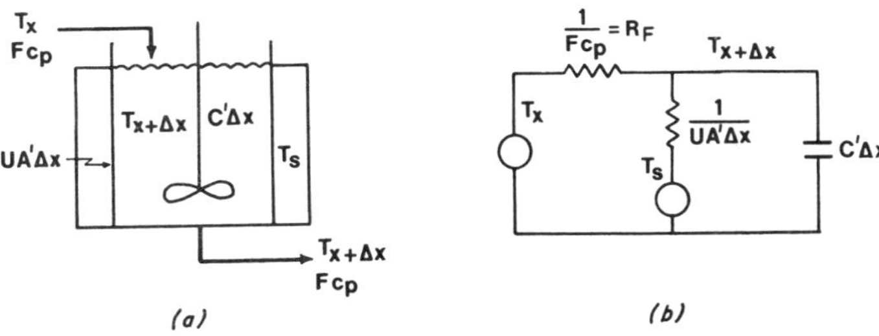

Figure 15.   Section of condenser tube modeled as a small stirred-tank heater. (*a*) Physical model. (*b*) Electrical analogy.

as a series of these tanks, each of differential length, *dx*, as indicated in Figure 16*a*. The temperature of the feed to the first tank is T_1 and the temperature of the leaving stream from the last is T_2. It is important to note that since axial conduction is assumed to be of negligible importance, the tanks are noninteracting; in other words, the behavior of the first tank is independent of the behavior of the second, and so on. The corresponding electrical analogy is shown in Figure 16*b*.

There is a temptation to combine the differential elements of Figure 16*b* into the form shown in Figure 17*a*. This analogy is incorrect since it

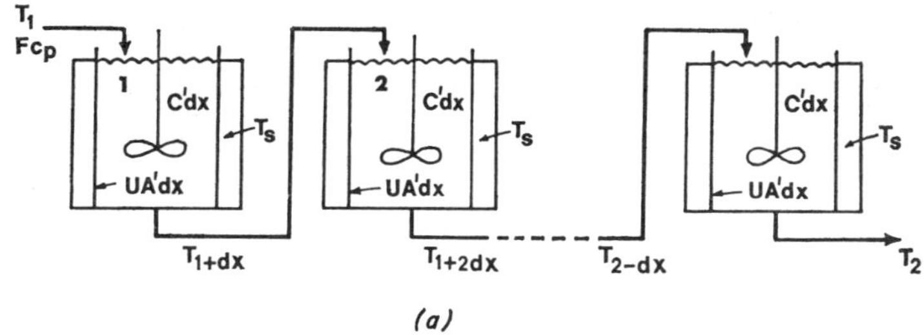

(*a*)

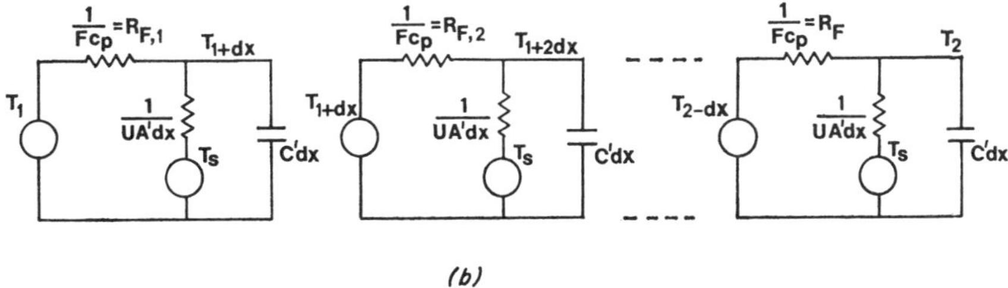

(*b*)

Figure 16.   Tube modeled as a series of stirred-tank heaters. (*a*) Physical model. (*b*) Electrical analogy.

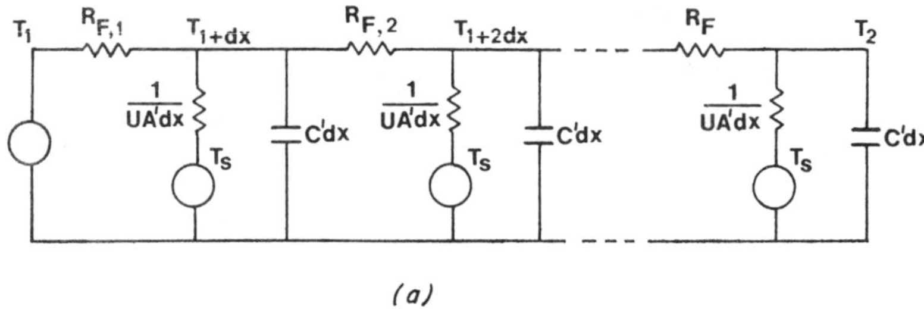

(a)

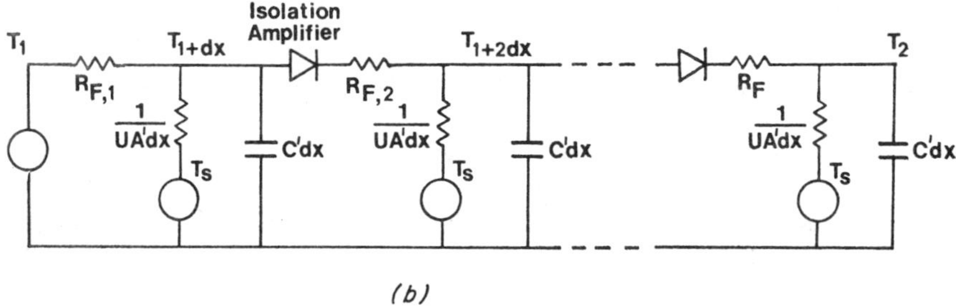

(b)

Figure 17.   Incorrect and correct electrical analogies for tube. (a) Incorrect analogy (b) Correct analogy.

indicates that a change in the temperature, $T_{1+2\,dx}$, would affect the temperature, T_{1+dx}, through the resistance, $R_{F,2}$; this is clearly not consistent with the physical situation as shown in Figure 16a. The correct analogy is presented in Figure 17b where the necessary isolation of each differential unit from the next is accomplished by the insertion of an "isolation amplifier."

It can be seen from the analog diagram of Figure 17b that the three input variables of interest will have markedly different effects on the output temperature, T_2. When there is a change in the entrance temperature, T_1, the response of T_2 will be relatively slow because the effect of the change must travel through each of the differential elements in series. Since a change in T_1 occurs at one point in the exchanger, it is an example of "localized forcing." Physically, a change in T_1 only affects the driving force for heat transfer if the overall coefficient is independent of temperature. Usually this is a reasonable assumption.

A change in steam temperature, T_s, is an example of "distributed forcing" since each differential element is simultaneously affected by the change. Hence, it would be expected that a change in T_s would be reflected in a relatively rapid change in T_2. As with T_1, a change in T_s only affects the driving force for heat transfer when the change has a negligible effect upon the overall coefficient.

The change in velocity, v, acts in an entirely different way from the previous two disturbances. A change in this variable simultaneously affects each differential element by changing R_F, which is a parameter of the exchanger. Therefore, this is an example of "parametric forcing." A change in velocity affects the overall driving force for heat transfer since, for example, an increase in flow rate would lower the average temperature of the fluid in the tube. But there can also be an additional effect if the inside film coefficient of heat transfer is controlling. When this is the case, a change in flow rate also changes the coefficient, U, and this has a distributed effect. It would be expected that T_2 would respond rapidly to changes in velocity.

b. TRANSFER FUNCTIONS

The transfer functions, T_2/T_1 and T_2/T_s, are fairly easily derived since Equation 66 is linear in the variables, T and T_s, when the velocity is constant. The overall coefficient, U, is only slightly dependent on temperature and can be considered constant. Dividing Equation 66 through by Δx and taking the limit at Δx approaches zero, the following partial differential equation results:

$$-v\frac{\rho c_p S}{UA'}\frac{\partial T}{\partial x} + T_s = T + \frac{S\rho c_p}{UA'}\frac{\partial T}{\partial t} \tag{68}$$

But

$$S = \frac{A'D}{4}$$

Hence, Equation 68 can be written

$$-v\frac{\rho c_p D}{4U}\frac{\partial T}{\partial x} + T_s = T + \frac{\rho c_p D}{4U}\frac{\partial T}{\partial t} \tag{69}$$

The grouping of the parameters, $\rho c_p D/4U$, has the units of time and it is convenient to refer to this as the "thermal time constant, τ_H." Thus, the final form of Equation 69 is:

$$-v\tau_H\frac{\partial T}{\partial x} + T_s = T + \tau_H\frac{\partial T}{\partial t} \tag{70}$$

where

$$\tau_H = \frac{\rho c_p D}{4U} \tag{71}$$

Because of the linearity of Equation 70 in T and T_s, this can be regarded as a deviation equation in these variables. Taking the initial deviations to be zero at time zero, the Laplace transformation of Equation 70 is

$$-v\tau_H \frac{dT}{dx} + T_s = T + \tau_H s T \tag{72}$$

where T is now the transform, namely $T(x, s)$, and T_s is the transform, $T_s(s)$. T_s is not a function of x since the steam temperature is the same at all points along the tube. This is an ordinary first-order differential equation in x since s appears only parametrically. The boundary condition is the deviation in entrance temperature at $x = 0$:

$$T(0, s) = T_1 \tag{73}$$

With this boundary condition, the solution of the differential equation is:

$$T_2 = \exp\left[-\frac{L}{v}\left(\frac{1}{\tau_H} + s\right)\right] T_1 + \frac{1}{\tau_H s + 1}\left\{1 - \exp\left[-\frac{L}{v}\left(\frac{1}{\tau_H} + s\right)\right]\right\} T_s \tag{74}$$

The transfer function, T_2/T_1, is found by setting the deviation, T_s, equal to zero; conversely, T_2/T_s is found by setting T_1 equal to zero. The ratio of the length, L, to the velocity, v, is the residence time of the fluid in the tube, τ_R. The ratio of this to the thermal time constant, τ_R/τ_H, arises in all transfer functions for exchangers and is given the symbol, P_o:

$$P_o = \frac{\tau_R}{\tau_H} \tag{75}$$

In terms of this parameter, the transfer functions are

$$\frac{T_2}{T_1} = e^{-P_o(\tau_H s + 1)} \tag{76}$$

and

$$\frac{T_2}{T_s} = \frac{1}{\tau_H s + 1}\left[1 - e^{-P_o(\tau_H s + 1)}\right] \tag{77}$$

The transfer function, T_2/v, is not as easily obtained as the previous two, so only the result will be presented here. The difficulty arises because both v and T are now variables in Equation 66, so the equation is no longer linear. An added complication, as mentioned earlier, is that the coefficient, U, may change significantly if the inside film coefficient is controlling. In the result to be presented, it is assumed that the derivative of the coefficient with respect to velocity can be approximated by the following equation:

$$\frac{dU}{dv} = b\,\frac{\overline{U}}{\overline{v}} \tag{78}$$

where $\overline{U}$ = normal value of the coefficient of heat transfer
$\quad\ \overline{v}$ = normal velocity

If b is zero, then the velocity has no effect on the coefficient. At the other extreme, if the inside film is completely controlling, then b could be estimated by assuming that the coefficient is proportional to the eight-tenths power of the velocity.

The transfer function is found by linearizing Equation 68 for small changes in v and T. Using Equation 78 to account for the change in the coefficient, the final result is

$$\frac{T_2}{v} = \frac{\delta_2(b-1)}{\overline{v}\tau_H}\left(\frac{1 - e^{-s\tau_R}}{s}\right) = \frac{\delta_2(b-1)P_o}{\overline{v}}\left(\frac{1 - e^{-s\tau_R}}{s\tau_R}\right) \tag{79}$$

where δ_2 is the normal driving force at the exit end of the exchanger. A derivation of this transfer function can be found in References 11ii, 13, and 14.

2. Interaction of Shell Fluid Dynamics with Tube Fluid Dynamics

The treatment in the previous section completely neglected the interaction between the temperature of the steam, T_s, and the thermal behavior of the water as it passed through the tube. There is obviously interaction since the rate of heat transfer at each point along the tube is determined by the difference in the temperatures of the two fluids at that point.

In this section, we shall assume that the outlet water temperature, T_2, is to be controlled by manipulating the pressure to the steam valve, P_V, as shown in Figure 13a. As in the example of Section III.D of Chapter 9, it will be assumed that critical flow exists in the valve and that the steam supply pressure is constant. Therefore, the flow rate of steam and the rate of heat transfer are under direct control and the steam temperature, T_s, is only indirectly manipulated. The important point is that T_s is not only a function of the steam rate, but also a function of the thermal behavior of the water in the tube. This contrasts sharply with the treatment of the previous section since by neglecting the interaction, T_s was considered to be independent of the water temperature.

a. DEVELOPMENT OF THE ELECTRICAL ANALOGY

The overall energy balance for the steam in the shell is

$$K_V \lambda P_V - UA'\int_0^L (T_s - T)\,dx = C_s\,\frac{dT_s}{dt} \tag{80}$$

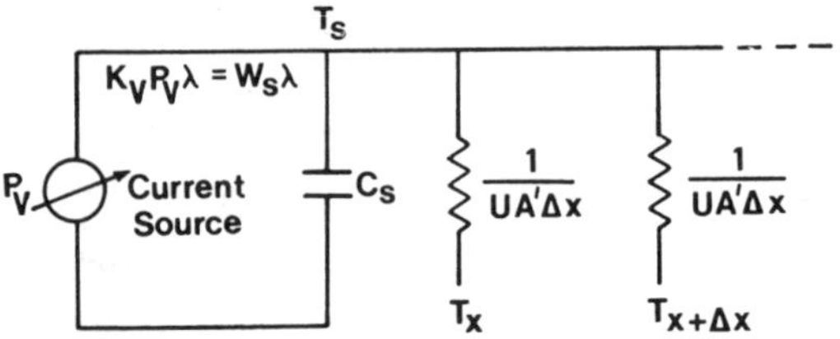

Figure 18.   Electrical analogy for the shell fluid.

where K_V = valve gain, lb steam/hr, psi

$\quad\quad \lambda$ = heat of condensation of steam, Btu/lb

$\quad\quad P_V$ = change in valve pressure, psi

$\quad\quad C_s = (\lambda m/P)(dP/dT)_s$ = shell capacitance as derived in Section III.D. of Chapter 9

Thus, the shell is treated as a lumped capacitance, but the entire heat transfer along the tube must be integrated because of the variation in the liquid temperature. The integral term, $UA' \int_0^L (T_s - T)\, dx$, can be thought of as a summation of rates of heat transfer to small tube sections of length, Δx:

$$UA' \int_0^L (T_s - T)\, dx \cong \sum \frac{T_s - T}{\dfrac{1}{UA'\, \Delta x}}$$

Thus, an electrical analogy for the shell is as shown in Figure 18. This diagram can be combined with that for the tube in Figure 17*b* to obtain the result shown in Figure 19. This is a current-source situation because critical flow has been assumed through the steam valve.

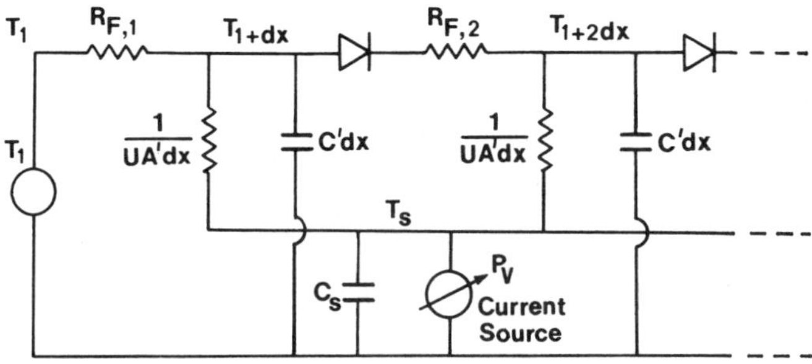

Figure 19.   Electrical analogy for two differential sections of condenser with large shell fluid capacitance.

b. TRANSFER FUNCTIONS

Just as three transfer functions were determined for the behavior of the tube alone, three could also be presented here. However, for brevity, only T_2/P_V will be developed here and that for T_2/T_1 is the subject of Problem 14 at the end of this chapter.

The derivation of the transfer function, T_2/P_V, is based on Equation 70 for the tube fluid and on Equation 80 for the shell fluid. The first step is to relate the temperature of the fluid in the tube at any point, $T(x)$, to T_s. The transfer function, T_2/T_s, given by Equation 77 can be used directly to obtain this relation if L is interpreted as x and T_2 is interpreted as $T(x)$. Thus, using Equation 75 in Equation 77 and rearranging, we obtain

$$T = \frac{T_s}{\tau_H s + 1}\left[1 - \exp\left(-\frac{\tau_H s + 1}{v\tau_H}x\right)\right] \tag{81}$$

Equation 80 is now transformed and Equation 81 substituted into it. After integration, the result is

$$C_s s T_s = K_V \lambda P_V - U A' T_s \left\{ L - \frac{L}{\tau_H s + 1} + \frac{v\tau_H}{[\tau_H s + 1]^2}\right.$$
$$\left. \times \left[1 - \exp\left(-\frac{\tau_H s + 1}{v\tau_H}L\right)\right]\right\} \tag{82}$$

This equation is then solved for T_s and substituted back into Equation 81, remembering that for $x = L$, $T = T_2$. After rearrangement, the desired transfer function is obtained:

$$\frac{T_2}{P_V} = \frac{K_1\left[1 - e^{-P_o(\tau_H s + 1)}\right]}{\tau_S \tau_H s^2 + (\tau_S + \tau_H)s + \dfrac{1 - e^{-P_o(\tau_H s + 1)}}{P_o(\tau_H s + 1)}} \tag{83}$$

where

$$K_1 = \frac{K_V \lambda}{U A' L} \tag{84}$$

$$\tau_S = \frac{C_s}{U A' L} \tag{85}$$

Comparing Equation 83 with that for T_2/T_s given in Equation 77, we see that accounting for the interaction considerably increases the complexity of the transfer function. Although the denominator of Equation 83 contains a second power of s, the transfer function cannot be classed as second order because of the complex exponential function of s.

B. Stirred Tanks

When an exothermic reaction is carried out in a stirred tank, temperature control can be achieved by passing a coolant through a jacket or coil. If the coolant is a boiling liquid, then a relatively simple, lumped parameter model will probably suffice. Usually, however, the coolant is a subcooled liquid. If the liquid passes through a jacket, then the exact flow pattern may be difficult to predict, and the model will be correspondingly approximate.

For the case of a liquid passing through a coil, at least the flow pattern is known and the assumption of plug flow is reasonable. The tank temperature can be controlled by either changing the inlet temperature of the coolant or by changing its flow rate. As in the case of the condenser, a change in the inlet temperature affects only the driving force for heat transfer, but a change in flow rate at least changes the driving force and may possibly change the coefficient.

It is interesting to note that the physical situation is almost the reverse of the condenser. This can be seen by referring to the physical picture of the reactor in Figure 20. The temperature of the reactants, T_s, can be considered

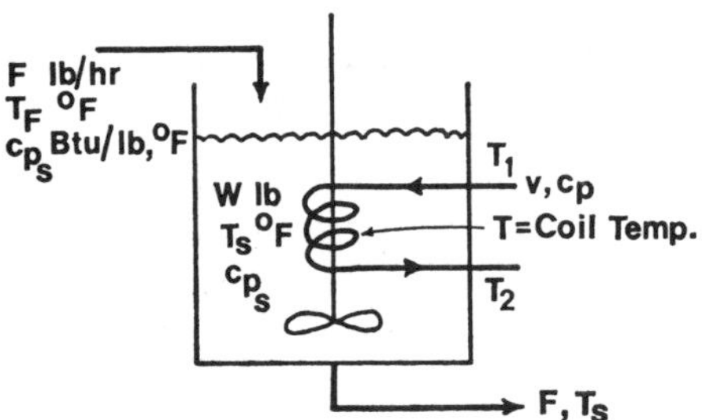

Figure 20. Continuous-flow, stirred-tank reactor with a cooling coil.

to be uniform and therefore their temperature is analogous to the steam temperature in the shell of the condenser. The liquid passing through the coil is, of course, analogous to that passing through the condenser tube. Thus, the temperature profiles shown in Figure 13b apply to the reactor as well. However, for the reactor, the transfer functions of interest are essentially in the opposite direction to those examined for the condenser; that is, for the reactor, we are interested in how the inlet temperature of the tube fluid and its velocity affect the temperature of the reactants.

The heat of reaction depends upon both the temperature and composition in the reactor. However, in one treatment given in Reference 15, it was assumed that the heat of reaction was constant, so the result basically describes the behavior of the reactor as a heat exchanger. The resulting transfer function is presented here:

$$\frac{T_s}{v} = \frac{\dfrac{\delta_1}{\bar{v}}\left\{\dfrac{(1-b)}{\tau_H s}\left[\dfrac{1 - e^{-P_o(\tau_H s + 1)}}{(\tau_H s + 1)}\right] - \left[\dfrac{(1-b)}{\tau_H s} + b\right](1 - e^{-P_o})\right\}}{\alpha + P_o + \beta s + \dfrac{1}{(\tau_H s + 1)^2}\left[1 - e^{-P_o(\tau_H s + 1)}\right] - \dfrac{P_o}{\tau_H s + 1}} \tag{86}$$

where $\bar{v}$ = normal velocity

$$\alpha = \frac{Fc_{p_s}}{S\rho c_p \bar{v}} \tag{87}$$

$$\beta = \frac{Wc_{p_s}}{S\rho c_p \bar{v}} \tag{88}$$

In the reference just cited, the results of some experimental tests were compared with those predicted from this relatively "exact" transfer function and there was very good agreement. However, it was found that several comparatively simple lumped parameter models were quite satisfactory approximations of the distributed model.

C. Cocurrent and Countercurrent Exchangers

The ultimate in complexity occurs when the temperatures of both fluids vary with time and position. Neglecting the possibility of multiple passes in the tubes and shell, a relatively complete analysis would include four partial differential equations—one for each fluid, one for the tube wall, and one for the shell wall.

The number of transfer functions of interest is quite large. Consider the drawing of a countercurrent exchanger shown in Figure 21. Possible inputs

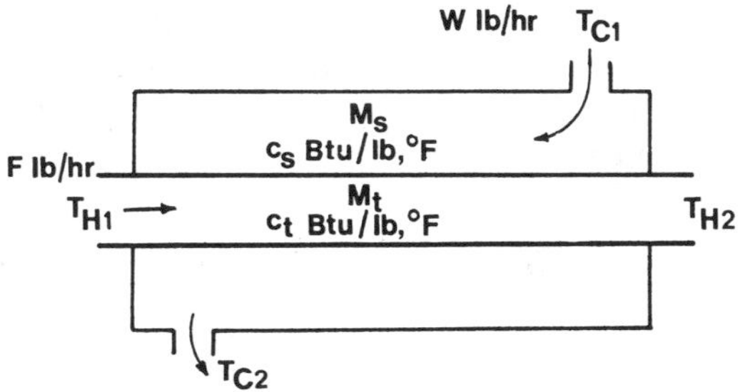

Figure 21.   Countercurrent heat exchanger.

are variations in T_{H1}, T_{C1}, F, and W, and possible outputs are T_{H2} and T_{C2}, giving a total of eight transfer functions. Then there would be an equal number for cocurrent flow. Takahashi (16) has treated the cases in which the inlet temperatures are inputs. If shell wall dynamics are neglected, only three partial differential equations are involved and this case is presented by Campbell (17) for both cocurrent and countercurrent flow. Needless to say, the resulting transfer functions are very complex and very difficult to use.

Most research effort has been directed toward developing lumped parameter models that satisfactorily describe measured dynamics. The simplest possible lumped parameter model is to treat the exchanger as two stirred tanks separated by a common area for heat transfer, as shown in Figure 22. It is assumed that the temperature of the liquid in each tank is

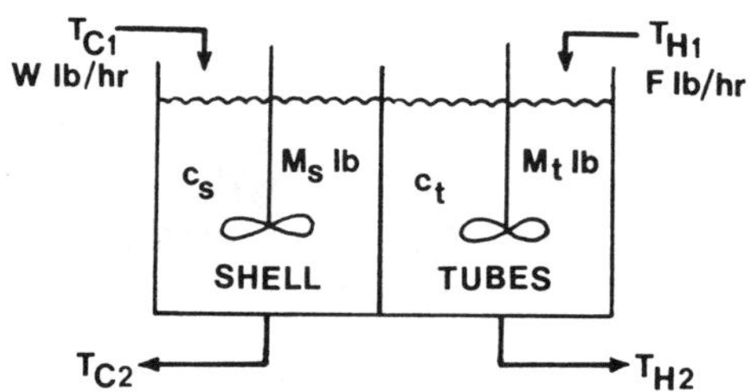

Figure 22. Very simple model of heat exchanger.

perfectly uniform. If the capacitances of the tube and shell walls are either neglected or lumped with those of the liquids, then a second-order model results. Mozley (18), in his "Model I," expressed the driving force for heat transfer as the difference between the exit temperatures of the two streams. In "Model II" he assumed the temperature in each tank to be the arithmetic average of the inlet and exit temperatures, and the driving force was the arithmetic average of the inlet and exit driving forces. Either of these models might be used for a jacketed tank or for a tank containing a coil. Model II was one of several used in the experimental study of reactor dynamics previously cited (15). It gave quite satisfactory results for the particular equipment employed. Neither of these models makes any distinction between cocurrent and countercurrent flow and therefore they have rather serious limitations.

SUMMARY

This chapter has been only a brief introduction to the topic of distributed parameter processes. Clearly, they represent an order-of-magnitude jump in complexity over that of lumped parameter processes. Even when the relatively exact distributed transfer functions can be derived, using them in a control analysis frequently presents computational and mathematical difficulties. Hence, the main thrust of research in process dynamics has been to find suitable lumped parameter approximations to them.

It is important to recognize what kinds of physical situations lead to distributed models. In all cases, the processes are described by partial differential equations and the transfer functions are characterized by exponentials in the operator, s. This leads to behavior similar to that displayed by a pure time delay. In the remaining chapters of this book, the presence of time delays will be found to be detrimental to control so it is important to be able to recognize distributed processes.

In closing our study of process dynamics, it should be noted that although the processes we have examined have usually been classified as first order, second order, or distributed, real processes can be of any order. Problem 4 of Chapter 10, for example, resulted in a transfer function of third order. The models and analogies used for some of the distributed processes in this chapter suggest the possibility of processes of much higher order than third.

Having developed some background in process dynamics, we are now ready to apply it in predicting the performance of process control systems. This analysis of systems occupies the remaining portion of this text.

REFERENCES

1. M. Bradner, Paper No. 48-4-2, ISA Conference, Philadelphia, Pa., September 1948.
2. N. B. Nichols, "The Linear Properties of Pneumatic Transmission Lines," *ISA Trans.*, **1,** 1 (1962), 5.
3. R. B. Bird, W. W. Stewart, and E. N. Lightfoot, *Transport Phenomena*, John Wiley and Sons, Inc., New York, 1960, (i) p. 83, (ii) p. 85.
4. J. C. Moise, "Pneumatic Transmission Lines," *ISA Journal*, **1,** 4 (1954), 35.
5. C. P. Rohmann and E. C. Grogan, "On the Dynamics of Pneumatic Transmission Lines," *Trans. ASME*, **79** (1957), 853.
6. G. H. Farrington, *Fundamentals of Automatic Control*, John Wiley and Sons, Inc., New York, 1951, (i) p. 141, (ii) pp. 130–131, (iii) pp. 138–140.
7. M. Bradner, "Pneumatic Transmission Lag," *Instruments*, **22,** 7 (1949), 618.
8. J. O. Hougen, *Experiences and Experiments with Process Dynamics*, C.E.P. Monograph Series, A.I.Ch.E., **60,** 4 (1964), 46.
9. J. O. Hougen, O. R. Martin, and R. A. Walsh, "Dynamics of Pneumatic Transmission Lines," *Control Eng.* (September 1963), 114.
10. C. B. Schuder and R. C. Binder, "The Response of Pneumatic Transmission Lines to Step Inputs," *Trans. ASME J. Basic Eng.*, **81** (1959), 578.
11. P. Harriott, *Process Control*, McGraw-Hill Book Co., Inc., New York, 1964, (i) p. 205, (ii) Chapter 11.

12. W. C. Cohen and E. F. Johnson, "Dynamic Characteristics of Double-pipe Heat Exchangers," *Ind. Eng. Chem.*, **48** (1956), 1031.
13. L. B. Koppel, "Dynamics of a Flow-forced Heat Exchanger," *Ind. Eng. Chem. Fund.*, **1**, 2 (1962), 131.
14. E. F. Johnson, *Automatic Process Control*, McGraw-Hill Book Co., Inc., New York, 1967, pp. 94–97.
15. T. W. Weber and P. Harriott, "Dynamics of Heat Removal From an Agitated Tank," *Ind. Eng. Chem. Fund.*, **4** (1965), 155.
16. Y. Takahashi and H. M. Paynter, ASME, Paper 55-SA-50.
17. D. P. Campbell, *Process Dynamics*, John Wiley and Sons, Inc., New York, 1958, pp. 191–204.
18. J. M. Mozley, "Predicting Dynamics of Concentric Pipe Heat Exchangers," *Ind. Eng. Chem.*, **48** (1956), 1035.

PROBLEMS

PROBLEMS CONCERNING HEAT TRANSFER IN A SLAB

Problem 1.   In Section II.B.1, a first-order lumped parameter approximation was presented in Equation 62 for a transmission line with a bellows termination. It was pointed out that this situation is completely analogous to heat transfer in a finite slab that is perfectly insulated at one face. The approximation calls for an effective time constant equal to one-half the product of the total capacitance and the total resistance. Hence, a possible electrical analogy is that shown in Figure 23*a*.

(a)   Why are these physical situations of pneumatic transmission and heat transfer analogous?

(b)   A second more sophisticated approximation is suggested in Figure 23*b*. For the case of heat transfer, this amounts to picturing the entire slab as a series of two half-slabs, each with half of the total resistance and half of the total capacitance.

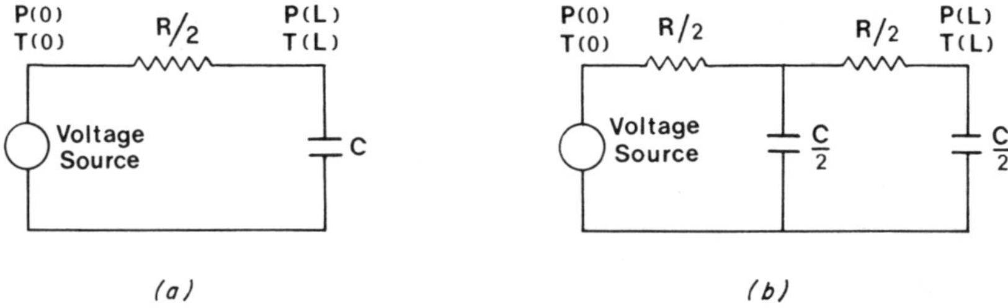

Figure 23.   Lumped parameter approximations for Problem 1. (*a*) First-order approximation. (*b*) Second-order approximation.

(1) Find the transfer function, $T(L)/T(0)$, in terms of:
   (i) R and C
   (ii) k = thermal conductivity
        ρ = density
        c_p = heat capacity
        L = thickness

(2) Calling the product, $RC = \tau$, find the effective time constants of this approximation. Is the response to a step input highly overdamped?

(3) For each of the approximations in Figure 23, find the response of $T(L)$ to a unit step input in $T(0)$. Plot the two responses on the same graph sheet versus normalized time in which $\tau = RC = 1$ time unit.

$$\textit{Answers:} \ \ (b)(1)(i) \ \ \frac{T(L)}{T(0)} = \frac{1}{\dfrac{R^2 C^2}{16} s^2 + \dfrac{3RC}{4} s + 1}$$

$$(b)(2) \quad 0.6545\tau, \ 0.0955\tau, \ \text{and} \ \zeta = 1.5$$

Problem 2.   Derive the transfer function of Section I.C.2.a but for a wall of infinite thickness. Compare the resulting transfer function with that in Equation 25. Under what physical conditions does the latter approach the transfer function you obtained? Does the result make sense?

$$\textit{Answer:} \ \ \frac{\theta}{\theta_F} = \frac{1}{\dfrac{M}{F} s + \dfrac{A\rho c_p s}{\sqrt{s/\alpha} \ Fc_1} + 1}$$

Problem 3.   This problem concerns some checks that can be made on the correctness of the transfer function given in Equation 25.

(a) Check the dimensionality of each of the terms.

(b) For a unit step in θ_F, apply the final-value theorem.

(c) Consider the situation in which the thermal conductivity of the wall approaches zero. (i) Derive the transfer functions directly without reference to Equation 25. (ii) Now check to see if Equation 25 approaches your predicted transfer function as the thermal conductivity approaches zero.

(d) What happens as the thermal conductivity of the wall approaches infinity? Do you obtain the expected result?

Problem 4.   A finite slab of thickness, L, is initially at a uniform temperature, T_0. At time zero, a change is made in the temperature at the face located at

$x = L$. The temperature at the face located at $x = 0$ is kept at T_0. Find the transfer function, $\overline{T}(x, s)/\overline{T}(L, s)$.

$$\textit{Answer:}\quad \frac{\overline{T}(x, s)}{\overline{T}(L, s)} = \frac{\sinh(\sqrt{s/\alpha}\, x)}{\sinh(\sqrt{s/\alpha}\, L)}$$

Problem 5. Draw the electrical analogy of the thick-walled tank in Section I.C.2.a. If the insulation around the tank were not perfect, how would this alter the analog?

Problem 6. Consider Equation 25 for the case in which the wall thickness, L, becomes very small. What form does the transfer function approach?

Problem 7. In Section I.C.2.a, the tank fluid portion of the process is essentially a lumped parameter situation with the capacitance represented by the fluid holdup: $C_1 = Mc_1$; the resistance is the reciprocal of the product of the flow rate times its thermal capacitance: $R_1 = 1/Fc_1$.

(a) If the wall were to be treated as a lumped parameter element, what would be its capacitance, C_2, in terms of its thickness and physical properties? What would be its resistance, R_2?

(b) Substitute your conclusions of part *a* together with the definitions of C_1 and R_1 given above into Equation 25 so that the transfer function is in terms of resistances and capacitances. Does the "typical" interaction factor appear in the result?

Problem 8. Solve Problem 12 of Chapter 10 and do the following:

(a) Compare the lumped parameter transfer function with that found for the distributed model (Equation 25). What are the major differences? What are the major similarities?

(b) Apply the various tests of Problem 3 of this chapter to the lumped model transfer function.

Problem 9. The transfer function for an insulated slab was given in Equation 17. Suppose that the flux at the uninsulated surface is given by Newton's law of cooling:

$$q = h(T_f - T_0)$$

where q = heat flux at wall, Btu/hr, ft^2
$\quad T_f$ = bulk fluid temperature, °F
$\quad T_0$ = wall temperature at fluid-solid interface, °F

(a) The flux, q, is related to the gradient in the slab at $x = 0$:

$$q = -k\,\frac{\partial T}{\partial x}\bigg|_{x=0}$$

Using this fact and other results of this chapter, derive the transfer function, q/T_f.

(b) Extend the result of part *a* to find the same transfer function for a semi-infinite slab.

$$\textit{Answers: (a)} \quad \frac{q}{T_f} = \frac{k\sqrt{s/\alpha}\ \tanh\ (\sqrt{s/\alpha}\ L)}{1 + \dfrac{k\sqrt{s/\alpha}}{h}\ \tanh\ (\sqrt{s/\alpha}\ L)}$$

$$\text{(b)}\quad \frac{q}{T_f} = \frac{k\sqrt{s/\alpha}}{1 + \dfrac{k\sqrt{s/\alpha}}{h}}$$

PROBLEM CONCERNING TRANSMISSION LINES

Problem 10. A No. 6 Foxboro valve motor has a volume of 25 in.^{3} at a pressure of 3 psig and a volume of 55 in.^{3} at 15 psig. If this valve is at the end of a line, $\frac{3}{16}$ in. in inside diameter and 500 ft long, find the values of the parameters in Equation 64.

$$\textit{Answers:} \quad \begin{aligned} I &= 6.1 \times 10^3 \ \text{lb}_f/\text{sec}^2/\text{ft}^5 \\ R &= 7.7 \times 10^4 \ \text{lb}_f/\text{sec}/\text{ft}^5 \\ C &= 4.53 \times 10^{-5} \ \text{ft}^5/\text{lb}_f \\ C_T &= 2.71 \times 10^{-5} \ \text{ft}^5/\text{lb}_f \end{aligned}$$

PROBLEMS CONCERNING HEAT EXCHANGERS

Problem 11. Commencing with Equation 69, find the transfer function, $T(L)/T(0)$, when there is no heat transfer between the tube fluid and the walls (i.e., $U \to 0$).

$$\textit{Answer:} \quad \frac{T(L)}{T(0)} = e^{-(L/V)s} = e^{-\tau_R s}$$

Problem 12. Invert each of the transfer functions given by Equations 76, 77, and 79, for unit step inputs. In each case, sketch the output and check the final value of the output using the final-value theorem.

Problem 13. Consider a fluid flowing through a thick-walled pipe surrounded by perfect insulation as illustrated in Figure 24. Using Equation 70, the partial differential equation for the fluid can be written

$$\frac{\partial T}{\partial t} + v\frac{\partial T}{\partial x} = K_1(T_w - T) \tag{A}$$

where $K_1 = 4h_1/\rho c_p D$

   h_1 = inside film coefficient of heat transfer, Btu/hr, ft^2, °F

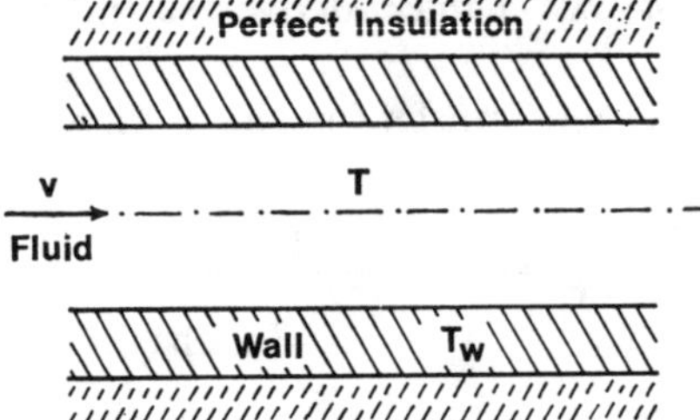

Figure 24. Thick-walled pipe surrounded by perfect insulation for Problem 13.

If there are no radial or axial temperature gradients in the wall, the partial differential equation describing its behavior is

$$\frac{\partial T_w}{\partial t} = K_2(T - T_w) \tag{B}$$

where $K_2 = A_1' h_1 / M_s' c_{p_w}$
$\quad\quad M_s' = $ mass of wall per foot of pipe, lb_m/ft
$\quad\quad c_{p_w} = $ heat capacity of wall, Btu/lb_m, °F

Other symbols have been defined in the text.

(a) Derive Equation B.
(b) Take the Laplace transforms of Equations A and B and solve them simultaneously to obtain the transfer function, $T(L)/T(0)$, where $T(L)$ is the leaving temperature of the stream at a distance L ft from the entrance, and $T(0)$ is the entering temperature of the stream.
(c) Test your transfer function of part b by examining what happens as:

   (1) The heat transfer coefficient, h_1, approaches zero. Does the result make sense?
   (2) The heat transfer coefficient approaches infinity. (*Hint:* Consider the parameter, s, to be finite as you take the limit.) Can you explain the result?

(d) Draw the electrical analogy for this process. Check its correctness by seeing what happens to it as the tests of part c are applied.

$$\textit{Answer: } \text{(b)} \quad \frac{T(L)}{T(0)} = \exp\left[-\frac{s^2 + s(K_1 + K_2)}{v(s + K_2)} L \right]$$

Problem 14. The transfer function, T_2/P_V, was presented in Equation 83 for the case in which there is an interaction between the shell and the tubes.

(a) Find the corresponding transfer function, T_2/T_1.
(b) What is the gain factor of the resulting transfer function?

(c)   What is the gain factor for the single-capacitance model given by Equation 76? Would you expect it to be the same as that which you found for the two-capacitance model?

$$\textit{Answers: (a)} \quad \frac{T_2}{T_1} = \frac{(\alpha^2 \tau_S s + \alpha^2 - \alpha)e^{(\alpha/\beta)L} + \dfrac{\gamma}{P_o}}{\alpha^2 \tau_S s + \alpha^2 - \alpha + \dfrac{\gamma}{P_o}}$$

$$\text{where } \alpha = 1 + \tau_H s$$
$$\beta = -v\tau_H$$
$$\gamma = e^{(\alpha/\beta)L} - 1$$

(b) Unity

(c) e^{-P_o}

PART III

Behavior of Controlled Systems

Part II was concerned with the modeling of processes and with the prediction of their dynamic behavior. Part III deals with the combination of these various processes to form a control system and with the response of the system to load disturbances and to changes in set point.

Part III is composed of three chapters. Chapter 12 is a synthesis of material that has been previously presented. It conveys some perspective and insight into the complexity of real processes and systems. Chapters 13 and 14 present two different ways of examining the unsteady-state behavior of controlled systems. Chapter 13 is concerned with the transient behavior of the system. A primary goal of the chapter is to present some of the background and philosophy behind the reaction curve method for controller tuning. Chapter 14 deals with frequency response, which is a powerful tool in the analysis of feedback systems. This leads to some fundamental understanding of the continuous cycling method for controller tuning.

CHAPTER XII

Introduction to Closed-Loop Analysis

Having developed transfer functions for a large variety of processes, it is instructive to see how a number of them might fit together to provide a model of an entire control system. Also, an example will serve as a good introduction to the topic of transient response.

I. PHYSICAL DESCRIPTION OF EXAMPLE

Let us suppose that an exothermic reaction is being carried out in a continuous-flow, stirred-tank reactor. The temperature is controlled by manipulating the flow rate of coolant through a coil and is sensed by a gas-filled bulb that is protected by a well. The bulb in turn is part of a pneumatic temperature transmitter. The controller is a three-mode unit having proportional, integral, and derivative action. The output of the controller is connected to a valve by pneumatic tubing, $\frac{1}{4}$ in. in outside diameter and 100 ft long. The only significant load variables are the entering coolant and

313

entering feed temperatures. The objective is to develop a detailed block diagram of the control system.

II. QUALITATIVE BLOCK DIAGRAM

Although a quantitative block diagram is the ultimate goal, we shall begin with a qualitative one as a guide to the development of the quantitative one. The introductory paragraphs of Chapter 5 noted that it is useful to think of the control system as consisting of two parts, the process itself and the control components. Accordingly, the block diagram might be drawn with two blocks, one for each of these parts. However, analyses of various processes presented in the previous four chapters indicate that each of these parts must be broken down into several others whose dynamics can be expressed in terms of transfer functions.

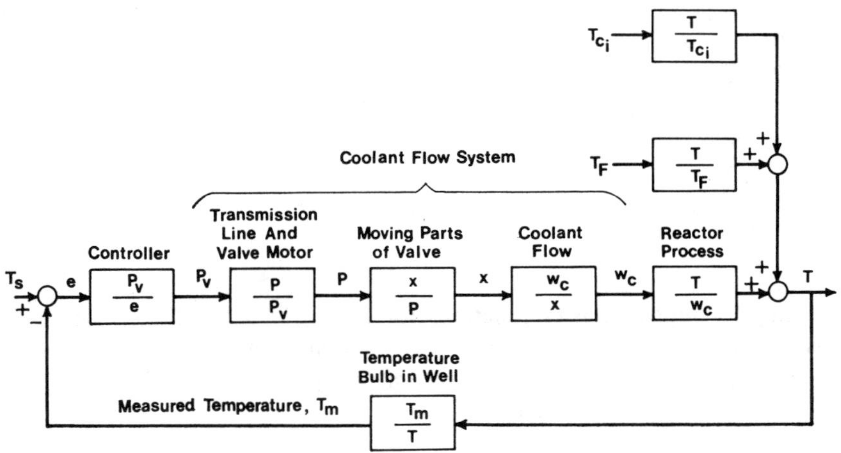

Figure 1.   Qualitative diagram for the reactor control system.

A *possible* block diagram is shown in Figure 1. The word "possible" is stressed, as others would be equally valid. The particular breakdown of the blocks basically depends on the transfer functions that are to be used to describe the various elements of the system. Thus, the blocks shown here were selected for the purpose of utilizing some of the transfer functions developed earlier.

The reactor block relates the temperature in the tank, T, to the coolant flow rate, w_c. The two main load variables would probably be the reactor feed temperature, T_F, and the inlet temperature of the coolant, T_{c_i}. Therefore, a block is shown for each of these.

Chapter 5 noted that the control components include the controller itself, the transmitter, and the valve. Blocks have been shown for the controller and the temperature transmitter. However, broadly speaking, the valve includes all of the elements of the coolant flow system and must account for the dynamic behavior between the output pressure of the controller and the actual flow rate of the coolant. Accordingly, the coolant flow system has been broken down into three parts. The first block relates the pressure in the valve motor, P, to the output pressure of the controller, P_V. The second block relates the stem position of the valve, x, to the pressure in the valve motor. Finally, the third block describes the response of the coolant flow rate to changes in the stem position.

With this background, we are now ready to examine each of these blocks in detail.

III. DEVELOPMENT OF TRANSFER FUNCTIONS

A. Process

Referring to Figure 1, we see that the required transfer functions for the process relate the tank temperature, T, to each of the variables, w_c, T_F, and T_{c_i}. In Section III.C of Chapter 9, a first-order model for a continuous-flow reactor was discussed and transfer functions for T/T_C and T/T_F were presented. This model neglects the dynamics of the coil and coolant. The temperature of the coolant, T_C, is its effective value and it would be necessary to relate this to the flow rate, w_c, in order to obtain the needed transfer function, T/w_c. Considering the distributed nature of the coil, this would be difficult, if perhaps not impossible to do.

A more "exact" approach to the coil-and-tank combination was that presented in Section III.B of Chapter 11, in which the transfer function relating tank temperature to coolant velocity was given in Equation 86. That transfer function neglected the effect of temperature on the reaction rate, but this factor could be incorporated into the derivation in the same way as it was in the first-order model just discussed. But then it would be necessary to derive transfer functions for the effects of T_F and T_{c_i}. The derivations would be algebraically complex and the results cumbersome and difficult to use.

Thus, for the purposes of this illustrative example, the first-order model of Chapter 9 is too simple, and the distributed model of Chapter 11, when modified for the effect of the heat of reaction, would be too complex. Hence, we shall take a middle position and derive a second-order model, accounting for the capacitance of the coolant in the coil and that of the reactants in the tank. As discussed in Chapter 9, although the effect of the change in the heat of reaction with temperature, $\partial Q_G/\partial T$, can have a very important effect on the dynamics, we shall neglect the effect in the derivation here so as to simplify the treatment slightly, but more importantly to enable us to use some experimental results of the author to illustrate several points.

The experimental studies described in References 1 and 2 concerned the dynamics of a continuous-flow, stirred-tank exchanger in which the temperature of the liquid in the tank was controlled by manipulating the flow rate through the coil. The distributed model described in Section III.B of Chapter 11 was tested together with several lumped parameter models. One of these was a second-order model, which will be used in this chapter.

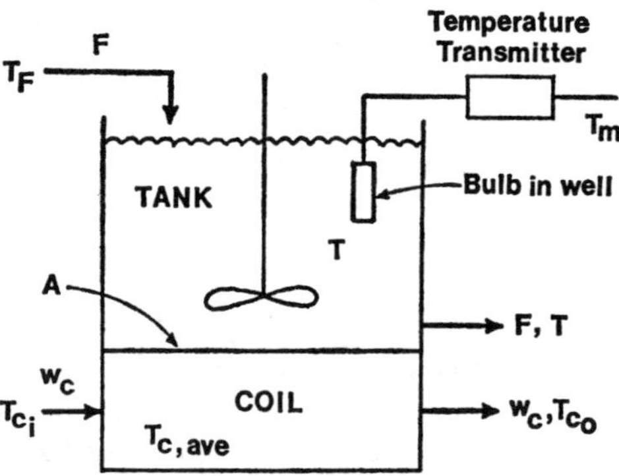

Figure 2.   Representation of second-order model for the reactor.

A representation of the model is shown in Figure 2. Basically, the model pictures the reactor and coil as two stirred tanks that are separated by the heat transfer area, A. The reactor liquid is assumed to be thoroughly mixed with temperature, T. The coolant is also assumed to be thoroughly mixed and therefore is treated as a lumped parameter element. However, its temperature is taken as the average of the inlet and outlet temperatures, and furthermore, the driving force for heat transfer is expressed as the average of the inlet and outlet driving forces. Hence, the distributed nature of the coolant in the coil is partially accounted for.

For brevity, detailed derivations of the transfer functions will not be carried out here. However, the energy equations for both the coolant and the reactor contents will be presented to clarify the exact nature of the model and to introduce and define the resistive and capacitive elements that appear in the transfer functions.

Turning first to the coolant, we can write the following energy balance:

$$w_c c_{p_c} T_{c_i} + UA(T - T_{c,\text{ave}}) - w_c c_{p_c} T_{c_o} = M_c c_{p_c} \frac{dT_{c,\text{ave}}}{dt} \tag{1}$$

where

$$T_{c,\text{ave}} = \frac{T_{c_i} + T_{c_o}}{2} \tag{2}$$

Using Equation 2, T_{c_o} can be eliminated from Equation 1 to yield

$$2w_c c_{p_c}(T_{c_i} - T_{c,\text{ave}}) + UA(T - T_{c,\text{ave}}) = M_c c_{p_c} \frac{dT_{c,\text{ave}}}{dt} \tag{3}$$

This equation can be written in terms of two resistances and a capacitance as follows:

$$\frac{T_{c_i} - T_{c,\text{ave}}}{R_1} + \frac{T - T_{c,\text{ave}}}{R_2} = C_1 \frac{dT_{c,\text{ave}}}{dt} \tag{4}$$

where

$$R_1 = \frac{1}{2w_c c_{p_c}} \tag{5}$$

$$R_2 = \frac{1}{UA} \tag{6}$$

$$C_1 = M_c c_{p_c} \tag{7}$$

The energy balance for the reactants is

$$F c_{p_t}(T_F - T) - UA(T - T_{c,\text{ave}}) = M_t c_{p_t} \frac{dT}{dt} \tag{8}$$

As noted earlier, the effect of the change in the heat of reaction is neglected in this analysis. It can be included rather easily and this modification is the subject of Problem 1 at the end of this chapter.

Equation 8 can be rewritten as follows:

$$\frac{T_F - T}{R_3} - \frac{T - T_{c,\text{ave}}}{R_2} = C_2 \frac{dT}{dt} \tag{9}$$

where

$$R_3 = \frac{1}{F c_{p_t}} \tag{10}$$

$$C_2 = M_t c_{p_t} \tag{11}$$

For the transfer functions, T/T_{c_i} and T/T_F, only the temperatures in Equations 3 and 8 are variables. Therefore, these equations are linear in the variables and they can be interpreted as deviation equations. Equations 4 and 9 are transformed and solved simultaneously to eliminate $T_{c,\text{ave}}$. After some algebraic manipulation, the desired transfer functions are obtained:

$$\frac{T}{T_{c_i}} = \frac{\dfrac{R_3}{R_1 + R_2 + R_3}}{as^2 + bs + 1} \tag{12}$$

$$\frac{T}{T_F} = \frac{\dfrac{R_1 + R_2}{R_1 + R_2 + R_3}\left(\dfrac{R_1 R_2 C_1}{R_1 + R_2}\, s + 1\right)}{as^2 + bs + 1} \tag{13}$$

where

$$a = \frac{R_1 R_2 R_3 C_1 C_2}{R_1 + R_2 + R_3} \tag{14}$$

and

$$b = \frac{C_1 R_1 R_3 + C_1 R_1 R_2 + C_2 R_1 R_3 + C_2 R_2 R_3}{R_1 + R_2 + R_3} \tag{15}$$

The denominators of the two transfer functions are identical, but it is interesting to note that a first-order form appears in the numerator of Equation 13. The presence of this is typical of processes having a "side capacitance." From the standpoint of the fluid in the reactor, the fluid in the coil acts as a side capacitance for thermal energy. For example, suppose that the entering feed temperature suddenly increased for a few seconds and then returned to its original value. The increase represents an increment of thermal energy added to the reactor fluid. Some of this energy would be transferred to the coil fluid. Then when the entering temperature returned to its former value, some of this stored energy in the coil would be returned back to the reactor fluid.

The derivation of the transfer function, T/w_c, is not quite so straight-forward. With w_c and T_{c_o} both variable, Equation 1 is no longer linear. The equation can be linearized by expressing each variable in terms of its original steady-state value plus a deviation, as follows:

$$(\overline{w}_c + \Delta w_c)c_{p_c}\overline{T}_{c_i} + UA(\overline{T} + \Delta T - \overline{T}_{c,\text{ave}} - \Delta T_{c,\text{ave}})$$

$$- (\overline{w}_c + \Delta w_c)c_{p_c}(\overline{T}_{c_o} + \Delta T_{c_o}) = C_1\frac{d\,\Delta T_{c,\text{ave}}}{dt} \tag{16}$$

For this transfer function, the inlet temperature is not a variable, so ΔT_{c_i} does not appear. Thus, it follows that the change in the average temperature is just half the change in the outlet temperature:

$$\Delta T_{c,\text{ave}} = \frac{\Delta T_{c_o}}{2} \tag{17}$$

Furthermore, the initial steady-state equation is

$$\bar{w}_c c_{p_c} \bar{T}_{c_i} + UA(\bar{T} - \bar{T}_{c,\text{ave}}) - \bar{w}_c c_{p_c} \bar{T}_{c_o} = 0 \qquad (18)$$

Subtracting Equation 18 from Equation 16 and using Equation 17, the deviation equation becomes

$$(\bar{T}_{c_i} - \bar{T}_{c_o})c_{p_c}\, \Delta w_c + UA\, \Delta T - (UA + 2\bar{w}_c c_{p_c})\, \Delta T_{c,\text{ave}}$$

$$= M_c c_{p_c} \frac{d\, \Delta T_{c,\text{ave}}}{dt} \qquad (19)$$

Using Equations 5 and 6 and dropping the Δ notation for convenience, Equation 19 can be written

$$(\bar{T}_{c_i} - \bar{T}_{c_o})c_{p_c} w_c + \frac{T}{R_2} - \left(\frac{1}{R_2} + \frac{1}{R_1}\right) T_{c,\text{ave}} = C_1 \frac{dT_{c,\text{ave}}}{dt} \qquad (20)$$

Equations 9 and 20 can be transformed and solved simultaneously to obtain the following transfer function:

$$\frac{T}{w_c} = \frac{\dfrac{(\bar{T}_{c_i} - \bar{T}_{c_o})c_{p_c}R_1 R_3}{R_1 + R_2 + R_3}}{as^2 + bs + 1} \qquad (21)$$

where $\bar{T}_{c_i}$ and $\bar{T}_{c_o}$ are the normal inlet and outlet coil temperatures, °F and a and b are given by Equations 14 and 15.

The transfer functions given by Equations 12, 13, and 21 are rather abstract. We shall examine the relative magnitudes of the resistances and capacitances as they might appear in a reactor. Some data, given in Reference 1, are presented here for illustrative purposes. The "reactor" was an agitated tank 2 ft in diameter and about 4 ft deep. The feed stream to the tank was hot water which was cooled in the tank by cold water passing through a copper coil.

The tank holdup was about 500 lb of water. The cooling coil was wound from a 60-ft length of Type K, $\frac{3}{4}$-in. copper water tubing. These and other physical parameters of the system are summarized below, along with operating conditions of a typical run:

M_t = tank holdup = 500 lb water (60 gal)
M_c = coil holdup = 11.4 lb water
A = area for heat transfer = 13.9 ft^2
$c_{p_c} = c_{p_t}$ = heat capacity of water = 1 Btu/lb, °F
$\bar{w}_c$ = normal cooling water rate = 1845 lb/hr
$\bar{F}$ = normal feed rate to tank = 4245 lb/hr
$\bar{T}_{c_i}$ = normal cooling water inlet temperature = 73.4°F
$\bar{T}$ = normal tank temperature = 123.7°F
$\bar{T}_{c_o}$ = normal cooling water outlet temperature = 117.1°F

The calculations for the transfer functions proceed as follows:

$$\bar{Q} = \text{normal rate of heat transfer, Btu/hr}$$

$$\bar{Q} = w_c c_{p_c}(\bar{T}_{c_o} - \bar{T}_{c_i}) = UA \, \Delta\bar{T}_{\text{ave}}$$

$$\Delta\bar{T}_{\text{ave}} = \text{average driving force} = [(\bar{T} - \bar{T}_{c_i}) + (\bar{T} - \bar{T}_{c_o})]/2$$

$$\bar{Q} = 1845(1)(117.1 - 73.4) = 80{,}600 \text{ Btu/hr}$$

$$\Delta\bar{T}_{\text{ave}} = 28.5°\text{F}$$

$$U = 205 \text{ Btu/hr, ft}^2, °\text{F}$$

$$R_1 = 1/2w_c c_{p_c} = 1/2(1845)1 = 0.271 \times 10^{-3} \text{ hr, }°\text{F/Btu}$$

$$R_2 = 1/UA = 1/205(13.9) = 0.353 \times 10^{-3} \text{ hr, }°\text{F/Btu}$$

$$R_3 = 1/Fc_{p_t} = 1/4245(1) = 0.236 \times 10^{-3} \text{ hr, }°\text{F/Btu}$$

$$C_1 = M_c c_{p_c} = 11.4(1) = 11.4 \text{ Btu/}°\text{F}$$

$$C_2 = M_t c_{p_t} = 500(1) = 500 \text{ Btu/}°\text{F}$$

It is interesting to note that the three resistances are of roughly the same magnitude but the capacitances are quite different. The resulting transfer functions are

$$\frac{T}{T_{c_i}} = \frac{0.274 \; °\text{F/}°\text{F}}{1940s^2 + 315s + 1} \tag{22}$$

$$\frac{T}{T_F} = \frac{0.726(6.67s + 1) \; °\text{F/}°\text{F}}{1940s^2 + 315s + 1} \tag{23}$$

$$\frac{T}{w_c} = \frac{3.25 \times 10^{-3} \; °\text{F/lb/hr}}{1940s^2 + 315s + 1} \tag{24}$$

The factors 1940 and 315 in the denominators are in terms of seconds rather than hours since these will prove to be more convenient units later on.

The denominators are overdamped, second-order forms that can be factored to give two effective time constants. Thus, the three transfer functions can be expressed as follows:

$$\frac{T}{T_{c_i}} = \frac{0.274}{(6.28s + 1)(309s + 1)} \tag{25}$$

$$\frac{T}{T_F} = \frac{0.726(6.67s + 1)}{(6.28s + 1)(309s + 1)} \tag{26}$$

$$\frac{T}{w_c} = \frac{3.25 \times 10^{-3}}{(6.28s + 1)(309s + 1)} \tag{27}$$

Because of these particular forms, each transfer function can be decomposed into two blocks, as shown in Figure 3. This breakdown is permissible since consecutive transfer functions multiply one another.

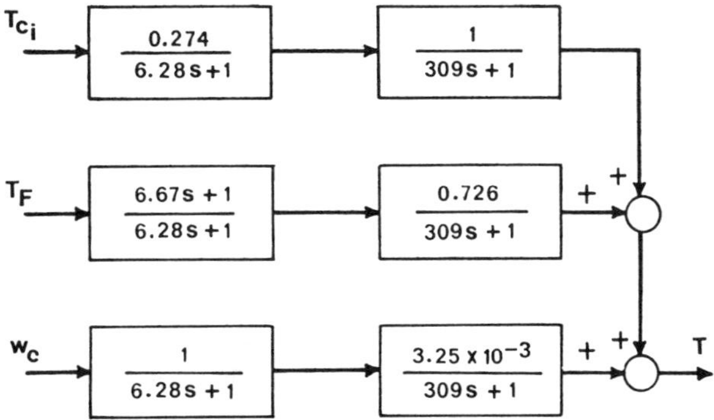

Figure 3.   Block diagram of reactor transfer functions.

Consider the first of the two blocks comprising T/T_F in Figure 3. Calling the output of this block, T', this pseudotransfer function is

$$\frac{T'}{T_F} = \frac{6.67s + 1}{6.28s + 1}$$

Although the output has the units of temperature, it cannot be given any physical interpretation. If the numerator and denominator were identical, they could be canceled, in which case the block would merely have a gain of one. Since they are almost the same, one suspects that their combined dynamic effect will be small. This reasoning can be partially substantiated in the following way. Suppose there is a unit step change in T_F, then:

$$T' = \frac{1}{s}\frac{6.67s + 1}{6.28s + 1}$$

This transform can be inverted to yield

$$\begin{aligned}T' &= 1 + [(6.67/6.28) - 1]e^{-t/6.28}\\ &= 1 + 0.06e^{-t/6.28}\end{aligned}$$

The corresponding response is shown in Figure 4. It deviates only slightly from that which would result from a pure gain of one, and furthermore, since the time constant of 6.28 is much less than the following one of 309, this transfer function can be neglected.

The two-block representation of T/T_{c_i} in Figure 3 is repeated in Figure 5a. The overall transfer function is unchanged if the gain of the first transfer function is divided by 3.25×10^{-3} while the gain of the second is multiplied by the same factor, as shown in Figure 5b. Again, the intermediate signal, labeled T'_c, has no real physical significance, although it must be closely associated with the effective coolant temperature.

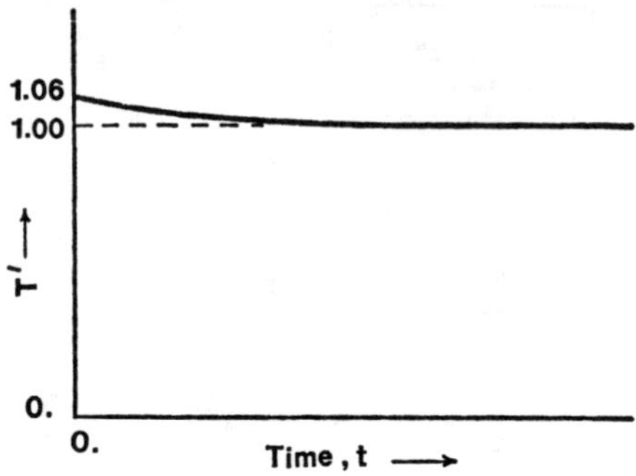

Figure 4. Step response for pseudo-transfer function, T'/T_F.

Based on the preceding discussion, the block arrangement of Figure 3 can be redrawn as shown in Figure 6. Looking at the process physically, the smallest time constant, 6.28, is closely associated with the coil and the largest, 309, with the tank itself, although these are not the time constants of these two parts because they are interacting. Considering the process to consist of two elements in series—the coil followed by the tank—then the smaller of the two time constants is associated with the first element, and the larger with the second element. Referring to Figure 6 as well as to the physical setup, changes in cooling water temperature occur before the first element, and changes in feed temperature occur before the second. Because of these differences in location, one would expect the response in tank temperature to be quite different for the two disturbances.

The gain factors in Equations 12, 13, and 21 are all in error because they are based on the use of an average driving force for the rate of heat

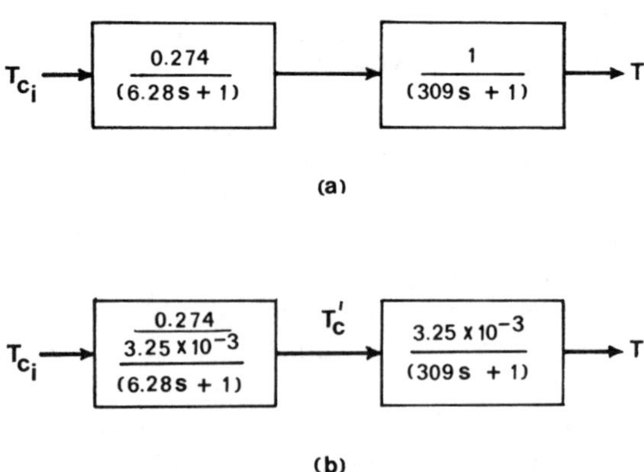

Figure 5. Equivalent representations of T/T_{c_i}.

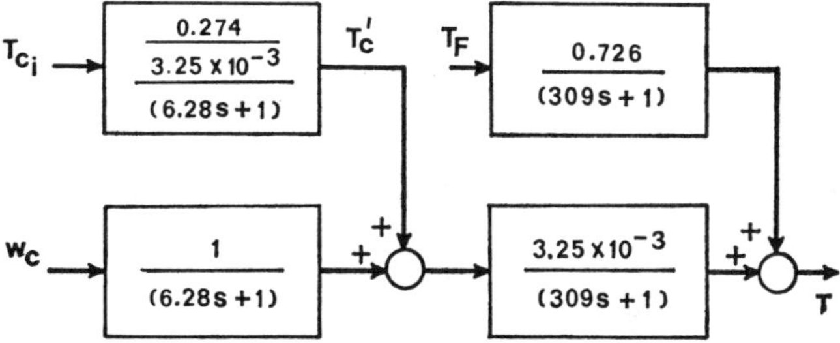

Figure 6.   Modified block diagram of reactor transfer functions.

transfer. Since the overall coefficient is probably about constant along the coil, the logarithmic mean driving force should be used. Thus, under steady-state conditions, the following equations must hold:

$$Fc_{p_t}(T_F - T) = w_c c_{p_c}(T_{c_o} - T_{c_i}) \qquad \text{energy balance} \qquad (28)$$

$$w_c c_{p_c}(T_{c_o} - T_{c_i}) = U_{\mathrm{lm}}A \, \frac{(T - T_{c_i}) - (T - T_{c_o})}{\ln\left(\dfrac{T - T_{c_i}}{T - T_{c_o}}\right)} \qquad (29)$$

where U_{lm} is the overall coefficient based upon a logarithmic mean driving force.

The actual values of the gain factors could be found by trial and error using Equations 28 and 29. Suppose, for example, that we wished to calculate the gain factor for T/T_{c_i}. The original values of each of the variables and parameters are known. Then a small change in T_{c_i} could be assumed and the resulting new value of T could be found by trial and error. The change in T divided by the assumed change in T_{c_i} would give the gain for the transfer function, T/T_{c_i}. Then the same procedure could be used to find the gain factors for T/T_F and T/w_c. In the case of the latter, allowance could also be made for the change in the coefficient with flow rate if necessary.

Actually, for T/w_c, an analytical expression for the gain can be obtained from the transfer function for the distributed model, given by Equation 86 of Chapter 11. The gain is found by assuming a unit step increase in flow rate followed by the application of the final-value theorem; in effect, this simply amounts to taking the limit of the transfer function as s approaches zero. The result is

$$\text{gain for } \frac{T}{w_c} = \frac{(\bar{T} - \bar{T}_{c_i})[(1 - b)P_o e^{-P_o} + (e^{-P_o} - 1)]}{\bar{w}_c(\alpha + 1 - e^{-P_o})} \qquad (30)$$

where $\bar{T}$ = normal tank temperature, °F

$\quad\quad \bar{w}_c$ = normal cooling water flow rate, lb/hr. (In Chapter 11, the velocity, v, was used instead of w_c, but these are proportional to each other.)

Other symbols have been previously defined in this and the preceding chapter.

For the conditions of the run, Equation 30 gives a gain of 5.86×10^{-3} °F/lb/sec, which is roughly twice the value given by the lumped parameter model. From measurements of the change in coefficient with flow rate, the parameter, b, had a value of 0.29, and this value was used in the above computation. If b were zero, the gain from Equation 30 would be 5.20×10^{-3}, which is about 10% less; in other words, the inside film resistance to heat transfer is not very controlling here.

B. Coolant Flow System

Referring back to Figure 1, we see that the coolant flow section of the diagram consists of the transmission line, valve motor, moving parts of the valve, and the dynamic behavior of the coolant flow to changes in valve position. In Problem 4 of Chapter 10, it is found that the motor and moving parts of the valve might be modeled by a third-order transfer function. Since these elements terminate the transmission line, which is a distributed process, we see that an exact analytical treatment for the transfer function, x/P_V, would result in a very complex expression.

From physical intuition as well as from our background developed in Chapters 8 to 11, we might expect that the elements of the coolant flow system could be characterized by relatively small time constants compared with those of the process. Hence, approximations will be made in estimating the transfer functions for these elements.

The first simplification is to decompose the transfer function, x/P_V, into the blocks shown in Figure 1. This is done for convenience since the data of Bradner (3) were given in Figure 12 of Chapter 11 for P/P_V. We shall assume that the valve is terminated by the bellows of a valve positioner. The figure indicates that for a 100-ft line, the time constant is less than 1 sec, so we shall use 0.8 sec. Very little information is available in the literature concerning the dynamics of the moving parts of valves. We shall assume that the valve displays overdamped behavior and can be characterized by a single time constant of 0.5 sec.

The last transfer function relating flow rate to stem position was discussed in Section IV.A of Chapter 9. A detailed estimate of the time constant for changes in stem position would require knowledge of the piping layout, line sizes, and so forth. The example of that chapter indicates that this time constant is likely to be much less than a second, so we shall assume a value of 0.1 sec.

The gain factor of the coolant flow system could be calculated from the valve characteristic and the line sizes. As in some earlier chapters, this gain will be given the symbol, K_V. Hence with the approximations of the time constants given above, the overall transfer function for the coolant system is

$$\frac{w_c}{P_V} = \frac{K_V}{(0.8s + 1)(0.5s + 1)(0.1s + 1)} \tag{31}$$

C. Temperature Bulb in Well

As developed in Section III.B of Chapter 10, the combination of a bulb in a well is interacting second order. The response depends on the coefficient of heat transfer between the well and water in the tank. In the studies of the author previously described (2), the estimated effective time constants for this sensor were 6 and 26 sec. When a bare bulb was used, the dynamics of the bulb were adequately characterized by a single time constant of 2.2 sec. Therefore, the transfer function for the transmitter-sensor combination is:

$$\frac{T_m}{T} = \frac{1}{(26s + 1)(6s + 1)} \qquad \text{bulb in a well} \tag{32}$$

$$\frac{T_m}{T} = \frac{1}{2.2s + 1} \qquad \text{bare bulb} \tag{33}$$

D. Controller

The last element to be considered is the controller. In Chapter 6, the behavior of an ideal three-mode controller was given by the equation

$$P_V = K_C \left(e + \frac{1}{T_R} \int edt + T_D \frac{de}{dt} \right) \tag{34}$$

In Chapter 6, the symbol P was used for the controller output pressure. In this chapter the corresponding symbol is P_V, while P is the pressure in the valve motor at the end of the pneumatic transmission line. Taking the Laplace transform, we obtain the transfer function

$$\frac{P_V}{e} = K_C \left(1 + \frac{1}{T_R s} + T_D s \right) = \frac{K_C(T_R T_D s^2 + T_R s + 1)}{T_R s} \tag{35}$$

At this point we cannot be quantitative about the magnitudes of the parameters, K_C, T_R, and T_D; these must be chosen to give a desirable response of the temperature, T, to changes in set point, T_s, and in the load variables, T_{c_i} and T_F.

IV. QUANTITATIVE BLOCK DIAGRAM

We are now ready to put the actual transfer functions in the block diagram of Figure 1. The result is shown in Figure 7. It is particularly interesting to note the wide range in the magnitudes of the time constants. Surprisingly, perhaps, the second largest time constant is one of the effective ones of the bulb in the well. The other effective time constant for the bulb in the well is essentially the same size as the one closely associated with the coil.

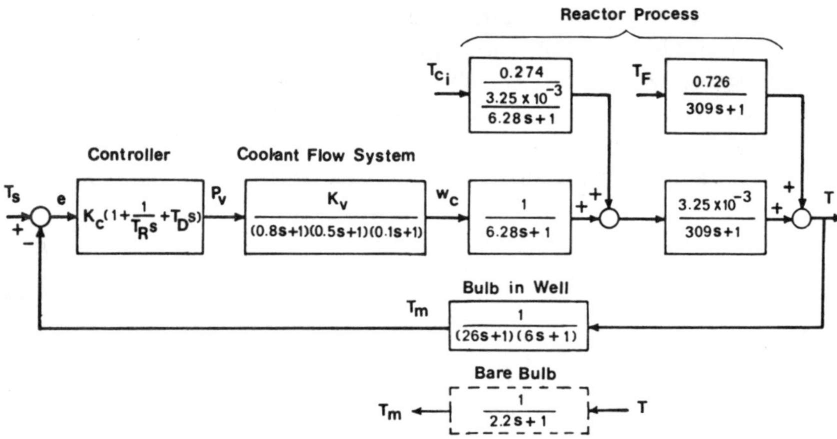

Figure 7.   Final block diagram of reactor control system.

Qualitatively, one might suspect that the response of the system will mainly depend on the relatively large time constants—6, 6.28, 26, and 309 sec—and will not be too dependent on the smaller ones—0.1, 0.5, and 0.8 sec. Furthermore, the fact that the second largest time constant is in the sensor implies that at times, the measured temperature, T_m, may be significantly different from the actual temperature, T, and this could be undesirable. Since the bulb was in a dry well, some improvement in its dynamics would be achieved if a tight-fitting sleeve were placed between the bulb and the well. A substantial improvement in system response might be expected if the bare bulb were used.

SUMMARY

This chapter has illustrated the kind of piece-by-piece analysis one might make in developing a quantitative block diagram. Physical intuition and experience will frequently be helpful in separating the significant elements of a system from the insignificant ones. Thus, ordinarily, a quantitative diagram would not contain more than three or four transfer functions.

Ultimately, one would be interested in finding optimum or acceptable controller settings. This subject is covered in Chapters 13 and 14.

REFERENCES

1. T. W. Weber and P. Harriott, "Dynamics of Heat Removal from an Agitated Tank," *Ind. Eng. Chem. Fund.*, **4** (1965), 155.
2. T. W. Weber and P. Harriott, "Control of a Continuous-Flow Agitated-Tank Reactor," *Ind. Eng. Chem. Fund.*, **4** (1965), 264.
3. M. Bradner, "Pneumatic Transmission Lag," *Instruments*, **22** (1949), 618.

PROBLEMS

Problem 1.

(a) In this problem, the transfer functions given by Equations 12, 13, and 21 are to be modified to take into account the heat of reaction. The change in heat of reaction, $\partial Q_G/\partial T$, can be considered to be the equivalent of an inverse resistance:

$$\frac{\partial Q_G}{\partial T} = \frac{1}{R_4}$$

where Q_G is the rate of heat generation, Btu/hr. Considering Equation 9 to be a deviation equation for temperatures, this introduces a term, $+(1/R_4)T$, into the left-hand side.

If one examines the electrical analogy for the tank, it can be seen that the resistance, R_3, is in parallel with a *negative* resistance, R_4. This suggests defining a new resistance, R_5, which is the equivalent of this parallel combination:

$$\frac{1}{R_5} = \frac{1}{R_3} - \frac{1}{R_4}$$

Using this idea, modify the parameters, a and b, given by Equations 14 and 15, directly. This same technique can be used for the gain factors in Equations 12 and 21. However, the technique fails to succeed in the case of the gain factor for Equation 13. Find the modified gain factor for this transfer function by starting with Equation 4 and the modified form of Equation 9. Why does the simple modification technique fail in this case?

(b)   Suppose that the coil and tank are noninteracting. Find the resulting transfer function for changes in feed temperature, T/T_F. What happens to the gain factor and time constant, τ_{tank}, as the magnitude of $\partial Q_G/\partial T$ increases (R_4 correspondingly decreases)? In particular, suppose that $C_2 = 10$ and $R_2 = R_3 = 1$. (These values are very hypothetical and are used only for illustrative purposes). Now sketch a plot of the time constant, τ_{tank}, versus the value of R_4; begin with a value of four for R_4 and let R_4 approach zero. What does this indicate about the possible behavior of the reactor?

(c)   Now consider the transfer functions derived in part *a* for the interacting case. If there is no reaction, then R_4 is infinite and the denominator is overdamped second order. Therefore it can be factored into two effective time constants. The larger one is mostly related to the tank, and the smaller to the coil, as in the example of this chapter. Now, as R_4 decreases in magnitude from infinity, what probably happens to the larger of the two effective time constants? Show that for R_4 sufficiently small, the gain factor of T/T_F becomes negative at the same time that the character of the denominator indicates that the reactor will be inherently unstable.

Problem 2.   In the example of this chapter, the coil and tank are interacting. Suppose they were not interacting. What would be the time constants of these two parts? Compare the numerical values with the effective ones given in the discussion.

$$\textit{Answers:} \quad T_1 = 6.29 \text{ sec}$$
$$T_2 = 255 \text{ sec}$$

Problem 3.   A constant temperature bath in a laboratory consists of a round glass tank, 1.5 ft in inside diameter and 1.5 ft high. The glass walls are $\frac{1}{4}$-in. thick Pyrex, which has a thermal conductivity of 0.68 Btu/hr, ft^2, °F/ft. The heat transfer coefficient between the outside wall and the air is 5 Btu/hr, ft^2, °F and the same value applies to that between the exposed water and the air. The film coefficient between the water and the walls is 400 Btu/hr, ft^2, °F. The bottom is essentially insulated.

The heating element is a steel rod containing electrical heating wires with an output of 100 W. The rod is $\frac{3}{8}$-in. in diameter and 1 ft long. The film coefficient between the water and the rod is 400 Btu/hr, ft^2, °F.

When the bath becomes too warm, heat is removed by a cooling coil consisting of ten feet of $\frac{3}{8}$-in. O.D. (0.305 I.D.) copper tubing with cooling water passing through at a velocity of 5 ft/sec. The outside film coefficient is 400 Btu/hr, ft^2, °F.

The temperature is sensed by a thermocouple with a time constant of

0.2 sec. The controller is an on-off device with a differential gap. In one position, the cooling water is turned on by a solenoid valve and the heater is turned off; in the other position, the heater is turned on and the cooling water is turned off.

The purpose of this problem is to develop a detailed block diagram of the system and to obtain values for as many of the time constants as possible. It is best to consider the problem in two parts, as follows:

Part 1—Heater on, Coolant Off

(a) When the heater is on, the electrical heating wires inside of the heating element are analogous to a "current source" and the process is analogous to Figure 8*b* of Chapter 10. In this problem, R_2 consists of the parallel resistances to heat transfer through the wall of the tank and at the air-water interface. Set up a table listing all of the quantities shown in Figure 8*b* and name the corresponding analogs for this problem.

(b) Calculate numerical values of C_1, C_2, R_1, and R_2 of Figure 8*b*.

(c) Draw the electrical analogy for this part.

(d) Find the transfer function, T/Q, where T is the bath temperature and Q is the flow rate of thermal energy in Btu/hr. Find the numerical value of the gain factor and the values of the effective time constants.

(e) The main load disturbance will be the room temperature, T_r. Find an approximate value of the transfer function, T/T_r, with numerical values of the gain and time constant. The gain factor is not given in the transfer functions of Chapter 10 so it must be derived.

Part 2—Coolant On, Heater Off

(f) When the coolant is on, the treatment of this chapter for a stirred-tank reactor can be used with some modification by recasting the parameters of this problem in terms of those in the development of this chapter. Set up a table of analogs for this part similar to that done in part *a*.

(g) Calculate the numerical values of C_1, C_2, R_1, R_2, and R_3. Assume that the inside film coefficient of heat transfer for the water in the tubing is 1150 Btu/hr, ft^2, °F.

(h) Draw the electrical analogy for this part.

(i) Find the transfer function, T/w_c. Calculate the effective time constants. A numerical value of the gain cannot be calculated since the inlet and outlet temperatures of the cooling water have not been given.

(j) Find an approximate value of the transfer function, T/T_r. The gain factor can be calculated using one of the transfer functions in this chapter.

Answers: (b) $C_1 = 0.0450$ Btu/°F; $C_2 = 165$ Btu/°F

$R_1 = 0.0253$ hr, °F/Btu

$R_2 = 0.0250$ hr, °F/Btu

(d) $$\frac{T}{Q} = \frac{0.0250 \text{ °F/Btu/hr}}{(14,900s + 1)(4.09s + 1)}$$
(time in seconds)

(g) $C_1 = 0.316$ Btu/°F; $C_2 = 165$ Btu/°F

$R_1 = 8.79 \times 10^{-4}$ hr, °F/Btu

$R_2 = 0.00366$ hr, °F/Btu

$R_3 = 0.0250$ hr, °F/Btu

(i) $$\frac{T}{w_c} = \frac{K}{(2300s + 1)(0.806s + 1)}$$
(time in seconds)

CHAPTER XIII

Transient Response

In Chapter 12, a block diagram for a complete control system was developed to illustrate the complexity that even a relatively simple feedback system might display. A similarly detailed diagram could be developed for a feedforward system.

In this chapter, system transfer functions will be derived and examined. Most of the emphasis will be on feedback configurations, since this is the most common method of control. We shall then study the transient response of some processes under feedback control. Finally, the reaction curve method for tuning a controller in a feedback system will be explained and illustrated. The goal in tuning a controller is to find the best settings of K_C, T_R, and T_D.

I. SYSTEM TRANSFER FUNCTIONS

When we speak of a "system transfer function," we are referring to the transfer function relating an output of the system to one of its inputs. For example, one system transfer function relates the controlled variable to the set point, and another relates the controlled variable to one of the load variables. The relationship appropriate to feedforward control is particularly simple, so we shall turn to it first.

A. Feedforward Control System

Recall from Chapter 2 that the idea behind a feedforward control system is to compensate for a particular load disturbance so that no error results. A typical feedforward block diagram is shown in Figure 1. The block containing

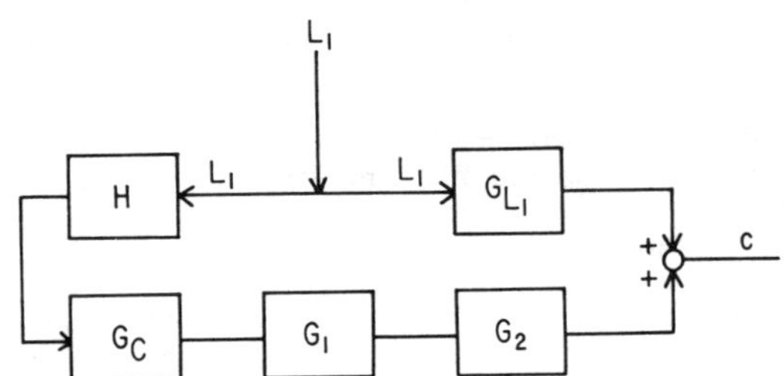

Figure 1.   General feedforward block diagram.

H represents a transmitter or measuring element that detects a change in the load variable, L_1. The feedforward computer is represented by G_C. The function, G_1, might characterize a valve, and G_2, the process.

From the diagram, it can be seen that

$$c = G_{L_1}L_1 + HG_CG_1G_2L_1 \tag{1}$$

For perfect control, c must be zero. Accordingly, the required computer transfer function is

$$G_C = \frac{-G_{L_1}}{HG_1G_2} \tag{2}$$

It can be seen from Equation 1 that in the event that the compensation is not perfect, then the system transfer function is

$$\frac{c}{L_1} = G_{L_1} + HG_CG_1G_2 \tag{3}$$

B. Feedback Control System

Most process control systems employ feedback, so the remainder of this chapter will be devoted to it. Generalizing from the example of Chapter 12, a typical feedback control system will have the configuration shown in Figure 2. A common convention is to designate the transfer functions in the forward

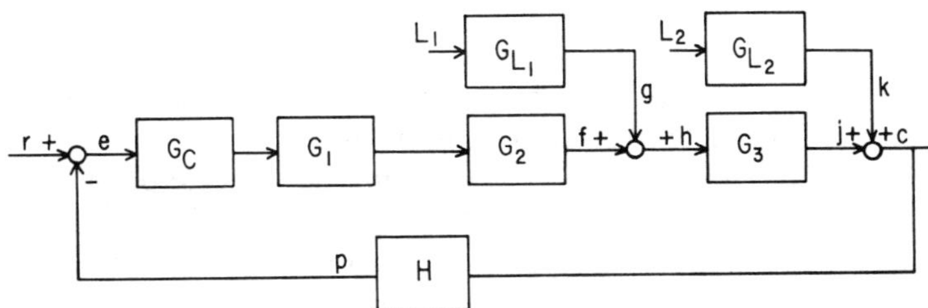

Figure 2.   General feedback block diagram.

part of the loop by G's and those in the feedback direction by H's. If this diagram is compared with Figure 7 of Chapter 12, it will be seen that G_1 could represent the collective transfer functions for the coolant flow system, G_2 could represent the cooling coil in the tank, and G_3 could represent the tank itself. In the general case, there would be n transfer functions representing n elements in the forward portion of the loop, and m transfer functions representing m elements in the feedback portion.

Two possible load variables, L_1 and L_2, are shown in the figure, each with its respective transfer function. As the example in Chapter 12 illustrated, it would be common to have several variables, each entering the loop at a different location.

The key point in the development of the system transfer functions is that for each block, the output is equal to the product of the input and the transfer function. Referring to Figure 2, the following equations can be written:

$$e = r - p \tag{4}$$

$$f = G_C G_1 G_2 e \tag{5}$$

$$g = G_{L_1} L_1 \tag{6}$$

$$h = f + g \tag{7}$$

$$j = G_3 h \tag{8}$$

$$k = G_{L_2} L_2 \tag{9}$$

$$c = j + k \tag{10}$$

$$p = Hc \tag{11}$$

Combining these relationships, we find

$$c = \frac{G_C G_1 G_2 G_3}{1 + G_C G_1 G_2 G_3 H}\, r + \frac{G_{L_1} G_3}{1 + G_C G_1 G_2 G_3 H}\, L_1 + \frac{G_{L_2}}{1 + G_C G_1 G_2 G_3 H}\, L_2 \tag{12}$$

If the response to a change in set point is to be determined, then the deviations, L_1 and L_2, are set equal to zero and only the first term on the right remains. Similarly, the response to a change in either load variable is found by setting r and the other load variable equal to zero. In the case of unity feedback the transfer function, H, is unity.

Comparing the numerators of the second and third terms on the right, it can be seen that an alternate representation of Figure 2 can be drawn as shown in Figure 3. The generalization of Equation 12 and Figures 2 and 3 to any number of transfer functions is evident by inspection.

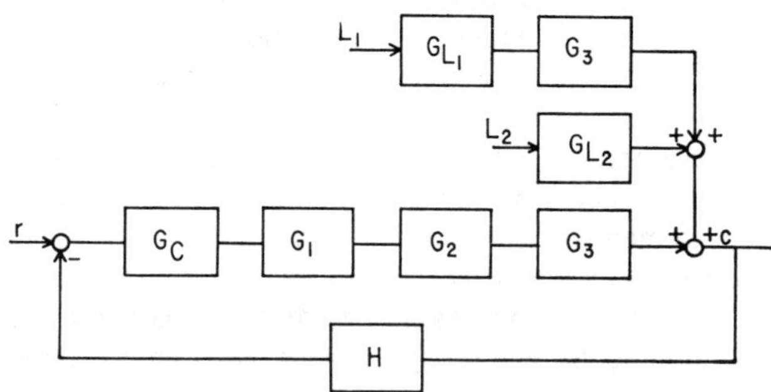

Figure 3.   Alternate representation of Figure 2.

C. Unity Feedback System

It was pointed out in Section I.B.5 of Chapter 4 that the block for the transmitter is usually moved from the feedback portion of the loop up into the forward portion in order to obtain the unity feedback configuration shown in Figure 4. The transfer function, G_C, now represents the product of the controller *and transmitter* transfer functions, and the transfer function, G, represents the product of all other transfer functions in the forward portion of the loop. The load transfer function, G_L, might be G_{L_2} or $G_{L_1}G_3$ of Figure 3, for example.

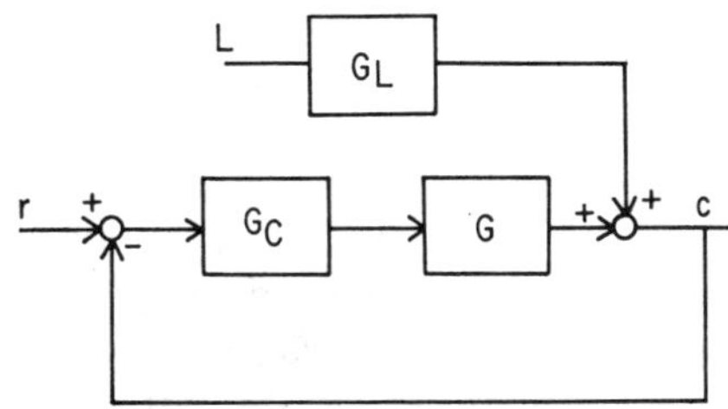

Figure 4.   Simple unity feedback control system.

Recall from Chapter 4 that the reason for moving the transmitter block up into the forward portion was so that the set point and error could be expressed in the units of the controlled variable and also because the chart of the controller is calibrated in these units. It should be apparent from the material covered in Chapters 8 to 11 that the dynamics of measuring elements can be significant in some control systems. In fact, this was the case in the example discussed in Chapter 12. When the temperature bulb was in a well, the second largest time constant in the loop was one associated with the bulb and well. Therefore, it is important to keep in mind that since a transmitter is always *physically* in the feedback portion of the loop, it is the *measured or observed* value of the controlled variable that is actually being controlled, and not the *actual or true* value. When the transmitter transfer function is placed in the forward portion, the diagram then implies that it is the actual value of the controlled variable that is being controlled. As noted in Chapter 4, this rearrangement is always permissible in a steady-state analysis and results in negligible error if the dynamics of the measuring element and transmitter are insignificant compared with those of the other elements in the loop.

Throughout the remainder of this text, most of the analyses will assume a unity feedback configuration. Furthermore, the gain of the transmitter will usually be assumed to be included in the controller gain, K_C. Hence, the units of K_C will be those of pressure divided by the units of the controlled variable.

1. Effect of Controller on Closed-Loop Response

For a unity feedback system, it can be seen from Equation 12 that the two system transfer functions are:

$$\frac{c}{r} = \frac{G_C G}{1 + G_C G} \tag{13}$$

$$\frac{c}{L} = \frac{G_L}{1 + G_C G} \tag{14}$$

In these two equations, G is the product of all of the transfer functions in the forward portion of the loop with the exception of that for the controller.

In servo operation, the transfer function of interest is c/r; the ultimate in performance is achieved if this function is unity. This implies that the product, $G_C G$, should be as large as possible. The transfer function, c/L, relates to the regulator operation of the system, and ideally it should be zero. This objective is approached for small values of G_L and large values of $G_C G$. The transfer functions, G and G_L, are basically determined by the demands or goals of the process, so if $G_C G$ is to be made large, G_C, and hence the corrective action, must be made large. As the controller action is increased, the system will tend to become unstable, so some compromise in the controller settings is usually necessary.

2. Stability Determined by the Open-Loop Transfer Function

The denominators of the transfer functions given in Equations 13 and 14 are identical, and it will now be shown that the stability of the system is determined by the open-loop transfer function, $G_C G$. Although a general proof of this could be given, it is perhaps a little easier to understand in the context of an example.

Consider the block diagram of the second-order process shown in Figure 5. The transfer function of a three-mode controller, as given in Equation 35

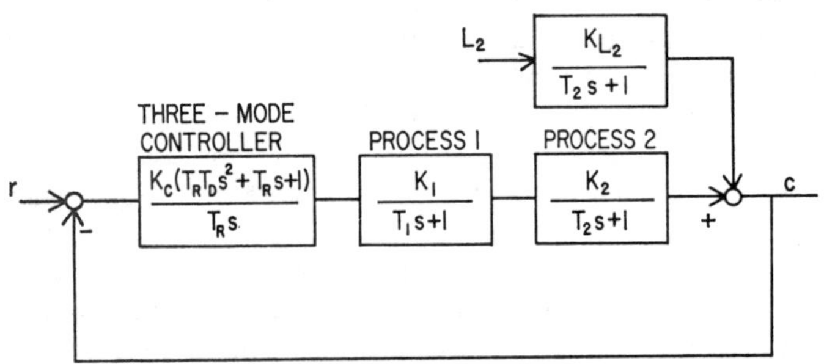

Figure 5. Second-order process under three-mode control.

of Chapter 12, is indicated. Thus, the open-loop transfer function is

$$G_C G = \frac{N}{D} = \frac{K_C K_1 K_2 (T_R T_D s^2 + T_R s + 1)}{(T_1 s + 1)(T_2 s + 1) T_R s} \tag{15}$$

where N denotes the numerator of the fraction and D, the denominator. In a similar way, we can write

$$G_L = \frac{N_L}{D_L} = \frac{K_{L_2}}{T_2 s + 1} \tag{16}$$

Inserting these relations into Equations 13 and 14 and simplifying, we obtain:

$$\frac{c}{r} = \frac{N}{D + N} \tag{17}$$

$$\frac{c}{L_2} = \frac{N_L D}{D_L (D + N)} \tag{18}$$

Referring to Equations 15, 16, and 18, the ratio of D to D_L is

$$D' = \frac{D}{D_L} = \frac{(T_1 s + 1)(T_2 s + 1) T_R s}{(T_2 s + 1)} = (T_1 s + 1) T_R s \tag{19}$$

Thus, Equation 18 becomes

$$\frac{c}{L_2} = \frac{N_L D'}{D + N} \tag{20}$$

Equations 17 and 20 have the same denominator, and this is the characteristic polynomial of the system. Section VII of Chapter 7 shows that the roots of this polynomial determine the stability of the system; furthermore, if the response is oscillatory, such properties as the damping coefficient and frequency of oscillation are also determined by these roots.

II. TRANSIENT RESPONSE OF CONTROLLED SYSTEMS

The problem of finding the best controller settings can be looked upon from two points of view. If the transfer functions of all of the process elements are known, then a theoretical analysis of some kind can be made to find a satisfactory set of controller settings. On the other hand, the usual situation is that the transfer functions are unknown and the controller settings are frequently found by some trial-and-error procedure in combination with empirical "rules of thumb"; such a procedure is referred to as "tuning the controller." Two of these empirical approaches will be discussed, one in this chapter and one in the next. However, in order to understand some of the reasoning behind these methods, we shall assume that the process transfer functions are known and study the responses of systems composed of these known functions.

A. Interaction of Controller Modes

Although some special controller actions are available, the following discussion will be confined to the standard controller actions of proportional, integral, and derivative. The problem then, is, given a process, what action or actions should be used, and how much of each? Qualitatively, it has been demonstrated before that proportional action is relatively fast but if used alone, will usually result in offset. Integral action is relatively slow but eliminates offset. Finally, derivative action is anticipatory and may be quite helpful in controlling certain types of processes.

From the standpoint of analysis, it would be desirable if the effects of these controller modes on the controlled variable were additive. Then it would be comparatively simple to develop some guidelines for determining satisfactory controller settings. Unfortunately, the various controller modes interact in a rather complex way. This is best shown by considering the block diagram presented in Figure 6.

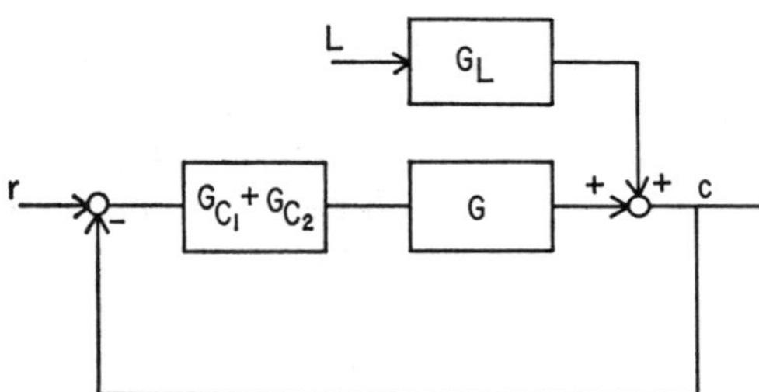

Figure 6.   Block diagram to illustrate interaction of controller modes.

Suppose, for example, that the controller is a proportional-integral unit that has the following transfer function:

$$\frac{P}{e} = K_C + \frac{K_C}{T_R s} \tag{21}$$

Thus, in terms of Figure 6, the proportional action is denoted by G_{C_1} and the integral action by G_{C_2}:

$$G_{C_1} = K_C \tag{22}$$

$$G_{C_2} = \frac{K_C}{T_R s} \tag{23}$$

The point to be made is illustrated by looking at the transfer function, c/r. If only the single controller action, G_{C_1}, is used, then

$$\frac{c}{r} = \frac{G_{C_1} G}{1 + G_{C_1} G} \tag{24}$$

Similarly, if only G_{C_2} is used, then

$$\frac{c}{r} = \frac{G_{C_2} G}{1 + G_{C_2} G} \tag{25}$$

But when both are used, the result is

$$\frac{c}{r} = \frac{G_{C_1} G + G_{C_2} G}{1 + G_{C_1} G + G_{C_2} G} \tag{26}$$

Since Equation 26 is not the sum of Equations 24 and 25, the responses resulting from the two controller actions are not additive.

B. Hypothetical Cases for Study

There is no such thing as a "typical" process since a process can be composed of any combination of lumped and distributed parameter elements and pure delays. Sometimes a process can be characterized by two or three lumped parameter elements; at other times, a combination of a pure delay plus one or two lumped parameter elements is satisfactory; ultimately, a distributed parameter model would be employed only in unusual situations where lumped parameter models were inadequate.

In Chapter 11 we found that the transfer functions of distributed processes usually have complex forms and that, whenever possible, these are approximated by lumped parameter expressions. It will be shown in this section that when a process consists of several lumped parameter elements, its response to a step input can be approximated to some degree by a first-order process plus a time delay. Therefore, we shall first study the behavior of several lumped parameter processes under various modes of control, and then make a similar study for a pure-delay process.

1. Closed-Loop Response of Lumped Parameter Processes

Many processes can be characterized by a single, relatively large time constant plus several smaller ones. This was the situation in the example of Chapter 12 in which the dominant time constant was 309 sec, followed by lesser time constants of 6.28 and 2.2 sec when a bare bulb was used, plus several relatively small time constants.

Apart from the action of the controller, it is reasonable to expect that the response of the entire system will be greatly influenced by the magnitude of the largest time constant. Suppose, for example, that process A has the following transfer function:

$$G_A = \frac{1}{(10s + 1)(s + 1)(0.5s + 1)}$$

and process B has the transfer function

$$G_B = \frac{1}{(20s + 1)(s + 1)(0.5s + 1)}$$

All other factors being equal, a control system for process A would probably respond about twice as fast as one for process B.

The smaller time constants of 1 and 0.5 in the two transfer functions would influence the allowable ranges of the controller parameters and the optimum controller settings. To assess the significance of these small time constants, we shall carry out a sequential study. The first process to be examined will have one time constant, the second will have two, and the third will have three. To reflect the typical spread in the time constants of a process, values of 10, 5, and 1 will be used. The unit of time is not important since it only affects the time scale of the response of a system. Summarizing then, the following three processes will be examined:

$$\text{process 1} \qquad G_1 = \frac{1}{10s + 1}$$

$$\text{process 2} \qquad G_2 = \frac{1}{(5s + 1)(10s + 1)}$$

$$\text{process 3} \qquad G_3 = \frac{1}{(s + 1)(5s + 1)(10s + 1)}$$

The process gain in all cases has been taken as unity for simplicity. Since the gain of the controller and the process multiply each other, all of the loop gain can be considered to be located at the controller.

Before examining the responses of the three processes under closed-loop control, it is worthwhile to compare their uncontrolled responses to step

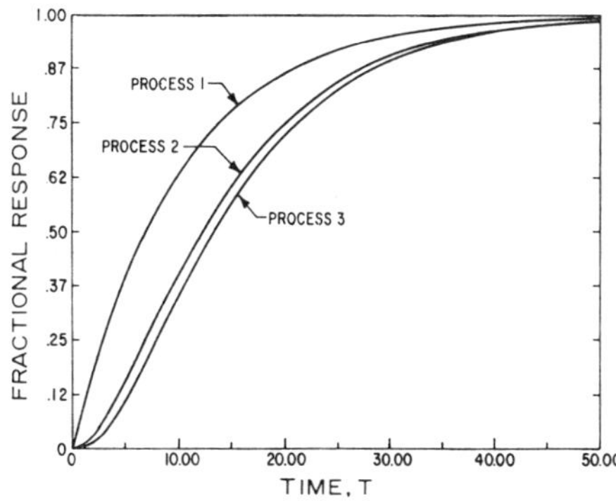

Figure 7.   Open-loop response of processes.

inputs—the so-called "open-loop responses." These are shown in Figure 7. As would be expected from the transfer functions, the response of process 2 is considerably retarded behind that of process 1. Since process 2 is second order, its slope is zero at time zero while the slope of process 1 is a maximum at time zero. The transfer functions of processes 2 and 3 differ very little, and their responses are very similar. The important point is, as more elements are added, the response becomes slower. All three responses are essentially complete by a time of 50 which is five times the largest time constant.

a. PROCESS 1

The block diagram for this case is shown in Figure 8. Note particularly that the load gain, K_L, has been set equal to unity. This gain factor changes the magnitude but not the form of the output and therefore this assumption is

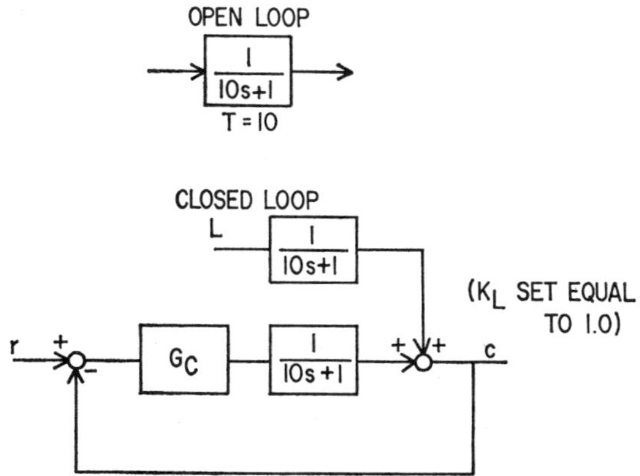

Figure 8.   Block diagram for process 1.

easily corrected for. The analysis will be confined to the following four controller actions.

Controller Action	*Controller Transfer Function*
Proportional, P	$G_C = K$
Integral, I	$G_C = \dfrac{K}{T_R s}$
Proportional-integral, P-I	$G_C = \dfrac{K(T_R s + 1)}{T_R s}$
Proportional-integral-derivative, P-I-D	$G_C = \dfrac{K(T_D T_R s^2 + T_R s + 1)}{T_R s}$

The resulting closed-loop transfer functions for changes in set point and load are presented in Table 1. These were obtained using Equations 13 and 14. For a unit step change in set point or in load, the input transform is $1/s$. For proportional action alone, the response to either a step change in set point or in load is first order. The system is stable for all loop gains, and an infinite value results in zero offset for changes in set point and in load.

Second-order behavior results for the remaining controller actions. For a given controller action, the denominators for changes in set point and load are identical. The resulting parameters of natural frequency and damping factor for each type of control are summarized below.

Controller Action

	I	P + I	P + I + D
Natural frequency, ω_n	$\sqrt{\dfrac{K}{T_R T}}$	$\sqrt{\dfrac{K}{T_R T}}$	$\sqrt{\dfrac{K}{T_R(T + T_D K)}}$
Damping factor, ζ	$\dfrac{1}{2}\sqrt{\dfrac{T_R}{KT}}$	$\dfrac{1 + K}{2}\sqrt{\dfrac{T_R}{KT}}$	$\dfrac{1 + K}{2}\sqrt{\dfrac{T_R}{K(T + KT_D)}}$

Responses can be overdamped, critically damped, or underdamped, depending upon the values of K and T_R. As K increases or T_R decreases, the response becomes faster. From Table 1 we see that ideal behavior occurs when T_R is zero, since c/r becomes unity and c/L becomes zero.

Since the system cannot be made unstable for any type of controller action at either infinite gain or zero reset time, ideal response is theoretically possible. Hence, to compare the response of the system under the various modes of control, moderate controller settings must be chosen arbitrarily.

TABLE 1

Open Loop

$$\frac{1}{Ts + 1}$$

Closed Loop

Controller Action	Set Point Transfer Function, c/r	Load Change Transfer Function, c/L
P	$\dfrac{\dfrac{K}{1 + K}}{\dfrac{T}{1 + K}\,s + 1}$	$\dfrac{\dfrac{K_L}{1 + K}}{\dfrac{T}{1 + K}\,s + 1}$
I	$\dfrac{1}{\dfrac{T_R T}{K}\,s^2 + \dfrac{T_R}{K}\,s + 1}$	$\dfrac{(K_L T_R/K)s}{\dfrac{T_R T}{K}\,s^2 + \dfrac{T_R}{K}\,s + 1}$
P + I	$\dfrac{T_R s + 1}{\dfrac{T_R T}{K}\,s^2 + T_R\left(\dfrac{1 + K}{K}\right)s + 1}$	$\dfrac{(K_L T_R/K)s}{\dfrac{T_R T}{K}\,s^2 + T_R\left(\dfrac{1 + K}{K}\right)s + 1}$
P + I + D	$\dfrac{T_D T_R s^2 + T_R s + 1}{T_R\left(\dfrac{T + KT_D}{K}\right)s^2 + T_R\left(\dfrac{1 + K}{K}\right)s + 1}$	$\dfrac{(K_L T_R/K)s}{T_R\left(\dfrac{T + KT_D}{K}\right)s^2 + T_R\left(\dfrac{1 + K}{K}\right)s + 1}$

A proportional gain of 3.0 was selected so as to yield an offset of 25% for proportional action alone. For integral action alone, a reset time of 5.81 was used to give a damping factor of 0.22, which corresponds to a decay ratio of 0.25. This ratio is sometimes considered a desirable response and more will be said about this later. The addition of derivative action to proportional-integral action only influences the natural frequency and the damping factor. Since both of these properties can just as well be set through adjustment of K and T_R, addition of derivative action here is somewhat trivial; however, to illustrate the effect, T_D was set at 3.0.

The responses of the system to unit step inputs are shown in Figure 9 and the corresponding transfer functions are presented in Table 2. In the

TABLE 2

	Open Loop	
	$\dfrac{1}{10s + 1}$	
	Closed Loop	
Controller Action	Set Point Transfer Function, c/r	Load Change Transfer Function, c/L
P $K = 3.0$	$\dfrac{0.75}{2.5s + 1}$	$\dfrac{0.25}{\leftarrow}$
I $K = 3.0$ $T_R = 5.81$	$\dfrac{1}{19.4s^2 + 1.94s + 1}$ $\omega = 0.227$ $\zeta = 0.220$	$\dfrac{1.94s}{\leftarrow}$
P + I $K = 3.0$ $T_R = 5.81$	$\dfrac{5.81s + 1}{19.4s^2 + 7.75s + 1}$ $\omega = 0.227$ $\zeta = 0.880$	$\dfrac{1.94s}{\leftarrow}$
P + I + D $K = 3.0$ $T_R = 5.81$ $T_D = 3.0$	$\dfrac{17.4s^2 + 5.81s + 1}{36.8s^2 + 7.75s + 1}$ $\omega = 0.165$ $\zeta = 0.639$	$\dfrac{1.94s + 1}{\leftarrow}$

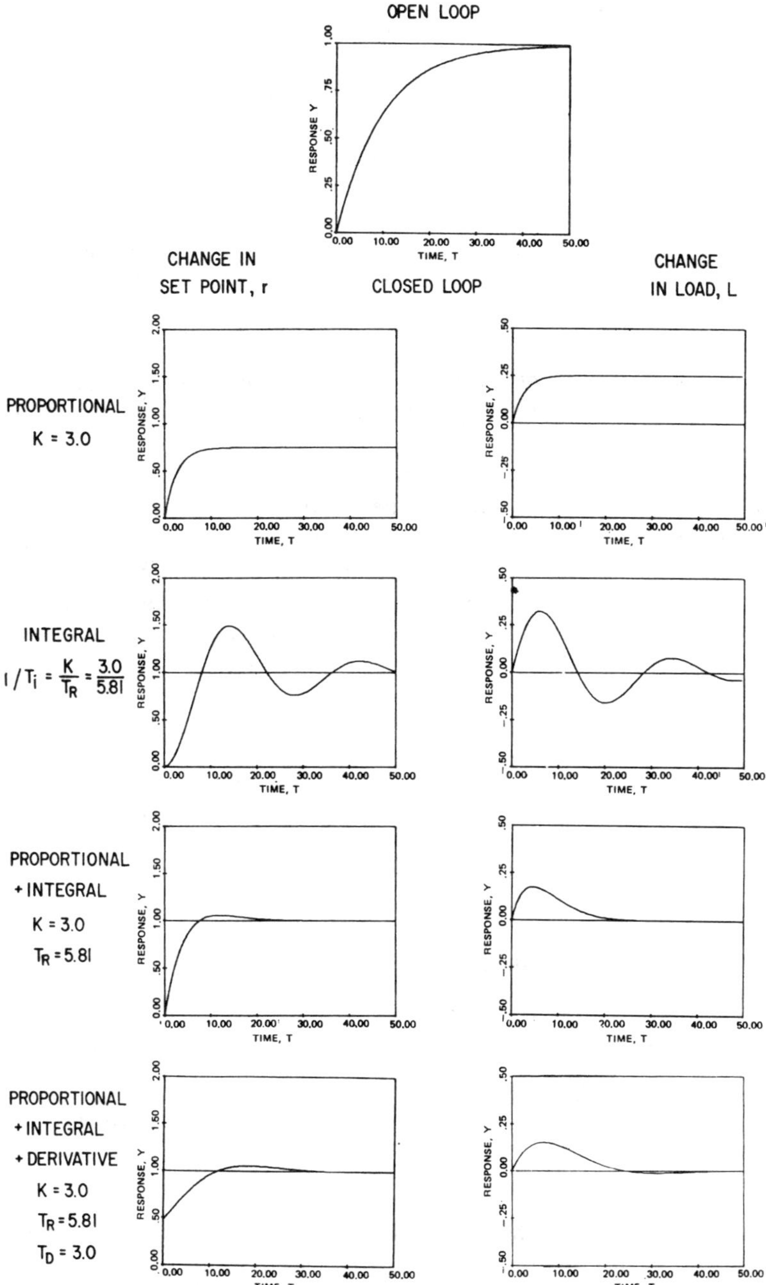

Figure 9. Response curves for process 1.

table, the damping factors and natural frequencies of the second-order forms in the denominators have been given. The numerator of the transfer function for a change in set point under proportional-integral-derivative control has a second-order form. This affects the magnitude and phase angle of the response, but neither the damping factor nor the frequency of oscillation. From Figure 9, it can be seen that proportional action alone is relatively fast but results in offset. Integral action alone is quite oscillatory but will eventually yield zero offset. If proportional action is added to integral, the damping factor is increased and the response is faster. Further addition of derivative action decreases the damping factor.

The last case involving derivative action requires some explanation. For a change in set point, the response does not meet the zero initial condition that is always assumed in the derivation of a transfer function. Furthermore, when a step change in set point is made, one would expect the derivative action to result in a momentary infinite response at the instant the step is made. The curve shown for this case results from the method used to generate the curves for these illustrations. Each transfer function in Table 1 was multiplied by $1/s$, the transform of a unit step, and then inverted to yield the time response. This method has the effect of treating the step input as the limiting case of a terminated ramp, as shown in Figure 10. As α approaches

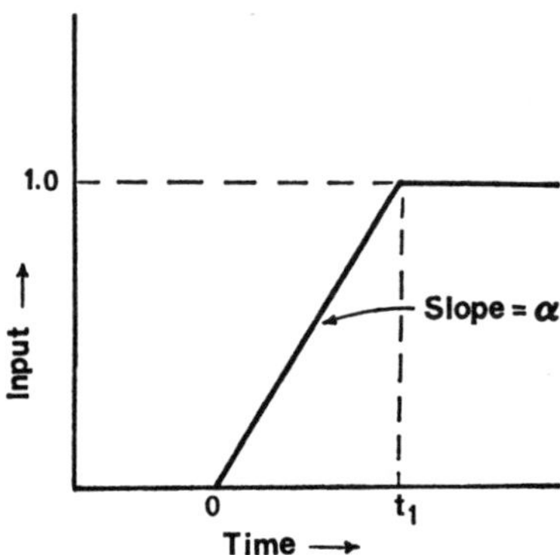

Figure 10. Terminated ramp function.

infinity, the input approaches a unit step, and the response approaches that presented in Figure 9 for each case. This subtle distinction between the limiting case and an actual step takes on significance only when derivative action is present and a change in set point is assumed. A closer look into this interesting anomaly forms the basis of Problem 9 at the end of this chapter.

b. PROCESS 2

The block diagram for this process is presented in Figure 11. The transfer functions in general form are given in Table 3. Since integral action is normally used in combination with proportional, the integral-only case has been omitted from this illustration.

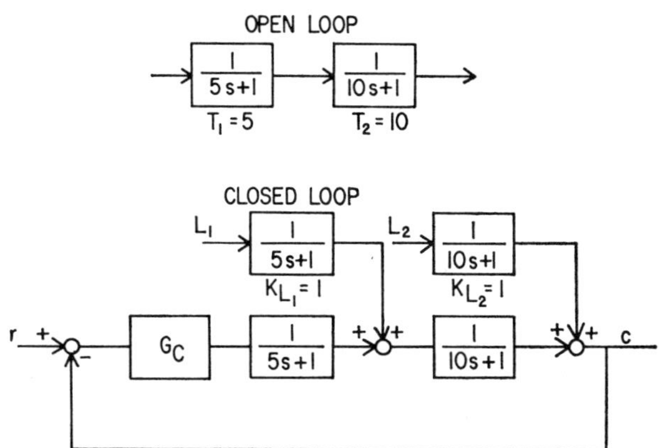

Figure 11.   Block diagram for process 2.

For proportional control, Table 3 indicates that the response will be second order. The response can range from overdamped to underdamped. As the gain is increased, the response becomes faster and more underdamped. At infinite gain, ideal behavior is achieved and the system is on the verge of instability.

When additional controller actions are added to proportional, the denominators of the resulting transfer functions are third order. At low gains and large reset times, the three roots of the characteristic equation are all real and the behavior is similar to that of an overdamped second-order system. As the gain is increased and the reset time is reduced, two of the three roots become complex conjugates and the behavior is oscillatory and underdamped. At high gains and low reset times, the system becomes unstable.

For proportional-integral control, the Routh criterion can be applied to yield the following lower limit on the reset time:

$$T_{R,\min} = \left(\frac{K}{1 + K}\right)\frac{T_1 T_2}{T_1 + T_2} \tag{27}$$

It will be shown shortly that the loop gain is usually greater than one so the minimum reset time is less than the smaller of the two time constants. This is

TABLE 3

Open Loop

$$\frac{K_1 K_2}{(T_1 s + 1)(T_2 s + 1)}$$

Closed Loop

Controller Action	Set Point Transfer Function, c/r	Load Change Transfer Function, c/L_1	Load Change Transfer Function, c/L_2
P	$\dfrac{K}{1 + K} \rightarrow$	$\dfrac{\dfrac{K_{L_1} K_2}{1 + K}}{\dfrac{T_1 T_2}{1 + K} s^2 + \dfrac{T_1 + T_2}{1 + K} s + 1}$	$\dfrac{\left(\dfrac{K_{L_2}}{1 + K}\right)(T_1 s + 1)}{\leftarrow}$
P + I	$\dfrac{T_R s + 1}{\rightarrow}$	$\dfrac{(K_{L_1} K_2 T_R / K)s}{\dfrac{T_1 T_2 T_R}{K} s^3 + T_R \left(\dfrac{T_1 + T_2}{K}\right) s^2 + T_R \left(\dfrac{1 + K}{K}\right) s + 1}$	$\dfrac{(K_{L_2} T_R / K)(T_1 s + 1)s}{\leftarrow}$
P + I + D	$\dfrac{T_R T_D s^2 + T_R s + 1}{\rightarrow}$	$\dfrac{(K_{L_1} K_2 T_R / K)s}{\dfrac{T_1 T_2 T_R}{K} s^3 + T_R \left(\dfrac{T_1 + T_2 + K T_D}{K}\right) s^2 + T_R \left(\dfrac{1 + K}{K}\right) s + 1}$	$\dfrac{(K_{L_2} T_R / K)(T_1 s + 1)s}{\leftarrow}$

K_{L_1} is the gain factor for L_1.

K is the product of the controller gain and $K_1 K_2$.

the subject of Problem 4 at the end of this chapter. When derivative action is added, the minimum reset time is decreased:

$$T_{R,\mathrm{min}} = \left(\frac{K}{1 + K}\right)\frac{T_1 T_2}{T_1 + T_2 + KT_D} \tag{28}$$

In effect, the derivative action makes the system more stable and this permits a higher value of gain and/or a lower value of reset time.

Since ideal behavior can theoretically be achieved with proportional action, it is necessary to arbitrarily select a moderate value of the gain in order to demonstrate its effect on the system. Similarly, arbitrary values of reset and derivative times must be assumed to illustrate their effects when added to proportional action. These controller parameters were selected using the Ziegler-Nichols reaction curve method (1), which will be discussed in some detail later on in this chapter. The resulting transfer functions are presented in Table 4 and are shown in their original forms as well as in their

TABLE 4

	Open Loop	
	$\dfrac{1}{(5s + 1)(10s + 1)}$	

	Closed Loop	

Controller Action	Set Point Transfer Function, c/r	Load Change Transfer Function, c/L_1	Load Change Transfer Function, c/L_2
P $K = 10.4$	$\dfrac{0.912}{\rightarrow}$	$\dfrac{0.0880}{4.40s^2 + 1.32s + 1}$ $\omega = 0.477$ $\zeta = 0.315$	$\dfrac{0.0880(5s + 1)}{\leftarrow}$
P + I $K = 9.33$ $T_R = 6.43$	$\dfrac{6.43s + 1}{\rightarrow}$	$\dfrac{0.690s}{34.5s^3 + 10.3s^2 + 7.12s + 1}$ $(6.35s + 1)(5.43s^2 + 0.774s + 1)$ $\omega = 0.429$ $\zeta = 0.166$	$\dfrac{0.690(5s + 1)s}{\leftarrow}$
P + I + D $K = 12.4$ $T_R = 3.86$ $T_D = 0.965$	$\dfrac{(2s + 1)(1.86s + 1)}{3.72s^2 + 3.86s + 1}$ $\rightarrow$	$\dfrac{0.690s}{15.5s^3 + 8.38s^2 + 4.17s + 1}$ $(3.08s + 1)(5.03s^2 + 1.09s + 1)$ $\omega = 0.446$ $\zeta = 0.242$	$\dfrac{0.690(5s + 1)s}{\leftarrow}$

factored forms to indicate the nature of the roots of the denominator polynomials. All of the second-order forms shown are underdamped and the corresponding natural frequency and damping factor are presented.

The responses of the system to unit steps in set point and load variables are shown in Figure 12. A number of interesting comparisons can be made, both horizontally and vertically. Looking across any given row, we see that, generally speaking, the shapes of the curves are the same. This is a direct result of the denominator polynomials being identical. It is important to note

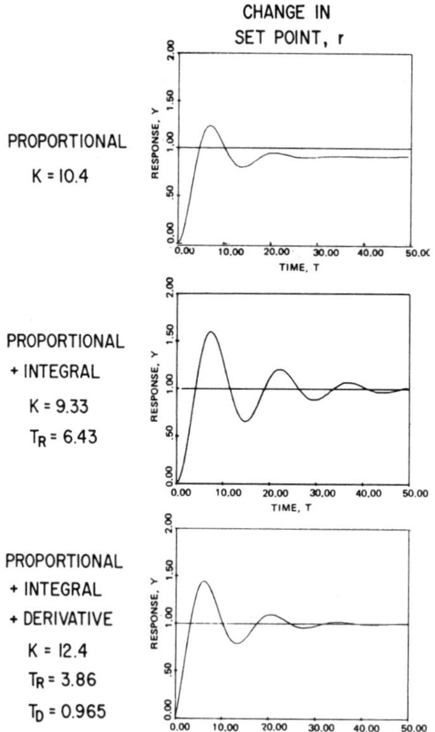

Figure 12.   Response curves for process 2.

that the ordinate scales are different; in particular, the scale for a change in L_2 is exactly twice that for a change in L_1. Thus, for comparable changes in load variables, a change in L_2 will have a more significant effect on the output than one in L_1. This is to be expected.

Comparisons made vertically indicate the effects of the several controller actions. Proportional action alone gives about 8.8% offset at the setting suggested by the Ziegler-Nichols method. Addition of the recommended amount of reset action results in considerably increased oscillation but the

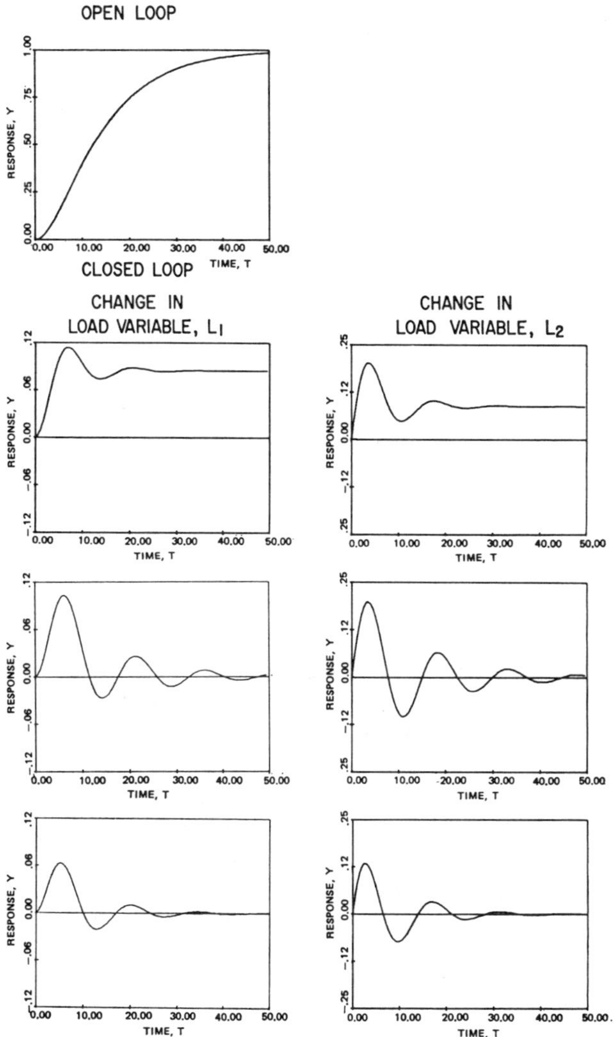

offset is removed. Finally, when derivative action is added, there is a significant improvement in the response, especially in the height of the first peak. This reduced height might be an important consideration in some processes.

c. PROCESS 3

The last lumped parameter process consists of three first-order elements in series. The block diagram is presented in Figure 13. Load disturbances can enter at the three locations indicated. However, we have seen for process 2

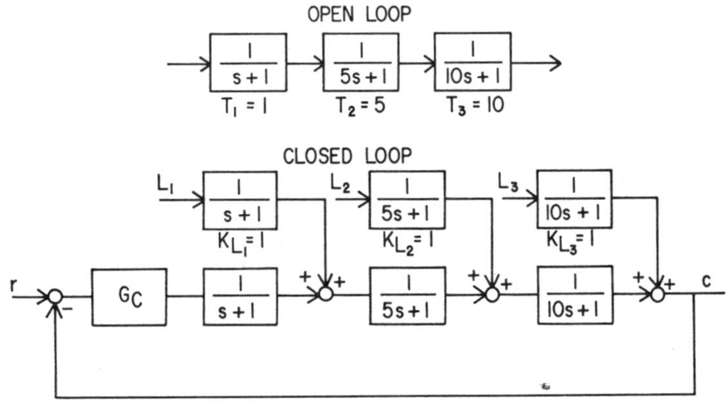

Figure 13.   Block diagram for process 3.

that for a given controller setting, the forms of the responses to these various disturbances will be similar. Therefore, for this illustration we shall only examine the responses to a change in set point and in the load variable, L_3. The general transfer functions are shown in Table 5.

For proportional action alone, the denominator polynomial is third order. At relatively low gains, the three roots of the characteristic equation are real and the response displays overdamped behavior. As the gain is increased, two of the roots become complex conjugates that are reflected by oscillatory, underdamped behavior. The maximum gain can be found using the Routh criterion and it was this case that was used as an example of this method in Section VIII of Chapter 7. The result was

$$K_{\max} = \left(1 + \frac{1}{R_2} + \frac{1}{R_3}\right)(1 + R_2 + R_3) - 1 \tag{29}$$

TABLE 5

Open Loop

$$\frac{K_1 K_2 K_3}{(T_1 s + 1)(T_2 s + 1)(T_3 s + 1)}$$

Closed Loop[a]

Controller Action	Set Point Transfer Function, c/r	Load Change Transfer Function, c/L_3
P	$$\dfrac{\dfrac{K}{1+K}}{\dfrac{T_1 T_2 T_3}{1+K}s^3 + \dfrac{T_1 T_2 + T_1 T_3 + T_2 T_3}{1+K}s^2 + \dfrac{T_1 + T_2 + T_3}{1+K}s + 1}$$	$$\dfrac{\left(\dfrac{K_L}{1+K}\right)(T_1 s + 1)(T_2 s + 1)}{\dfrac{T_1 T_2 T_3}{1+K}s^3 + \dfrac{T_1 T_2 + T_1 T_3 + T_2 T_3}{1+K}s^2 + \dfrac{T_1 + T_2 + T_3}{1+K}s + 1}$$
P + I	$$\dfrac{T_R s + 1}{\dfrac{T_R T_1 T_2 T_3}{K}s^4 + T_R\left(\dfrac{T_1 T_2 + T_1 T_3 + T_2 T_3}{K}\right)s^3 + T_R\left(\dfrac{T_1 + T_2 + T_3}{K}\right)s^2 + T_R\left(\dfrac{1+K}{K}\right)s + 1}$$	$$\dfrac{(K_L T_R / K)(T_1 s + 1)(T_2 s + 1)s}{\dfrac{T_R T_1 T_2 T_3}{K}s^4 + T_R\left(\dfrac{T_1 T_2 + T_1 T_3 + T_2 T_3}{K}\right)s^3 + T_R\left(\dfrac{T_1 + T_2 + T_3}{K}\right)s^2 + T_R\left(\dfrac{1+K}{K}\right)s + 1}$$
P + I + D	$$\dfrac{T_R T_D s^2 + T_R s + 1}{\dfrac{T_R T_1 T_2 T_3}{K}s^4 + T_R\left(\dfrac{T_1 T_2 + T_1 T_3 + T_2 T_3}{K}\right)s^3}$$	$$\dfrac{(K_L T_R / K)(T_1 s + 1)(T_2 s + 1)s}{T_R\left(\dfrac{T_1 + T_2 + T_3 + KT_D}{K}\right)s^2 + T_R\left(\dfrac{1+K}{K}\right)s + 1}$$

[a] Broken line indicates that the denominator in all three cases refers to two numerators, shown by their respective columns.

K_L is the gain factor for L_3.

K is the product of the controller gain and $K_1 K_2 K_3$.

where

$$R_2 = \frac{T_2}{T_1}$$

and

$$R_3 = \frac{T_3}{T_1}$$

This points out that the maximum gain is determined by the *ratios* of the time constants rather than by their individual magnitudes. For the three time constants of this process, the maximum gain is 19.8. Integral action tends to make a system oscillatory so the maximum proportional gain decreases away from 19.8 as integral action is added. Derivative action, on the other hand, has the opposite effect. The reasons for this will be better understood after frequency response is studied in Chapter 14.

When integral and derivative actions are added to proportional, the denominator polynomial becomes fourth order. The four roots of the resulting characteristic equation could all be real, two real and one pair of complex conjugates, or two pairs of complex conjugates. Thus, the behavior can range from overdamped to relatively underdamped with two frequencies of oscillation. General stability relationships similar to those in Equations 27, 28, and 29 are difficult to formulate, but the Routh criterion can be used to check if a given set of controller parameters will result in stable behavior.

The reaction curve method was used to obtain the controller parameters for the calculation of the transfer functions given in Table 6. The denominators are shown in both their original and factored forms. In all cases, one pair of complex conjugate roots occur, so there is some underdamped behavior.

Response curves have been plotted in Figure 14 for step changes in r and L_3. Comparisons across and down the figure can be made in the same way as was done for the second-order case of process 2. Across the figure, the similarity of curve forms is particularly striking and this results from the denominator polynomials being identical. Looking down the figure, we again see an improvement in response with addition, first of integral action, and then of derivative action. A comparison of Figures 12 and 14 indicates that three-mode control is somewhat more beneficial for the third-order process than for the second-order process.

In summary, these illustrations indicate the following.

1. Although proportional action alone results in some offset, this would not be serious if the time constants characterizing the process were well spread apart. In particular, a single time constant process places no limitation on the gain that can be used. Even with two time constants, the system is only on the verge of instability at infinite gain. However, the addition of a relatively small time constant to the second-order case

TABLE 6

Open Loop
$$\dfrac{1}{(s + 1)(5s + 1)(10s + 1)}$$

Closed Loop		
Controller Action	Set Point Transfer Function, c/r	Load Change Transfer Function, c/L_3
P $K = 7.08$	$\dfrac{0.876}{6.19s^3 + 8.05s^2 + 1.98s + 1}$ $(0.874s + 1)(7.08s^2 + 1.11s + 1)$ $\omega = 0.376$ $\zeta = 0.208$	$\dfrac{0.124(s + 1)(5s + 1)}{\leftarrow}$
P + I $K = 6.37$ $T_R = 9.53$	$\dfrac{9.53s + 1}{74.8s^4 + 97.3s^3 + 23.9s^2 + 11.0s + 1}$ $(0.891s + 1)(9.50s + 1)(8.84s^2 + 0.639s + 1)$ $\omega = 0.336$ $\zeta = 0.107$	$\dfrac{1.50(s + 1)(5s + 1)s}{\leftarrow}$
P + I + D $K = 8.50$ $T_R = 5.72$ $T_D = 1.43$	$\dfrac{\dfrac{(2.86s + 1)^2}{8.18s^2 + 5.72s + 1}}{33.7s^4 + 43.8s^3 + 19.0s^2 + 6.39s + 1}$ $(3.83s + 1)(1.17s + 1)(7.50s^2 + 1.39s + 1)$ $\omega = 0.365$ $\zeta = 0.254$	$\dfrac{1.50(s + 1)(5s + 1)s}{\leftarrow}$

resulted in an enormous reduction of the maximum gain to a value of 19.8. From a practical standpoint there are no truly first- or second-order processes, so there will always be some finite limitation on the maximum gain that can be used.

2. Integral action removes offset but tends to make the system less stable and to increase the period of oscillation. Therefore, when it is added to proportional action, it may be necessary to use a lower gain than would be used for purely proportional action.

3. Derivative action tends to reduce the peak heights of oscillatory responses. In particular, for a change in set point, the overshoot is reduced. In contrast to integral action, derivative action tends to stabilize a system and this permits the use of higher gains and lower reset times. It is

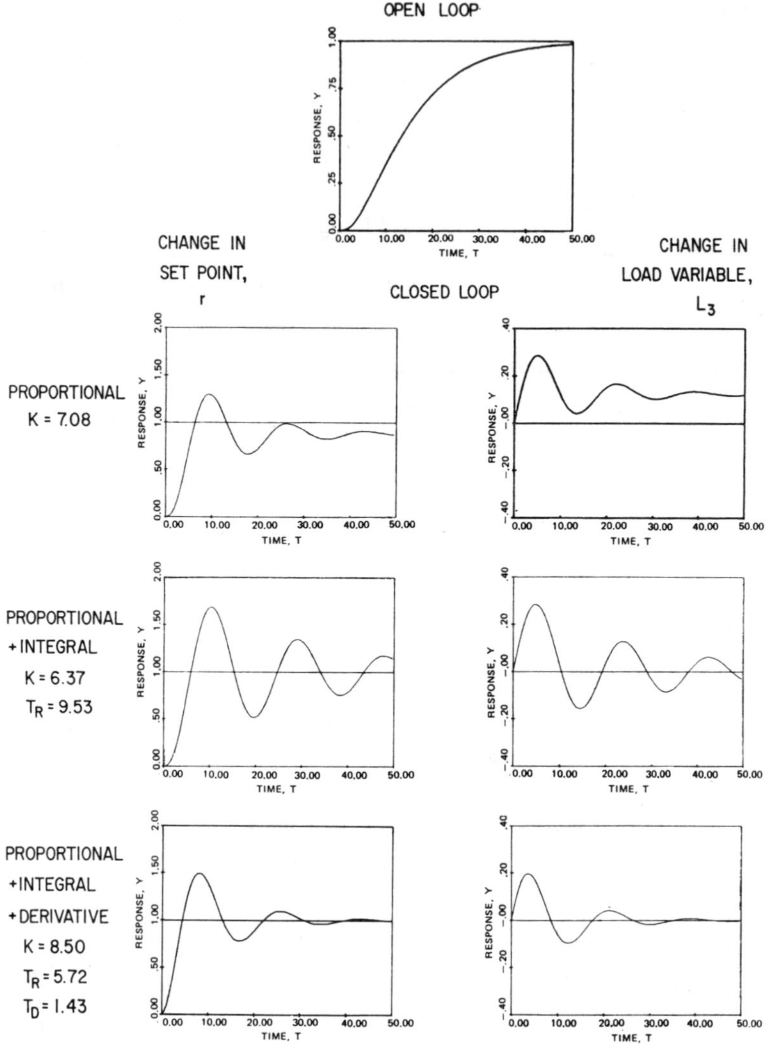

Figure 14.   Response curves for process 3.

important to realize, however, that these advantages of derivative action do not accrue to all types of processes. Processes characterized by relatively large time delays usually derive very little benefit from derivative action. It is also usually not advisable to use derivative action for noisy processes since the action amplifies the noise. Fast flow processes typically fall in this category.

2. *Closed-Loop Response of a Pure-Delay Process*

The addition of a relatively small time constant to a process roughly has the effect of delaying its response. The difference between processes 2 and 3 is a first-order element having a unit time constant. From Figure 7 we see that the response of process 3 is essentially the same as that of process 2, but delayed by about one time unit. Therefore, the transfer function of process 3 might be approximated as follows:

$$\text{process 3 transfer function} \cong \frac{e^{-s}}{(10s + 1)(5s + 1)}$$

The approximation suggested for process 3 was based upon an examination of the response curves of the two processes. It is interesting to note that there is some theoretical justification for approximating a series of non-interacting first-order elements by a pure delay. Suppose that a process consists of n first-order elements, each with a time constant of D/n, where D is some constant value of time. Then the overall transfer function for this process is as follows:

$$\text{overall transfer function} = \frac{1}{\left(\dfrac{D}{n}s + 1\right)^n}$$

Recalling the definition of the exponential, it can be seen that as n approaches infinity, the transfer function approaches e^{-Ds}, which is the transfer function of a pure delay.

These examples indicate that a pure delay may be *effectively* present in a process even though the process contains no pure delays as such. To assess the significance of a true or effective delay, we shall now examine the control of a pure delay under several modes of control. The block diagram for the system is shown in Figure 15.

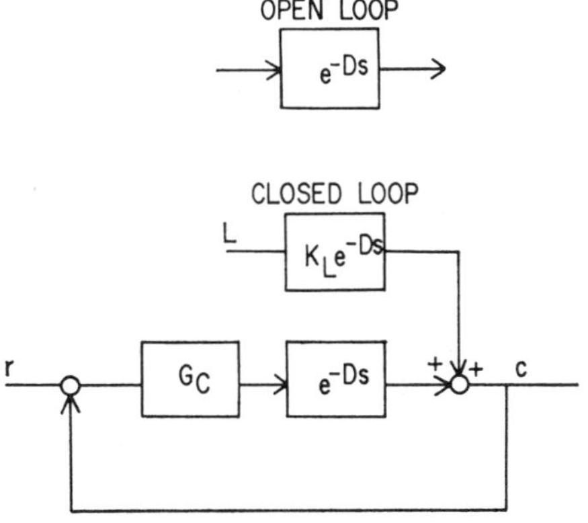

Figure 15. Block diagram for a pure-delay process.

Using Equation 14 for the case of proportional control, the transfer function for a change in the load variable is

$$\frac{c}{L} = \frac{K_L e^{-Ds}}{1 + K e^{-Ds}} \tag{30}$$

As in the illustrations for the lumped parameter processes, the load gain, K_L, will be taken as unity and a unit step change in L will be used. Since the time scale is arbitrary, the delay time, D, can be set at unity. The responses for three values of gain are shown in Figure 16. These graphs can be worked out

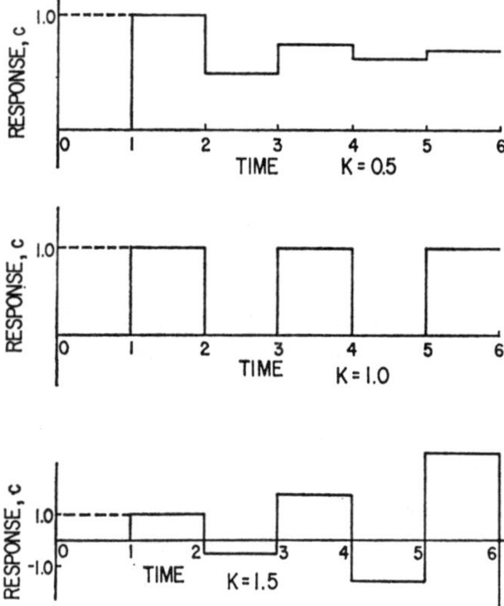

Figure 16.   Responses of a pure-delay process under proportional control.

by intuition or mathematically derived by one of several methods. The important point is that the maximum loop gain is unity and this will be the case for any pure-delay process. This is considerably lower than the maximum gain of 19.8 found for process 3, and processes 1 and 2 had no maximum gains at all.

If the pure-delay process is controlled by integral action alone, the resulting transfer function is

$$\frac{c}{L} = \frac{K_L T_i s e^{-Ds}}{T_i s + e^{-Ds}} \tag{31}$$

The nature of the denominator makes inversion of the transform difficult for a step change in load. However, the time response can be found numerically using a digital computer. There is a minimum value of integral time which is given by the following expression:

$$T_{i,\min} = \frac{2D}{\pi} = 0.637D \tag{32}$$

This equation can be obtained using methods of frequency response presented in Chapter 14. For this example having a delay of one time unit, the minimum integral time is 0.637. In Figure 17, three responses are shown for integral times of 0.5, 1.0, and 2.0. The lowest integral time, being less than the preceding minimum value, results in unstable behavior. At a value of 1.0, the response might be termed "underdamped" whereas a value of 2.0 results in a response that is on the verge of being "overdamped."

For the case of proportional-integral control, stability limits for gain and reset time are not so simply stated as for proportional or integral action alone. These limits can be found using frequency response techniques. Since the maximum gain without integral action is unity, and the minimum reset time without proportional action is $2D/\pi$, the addition of integral action to proportional will reduce the maximum gain from unity and increase the minimum reset time from $2D/\pi$.

A typical response under proportional-integral control is shown in Figure 18. The controller gain is 0.5 and the reset time, T_R, is 1.0. Since the integral time, T_i, is equal to T_R/K_C, it is 2.0. Therefore, this response can be compared with the first response in Figure 16 for purely proportional action, and with the third response in Figure 17 for purely integral action. The form of the response in Figure 18 is quite different from that displayed by lumped parameter elements.

It can be shown using a frequency response analysis that a control system for a pure-delay process will not benefit much by the addition of derivative action to proportional. Therefore, three-mode control will not be illustrated here.

These illustrations for a pure time delay indicate that this element is relatively more difficult to control than the lumped parameter processes we examined earlier. If a unit step input is assumed for L in the transfer function of Equation 30, application of the final-value theorem indicates that the offset is $K_L/(1 + K)$. Since the maximum gain for proportional action alone is 1.0, the offset for this type of action will always be at least $K_L/2$.

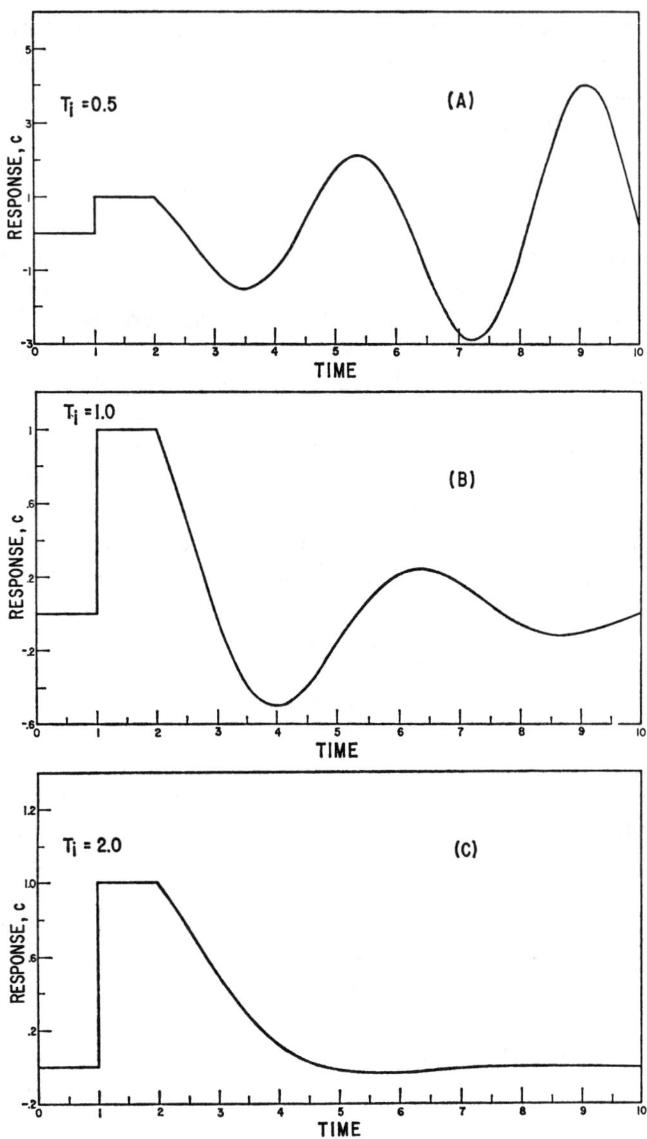

Figure 17. Responses of a pure-delay process under integral control.

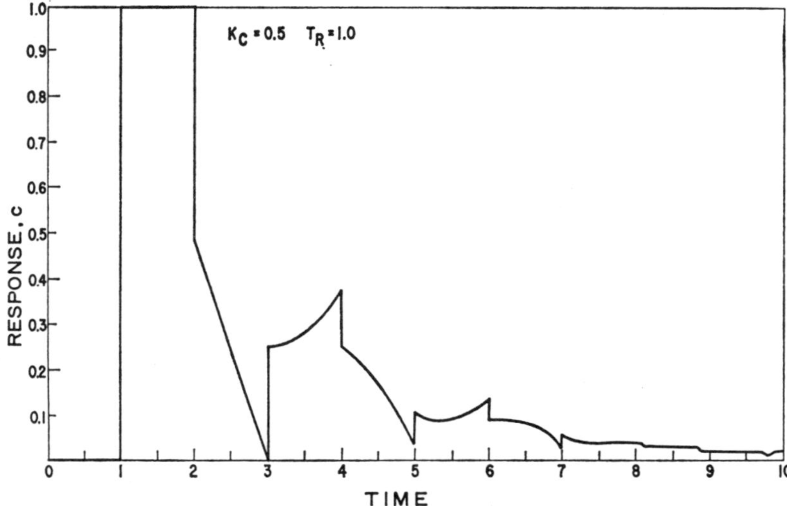

Figure 18.   Response of a pure-delay process under proportional-integral control.

III. TUNING BY THE REACTION CURVE METHOD

A. Basic Philosophy of the Method

As mentioned before, ordinarily one does not know the transfer functions of a process and satisfactory or optimum controller settings must be found empirically. The reaction curve method of tuning is based on the idea that an open-loop test of the process will reveal its dynamic properties and knowledge of these can be used to establish reasonably satisfactory controller settings.

Referring to Figure 7 again, we see that as first-order elements are progressively added to process 1, the response becomes slower and effectively more delayed. There is a corresponding decrease in the maximum allowable gain. A pure delay is an extreme example of slow response to an input and the maximum proportional gain turns out to be unity. This all suggests that the "effective" delay of a process would be one possible parameter for developing a correlation to predict satisfactory controller settings.

The correlating parameters of the Ziegler-Nichols method are indicated in Figure 19 for a typical open-loop response. This response curve is frequently referred to as a "reaction curve." One parameter is the maximum slope, S, and this tangency line is extended to the time axis. The time intercept is the effective time delay, L. It should be noted that the response is plotted

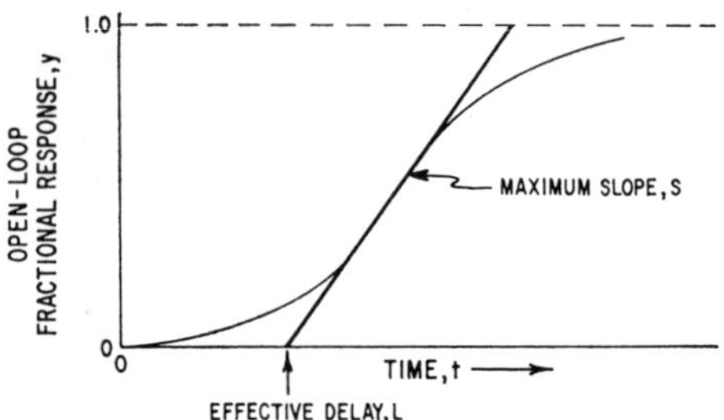

Figure 19.   Open-loop response or "reaction curve."

in terms of the fractional response and therefore it covers the range from zero to one.

By way of interpretation, suppose the process were a first-order element having a time constant, T, followed by a pure time delay, D. The open-loop response for this combination is shown in Figure 20. The maximum slope

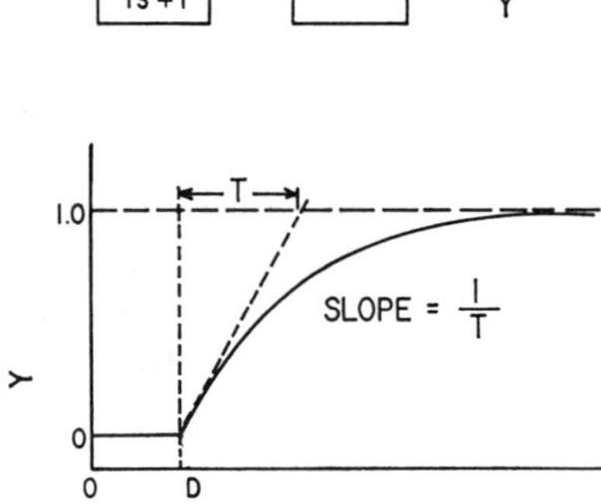

Figure 20.   Open-loop response of a first-order element plus a pure time delay.

occurs at the end of the delay interval at the instant the first-order response behavior begins, and this slope is the reciprocal of the time constant. In a sense, then, these parameters of maximum slope and effective delay characterize the true process as a first-order element followed by a pure time delay.

The suggested settings of Ziegler and Nichols are:

Proportional

$$K = \frac{1}{SL} \tag{33}$$

where S = maximum slope, reciprocal time
$\quad L$ = effective delay, time

Proportional-Integral

$$K = \frac{0.9}{SL} \tag{34}$$

$$T_R = \frac{L}{0.3} \tag{35}$$

Proportional-Integral-Derivative

$$K = \frac{1.2}{SL} \tag{36}$$

$$T_R = \frac{L}{0.5} \tag{37}$$

$$T_D = 0.5L \tag{38}$$

Comparing these three sets of settings, we see that they are consistent with the idea that addition of reset action tends to make the system more unstable while derivative action has the opposite tendency. Thus, for proportional-integral control, 10% less proportional action is recommended than for proportional action alone. On the other hand, addition of derivative action allows a 20% increase.

Application of these recommended settings to the control of a pure-delay process reveals that they are on the conservative side. The maximum slope of the reaction curve for a pure delay is infinite so the recommended gain for proportional control is zero. We know, however, that the maximum gain for a pure delay is unity. Therefore, the recommended value is overly conservative, and of course, meaningless. For proportional-integral control, Equation 34 is again meaningless. In the discussion of the control of a pure delay process in Section II.B.2, a pure delay of one time unit was used and Equation 35 leads to a recommended reset time of 3.3 time units. Looking at Figures 17 and 18, and recalling that T_i is equivalent to T_R/K_C, this value of 3.3 appears to be at least conservative if the proportional gain is not too high. As mentioned before, three-mode control was not examined for this process since derivative action is normally not beneficial in controlling a pure-delay process.

B. Application of the Method in Practice

Before the controller of a system can be tuned, the system must be brought to its design or normal operating condition. To achieve this, the controller might be placed in the manual mode. As indicated in Figure 21, this has the

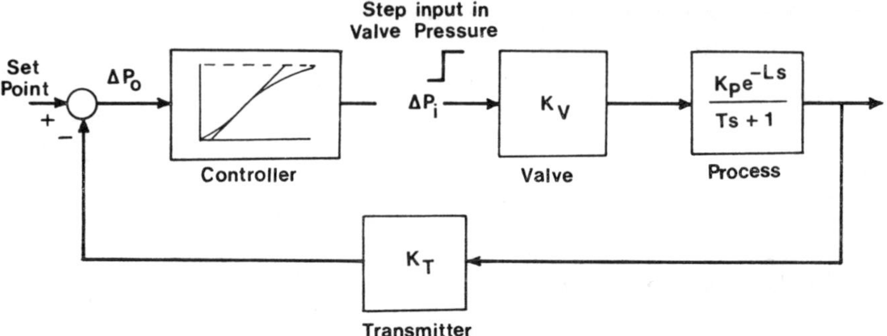

Figure 21.   Application of the reaction curve method in practice.

effect of breaking the loop between the controller and valve. The valve pressure can then be manipulated directly by adjusting the set point. When the design condition is reached, a step input in pressure is applied and the reaction curve subsequently appears on the chart of the controller. The controller parameters are then calculated from the reaction curve. The valve pressure is adjusted to return the controlled variable to its desired value. The controller is then placed in the automatic mode and the controller parameters are set at the calculated values.

The gain, K, in Equations 33, 34, and 36 is the overall gain, which is the product of the gains of the controller and other elements in the loop. In tuning the system, the controller gain, K_C, must be determined. This implies that it is necessary to explicitly know the remaining gains. However, this does not turn out to be the case, as shown by the following illustration.

Suppose a process consists of a first-order element and a pure delay, as shown in Figure 21. These are effectively the elements upon which the reaction curve method is based. The overall gain is the product of the various gains around the loop:

$$K = K_C K_V K_P K_T \tag{39}$$

Note that in this illustration, the transmitter gain, K_T, is shown separately and is not assumed to be included in the controller gain, K_C. If proportional action is to be used, then Equation 33 can be written as follows:

$$K_C = \frac{1}{K_V K_P K_T SL} \tag{40}$$

The reaction curve will have the form shown in Figure 20 with the delay equal to L. As indicated in the figure, the maximum slope is $1/T$ when the response is plotted in fractional form. Hence, Equation 40 can be written

$$K_C = \frac{T}{K_V K_P K_T L} \tag{41}$$

The delay time, L, is measured on the reaction curve. The only problem is to eliminate the factor, $T/K_V K_P K_T$, from Equation 41. This is accomplished using the measured slope on the controller chart in the following way. The change in input pressure to the valve, ΔP_i, results in the following response in the pressure input to the controller, ΔP_0:

$$\Delta P_0 = 0 \qquad 0 < t \le L \tag{42a}$$

$$\Delta P_0 = K_V K_P K_T (1 - e^{-(t-L)/T}) \, \Delta P_i \qquad t \ge L \tag{42b}$$

The negative sign of the comparator is not included here because the comparator mechanism is not connected to the recording chart mechanism of the controller. For a first-order process, the maximum slope occurs when the exponential increase begins. Thus, taking the derivative of Equation 42b and setting the time equal to L, we obtain

$$\left. \frac{d\,\Delta P_0}{dt} \right|_{\max} = \frac{K_V K_P K_T}{T} \, \Delta P_i \tag{43}$$

This slope, which has the units of psi per unit time, can now be made partially dimensionless by dividing by 12 psi, the pressure range of the recorder, giving the dimensionless slope, N:

$$N = \left. \frac{d(\Delta P_0/12)}{dt} \right|_{\max} = \frac{K_V K_P K_T}{T} \, \frac{\Delta P_i}{12} \tag{44}$$

or

$$\frac{T}{K_V K_P K_T} = \frac{\Delta P_i}{12N} \tag{45}$$

Substituting Equation 45 into Equation 41, we obtain

$$K_C = \frac{\Delta P_i/12}{NL} \tag{46}$$

In the actual test, the input pressure change, ΔP_i, is known. The measured value of the maximum slope, $(d\,\Delta P_0/dt)_{\max}$, is then divided by 12 to give the dimensionless slope, N. Equation 46 is then used to calculate the recommended controller gain. In accordance with Equations 34 and 36, factors of 0.9 and 1.2 are used in Equation 46 for proportional-integral and proportional-integral-derivative controllers, respectively.

C. Application of the Method to the Hypothetical Cases in Section II.B

The three reaction curves for processes 1, 2, and 3 are presented in Figure 22. For process 1, which has only a single first-order element, the effective delay is zero and the recommended proportional gain is infinite. It was for

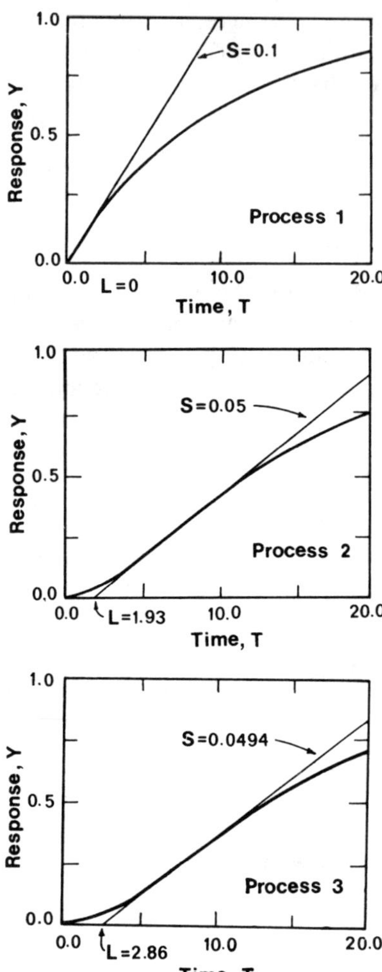

Figure 22. Process reaction curves for the three processes.

this reason that it was necessary to arbitrarily select the controller settings for the study cases in Figure 9. The maximum slopes and effective delays for processes 2 and 3 are indicated on the graphs and these were used in Equations 33 to 38 to obtain the settings for the study cases displayed in Figures 12 and 14.

As shown in Figure 22, the reaction curve for process 3 is characterized by a pure delay of 2.86 and the maximum slope is 0.0494. This implies that an approximation of the true response would be a delay of 2.86 followed by a first-order element having a time constant of 1/0.0494 or 20.2. As a matter of

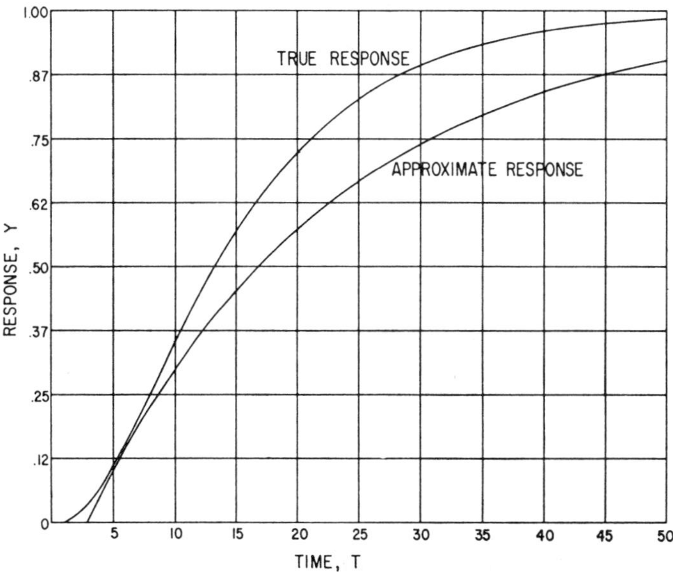

Figure 23.   Exact and approximate reaction curves for process 3.

interest, the true response is plotted with that of the approximation in Figure 23. The approximation is reasonably good for a fractional response less than about 0.5.

A review of Figures 12 and 14 indicates that the suggested settings result in underdamped behavior. Such behavior is ordinarily desirable since it results in relatively fast response compared to overdamped or critically damped behavior. Referring to Tables 4 and 6, note that purely second-order behavior is exhibited only for a second-order process under proportional action alone. Therefore, since the term, "decay ratio" only strictly applies to second-order, underdamped behavior, it has no precise meaning for the majority of the examples. However, a pseudo-decay ratio can be defined in terms of the first two peaks in a manner similar to that used for purely second-order behavior. Using this looser definition, the decay ratios were calculated and found to fall between 0.2 and 0.5, with the majority between 0.2 and 0.3. In developing the relationships in Equations 33 to 38, Ziegler and Nichols defined the optimum response as a one-quarter decay ratio, so it is not surprising that the responses for the examples here exhibit values close to that value.

A cautionary word is in order here regarding the reaction curve method since the settings dictated by it usually lead to underdamped behavior. In some systems, such "tight" tuning may be undesirable. For example, it was noted in Sections IV.B.1 and IV.C of Chapter 8 that in surge tanks, although the level or pressure is the controlled variable, the purpose of the control system is to prevent abrupt changes in the flow rates of the leaving streams. Therefore, this type of "averaging control" calls for "loose" tuning that is typified by relatively slow, overdamped behavior.

Under proportional action alone, only the third-order process can result in unstable oscillations; the maximum loop gain was previously given as 19.8. The reaction curve method suggests a loop gain of 7.08, which is about 35% of the maximum. As we shall see in Chapter 14, another tuning method involves finding the maximum gain and then setting the controller to yield half that value. Thus, the value of 35% appears to be conservative.

D. Application of the Method to the Example of Chapter 12

It is interesting to apply the reaction curve method to the example of Chapter 12. For convenience, the block diagram given in Figure 7 of Chapter 12 has been redrawn and rearranged in Figure 24. The gains have all been taken as

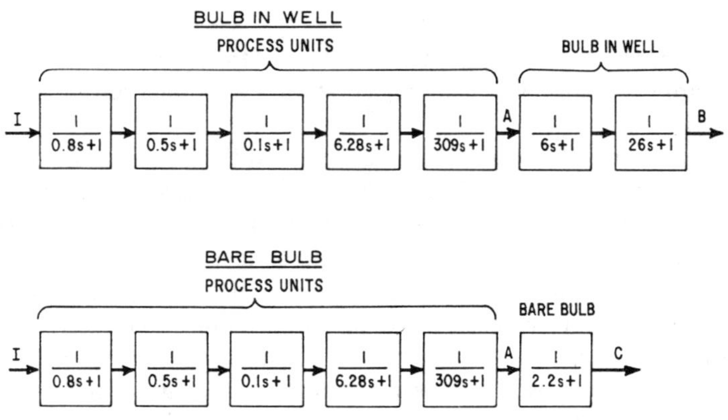

Figure 24. Open-loop block diagrams for example of Chapter 12.

unity since we are only interested in the loop gain. Using a digital computer, reaction curves and slopes were calculated for outputs at points A, B, and C. The maximum slopes and effective delays for these points are shown in the following tabular material.

Output Point	Maximum Slope, S	Effective Delay, L
A	0.002984	6.638
B	0.002563	30.715
C	0.002979	8.802

For proportional control alone, the suggested gain for each of these cases was calculated using Equation 33. The maximum gain for the corresponding components was determined using frequency response theory. The results of these calculations are shown in the following table.

Loop Consists of	K by Equation 33	K Maximum	Ratio of Gains
I to A plus proportional controller (no measuring element dynamics)	50.5	249	0.203
I to B plus proportional controller (bulb in well)	12.7	30.2	0.420
I to C plus proportional controller (bare bulb)	38.1	110	0.348

These results demonstrate the possible importance of the measuring element. Also, the recommended gain is quite conservative in all three cases.

SUMMARY

This chapter has been mainly devoted to the transient response of feedback systems. The presentation has been relatively qualitative and empirical. The effects of various controller actions have been illustrated for some simple processes and it should be evident that these effects will carry over to processes considerably more complex.

To develop a more fundamental understanding of the effects of various controller actions on system behavior, some knowledge of frequency response is helpful. This is the subject of Chapter 14.

REFERENCE

1. J. G. Ziegler and N. B. Nichols, "Optimum Settings for Automatic Controllers," *Trans. ASME*, **64** (1942), 759.

PROBLEMS

PROBLEMS CONCERNING SYSTEM TRANSFER FUNCTIONS

Problem 1. Consider the following cases involving feedforward control. In each, determine the nature and feasibility of the feedforward controller needed to achieve perfect control.

(a)

$$G_L = \frac{K_L}{T_P s + 1} \qquad \text{process}$$

$$G_P = \frac{K_P}{T_P s + 1} \qquad \text{process}$$

$$G_V = K_V \qquad\qquad \text{valve}$$

$$H = 1 \qquad\qquad \text{measuring element}$$

(b) Same as part *a* but $G_V = K_V/(T_V s + 1)$.
(c) Same as part *a* but $H = e^{-\tau_d s}$.

$$\text{Answers: (a) } G_C = -\frac{K_L}{K_V K_P}$$

$$\text{(b) } G_C = -\frac{K_L}{K_V K_P}(T_V s + 1)$$

$$\text{(c) } G_C = -\frac{K_L}{K_V K_P} e^{\tau_d s}$$

Problem 2. Equations 33 and 34 of Chapter 9 presented some transfer functions for a simplified model of a continuous-flow, stirred-tank reactor. It was pointed out that it is possible for both the gain factor and the time constant to be negative. Suppose the temperature of such a reactor is being controlled by manipulating the cooling water rate in a coil.

(a) The transfer function of the reactor is

$$\frac{T}{T_C} = \frac{-1}{-20s + 1}$$

where T = reactor temperature
T_C = effective coolant temperature

For simplicity, the magnitude of the gain factor has been taken as unity. This model basically accounts for the dynamics of the tank itself, neglecting that of the coil. If a step change in the effective coolant temperature, T_C, is made, what will be the open-loop transient response?

(b) Suppose now that the reactor is placed under proportional control. Assuming that the dynamics of the coil and valve are negligible, determine if it is possible to stabilize the system.

(c) Suppose the coil dynamics are characterized by a time constant of five. Then the reactor transfer function takes the following form:

$$\frac{T}{w_c} = \frac{-1}{(5s + 1)(-20s + 1)}$$

where w_c = coolant flow rate

Again, the gain has been taken as unity for simplicity. Investigate the stability of the system under proportional control.

(d) Finally, consider now the effect of valve dynamics. The process transfer function is

$$\frac{T}{P_V} = \frac{-1}{(s + 1)(5s + 1)(-20s + 1)}$$

Investigate the stability under proportional control.

$$\textit{Answers:} \quad \text{(a) } T = e^{t/20} - 1$$
$$\text{(b) Stable for } K > 1$$
$$\text{(c) Stable for } K > 1$$
$$\text{(d) Stable for } 1 < K < 17.1$$

PROBLEMS CONCERNING THE TRANSIENT RESPONSE OF CONTROLLED SYSTEMS

Problem 3. Consider the proportional-integral control of a first-order process. Suppose the reset time, T_R, is set exactly equal to the process time constant, T.

(a) Show that the closed-loop system is always either overdamped or critically damped.

(b) Find the effective time constants for the system.

(c) For a unit step change in set point, find the response of the controlled variable and sketch its behavior. The following transform pair may be helpful:

$$\text{Transform} = \frac{as + 1}{s(T_1 s + 1)(T_2 s + 1)}$$

$$\text{Time function} = 1 + \frac{T_1 - a}{T_2 - T_1} e^{-t/T_1} - \frac{T_2 - a}{T_2 - T_1} e^{-t/T_2}$$

(d) For a unit step change in load variable, find the response of the controlled variable and sketch its behavior.

Problem 4. Below Equation 27, the statement is made that the minimum reset time is less than the smaller of the two time constants when K is greater than one. Prove this.

Problem 5. Apply the Routh criterion for the case of a third-order process under proportional-integral control.

Answer: For stability:

$$T_R > \left(\frac{K}{1+K}\right)\frac{(T_1T_2 + T_1T_3 + T_2T_3)^2}{(T_1 + T_2 + T_3)(T_1T_2 + T_1T_3 + T_2T_3) - (1 + K)T_1T_2T_3}$$

Problem 6. A process consists of five first-order elements having time constants of 309, 6.28, 0.8, 0.5, and 0.1. If the last three are lumped together to yield a third-order approximation (309, 6.28, and 1.4), will the approximation result in a higher or a lower maximum gain than the fifth-order process? Explain.

Problem 7. Consider a process consisting of three first-order elements. What is the minimum possible maximum gain for such a process—that is, what set of three time constants results in the minimum value of the maximum overall gain? What is the maximum possible value of the maximum gain and under what conditions does it arise?

Problem 8. A block diagram of a control system is presented in Figure 25.

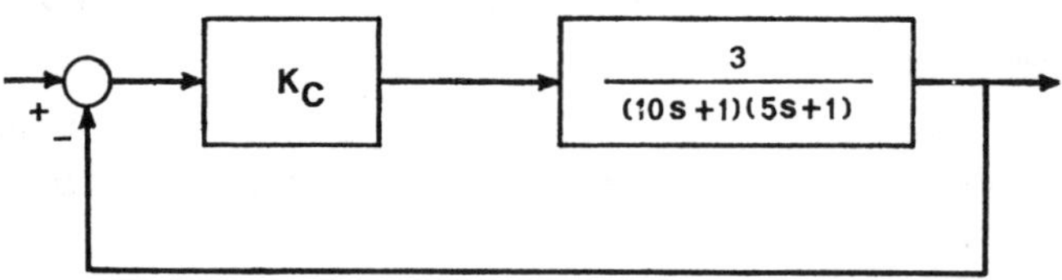

Figure 25. Block diagram for Problem 8.

(a) What value of proportional controller gain, K_C, will result in critically damped behavior?

(b) If the controller gain of part *a* is used, what offset will result from a unit change in set point?

(c) For a unit change in set point, what must the controller gain be to result in an offset of 10%?

(d) For the controller gain of part *c*, what is the resulting damping factor?

Answers: (a) $K_C = 0.0417$
(b) Offset $= 0.889$
(c) $K_C = 3.0$
(d) $\zeta = 0.336$

Problem 9.   A simple process is characterized only by a gain factor, K_P, and is under derivative action alone. A block diagram of the system is shown in Figure 26.

(a) For a unit step input in r, find the time behavior of c and make a rough sketch of it. At the instant the step is made, what should happen? Does your solution bear this out?

(b) Suppose the input, r, is the terminated ramp function of Figure 10. Find the time response and sketch it. What happens to this response as α approaches infinity?

$$\textit{Answers: (a) } c = e^{-t/K}$$
$$\text{where } K = K_D K_P$$
$$\text{(b) } c = \alpha K(1 - e^{-t/K}) \qquad 0 \le t \le t_1$$
$$c = \alpha K(e^{1/\alpha K} - 1)e^{-t/K} \qquad t \ge t_1$$

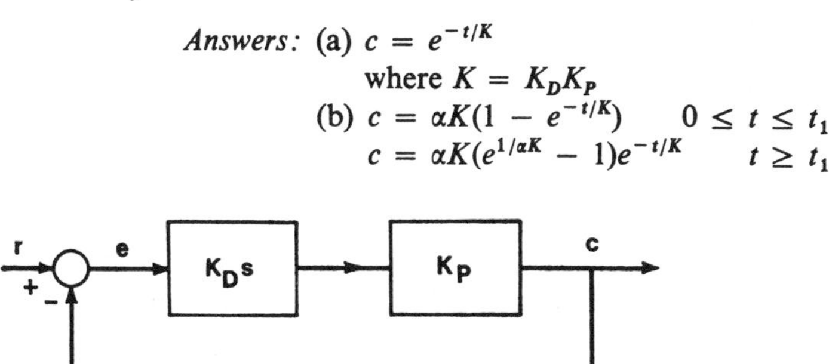

Figure 26.   Block diagram for Problem 9.

Problem 10.   A surge tank for a liquid stream is under level control. The manipulated variable is the entering flow rate and the liquid leaves through a constant displacement pump. The level is sensed by a float-transmitter combination whose dynamics are described by a first-order transfer function with a time constant of 6 sec and a gain of 1.2 psi/ft. The valve has linear trim with a maximum flow rate of 100 ft^3/min. The cross-sectional area of the tank is 10 ft^2. The liquid enters at the top of the tank as in Figure 3 on page 164.

(a) If the behavior of the system is to be critically damped under proportional control, what must the proportional band be set at?

(b) Will offset be a problem in this system if a change in set point is made?

$$\textit{Answer: (a) } 40\%$$

Problem 11.   This problem concerns the decay ratios of systems of second and higher order.

(a) The underdamped response of a second-order system was given in Equation 11a of Chapter 10. The decay ratio was defined in terms of the heights of the first two maxima relative to the ultimate steady-state value, which is K. Examining the form of this equation, it is apparent that the peaks will occur at times when the factor, $\sin(\omega_n\sqrt{1-\zeta^2}\,t-\psi)$ is equal to unity. Using this fact, show that the decay ratio for this system is that given by Equation 14 of Chapter 10.

(b)   For a second-order system, what is the decay ratio based on the ratio of the third peak to the second?

(c)   Based on your results for parts *a* and *b*, why is there some difficulty in defining the decay ratio for a third-order process having only one real root?

Problem 12.   Pure integral action is frequently used in flow control and ratio control applications. Flow systems are usually characterized by small time constants.

(a)   The integral time, T_i, is defined to be the ratio of the reset time, T_R, to the open-loop gain, K. Suppose that a flow process is described by a first-order transfer function. Over what range of integral times will the system be stable?

(b)   If the process is described by a second-order, overdamped transfer function, over what range of integral times will the system be stable?

$$\textit{Answers:} \text{ (a) Stable for all values of } T_i$$

$$\text{(b) } T_i > \frac{T_1 T_2}{T_1 + T_2}$$

Problem 13.   The flow rate, F, of a liquid through a line is controlled by manipulating a pneumatic valve. The flow rate responds relatively quickly to changes in supply pressure, P_S, so the transfer function, F/P_S, is a pure gain, K_L. The valve responds relatively slowly to changes in the controller output pressure, P_V, and the transfer function, F/P_V, is first order with a gain, K_P, and time constant, τ_V.

(a)   For proportional control, find the closed-loop transfer function, F/P_S. For a unit step increase in supply pressure, sketch the response. It may be helpful to use the initial- and final-value theorems.

(b)   Repeat part *a* for proportional-integral control. Sketch two possible responses that have basically different forms.

Problem 14.   Consider a surge tank for liquid in which the level is controlled by manipulating the leaving flow rate.

(a)   Suppose the controller action is purely proportional.
   (1)   Find the transfer function relating the level to the entering flow rate.
   (2)   Find the transfer function relating the leaving flow rate to the entering flow rate.

(3)   Sketch the responses of level and leaving flow rate to a step change in entering flow rate.

(b)   Repeat part *a* for a controller with proportional-integral action.

$$\text{Answers: (a)(1)} \quad \frac{h}{F_i} = \frac{K_L}{1 + K_o} \frac{1}{\dfrac{T}{1 + K_o} s + 1}$$

$$\text{(a)(2)} \quad \frac{F_o}{F_i} = \frac{1}{\dfrac{T}{1 + K_o} s + 1}$$

PROBLEMS CONCERNING CONTROLLER TUNING BY THE REACTION CURVE METHOD

Problem 15.   Water at 60°F is fed continuously at a rate of 20 lb/min to a continuous-flow, stirred-tank heater. The tank holdup is 200 lb of water and the temperature of the water in the tank is 160°F. Live steam is sparged into the tank to provide the heat and the steam has a latent heat of 1000 Btu/lb. The water leaves through a line that contains 12 lb of water between the tank and the temperature bulb. The bulb is connected through a transmitter to a proportional-integral controller having a chart span of 100°F. The steam control valve has a linear characteristic and the maximum steam rate is 6 lb/min. Draw the reaction curve for a 2-psi change in valve pressure and calculate the recommended settings. As an approximation, assume that the leaving flow rate of water is equal to the entering flow rate.

$$\text{Answers: } K_C = 5.50 \text{ psi/psi}$$
$$T_R = 1.82 \text{ min}$$

Problem 16.   The transfer function for two first-order elements in series is given by

$$\frac{Y}{X} = \frac{1}{(T_1 s + 1)(T_2 s + 1)}$$

The response for a unit step in X is

$$Y = 1 + \frac{T_1}{T_2 - T_1} e^{-t/T_1} - \frac{T_2}{T_2 - T_1} e^{-t/T_2}$$

The response has a "sigmoidal" shape such as those exhibited by processes 2 and 3 in Figure 22. The maximum slope is given by the expression

$$S = \frac{1}{T_2} \left(\frac{1}{R}\right)^{1/(R-1)}$$

where $R = T_2/T_1$

and occurs at a time, t_{max}, given by the following equation

$$t_{max} = T_2 \left(\frac{1}{R-1} \right) \ln(R)$$

where $\ln(R)$ is the natural logarithm of R.

The corresponding value of Y at t_{max} is denoted as $Y_{t_{max}}$ and is given by the following equation:

$$Y_{t_{max}} = 1 - (R+1)R^{-[R/(R-1)]}$$

The effective time delay is given by the expression

$$L = T_1 \left[1 + R - \left(\frac{1}{R} \right)^{R/(1-R)} + \frac{R}{R-1} \ln(R) \right]$$

(a) If $T_1 = 2.0$ and $T_2 = 10.0$, find the recommended proportional gain for the system.

(b) Suppose the process transfer function is

$$\frac{Y}{X} = \frac{e^{-5s}}{(2s+1)(10s+1)}$$

Find the recommended proportional gain.

Answers: (a) 14.0 psi/psi
(b) 2.46 psi/psi

Problem 17. A process consists of two first-order elements in series having time constants of 1.0 and 4.0.

(a) Under proportional control, what loop gain will result in a decay ratio of 0.25?

(b) If the gain recommended by the Ziegler-Nichols method is used, what will be the resulting decay ratio? Use the relations given in Problem 16 to simplify the calculations.

Answers: (a) 32.5
(b) 0.103

Problem 18. The transfer functions of three different processes are presented below:

$$\text{process 1} \qquad \frac{Y}{X} = \frac{1}{As}$$

$$\text{process 2} \qquad \frac{Y}{X} = \frac{1}{(T_1 s + 1)As}$$

$$\text{process 3} \qquad \frac{Y}{X} = \frac{1}{(T_1 s + 1)(T_2 s + 1)As}$$

(a) For each of these cases,
   (1) Sketch the reaction curve.
   (2) Find the recommended gain in accordance with the reaction curve method. In order to find the maximum slope, it may be helpful to utilize the fact that the transform of the slope is simply the transform of the reaction curve multiplied by s.
   (3) How does the recommended gain compare with the maximum gain?

(b) In retrospect, could the maximum slope in each case have been deduced by inspection or by physical intuition?

$$Answers: \text{(a)(2)}\ \ \text{process 1}\quad K = \infty$$
$$\text{process 2}\quad K = A/T_1$$
$$\text{process 3}\quad K = A/(T_1 + T_2)$$

Problem 19.   Water is to be heated in a shell-and-tube heat exchanger using steam at 220°F as the heating medium. Pertinent data are as follows:

Tube
$\frac{3}{4}$ in. in outside diameter
wall thickness 0.042 in.
length 16 ft
inside area for heat transfer 0.174 ft^2/ft of pipe
outside area for heat transfer 0.196 ft^2/ft of pipe
cross-sectional area for flow 0.00242 ft^2

Operating Conditions
water velocity 2 ft/sec
temperature of entering water 60°F
temperature of leaving water 182°F
steam temperature 220°F

It is proposed to control the leaving temperature of the water using a proportional-integral controller connected to a valve in the steam line. The measuring bulb will be located 2 ft from the tube exit. Using the reaction curve method, determine the suggested loop gain and controller reset time. Assume that the valve and other elements of the loop will not contribute significant delays.

$$Answers:\ K = 5.0$$
$$T_R = 3.33\ \text{sec}$$

CHAPTER   XIV

Frequency Response

Frequency response concerns the response of a process or a system to a sustained sinusoidal input. Usually one is interested in the variation of the response over a considerable range of frequencies. This information proves very useful in the theoretical analysis of systems and has some very important practical applications as well.

With respect to its theoretical significance, frequency response is a useful tool for determining the stability of feedback systems having certain types of controller actions and settings. The frequency response of a closed-loop feedback system can be used to predict its transient response. In research, frequency response is used as a means for comparing the dynamics of a process against those of various postulated models.

From a practical standpoint, frequency response analysis forms the basis of the continuous cycling method for tuning a controller. Frequency response theory can be useful in the design of dampers to remove pressure and flow pulsations in process systems. It can also be used to determine if a pneumatic transmission line is capable of conveying desired information from the transmitter to the controller.

I. ELEMENTARY CONCEPTS OF FREQUENCY RESPONSE

The basic idea in frequency response is that when the input to a linear element is a pure sine wave, the output is a pure sine wave of the same frequency and differs from the input only in magnitude and phase. These concepts are illustrated in Figure 1. The magnitude of the output is shown to be smaller

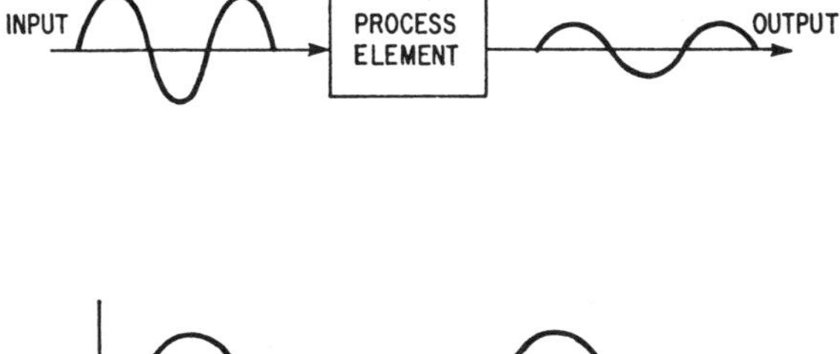

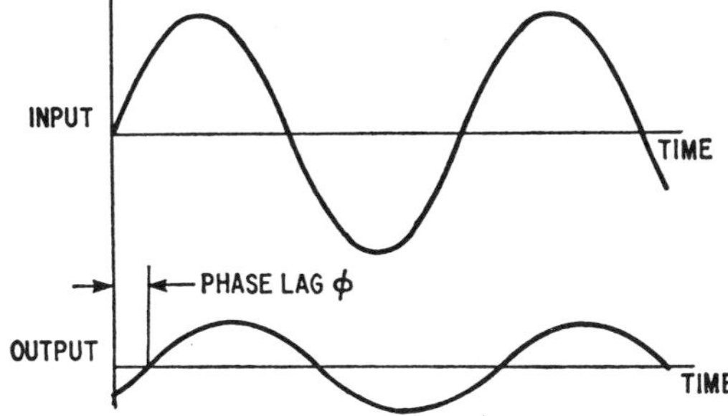

Figure 1.   Effect of a linear element on a pure sine wave input.

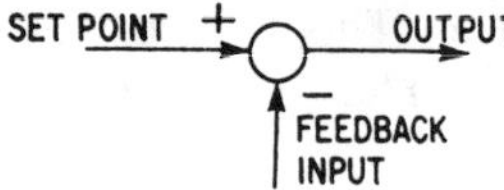

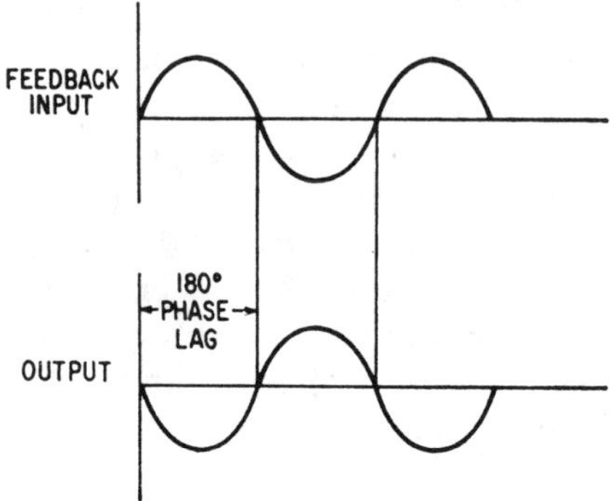

Figure 2. Effect of the comparator on a sinusoidal feedback input.

than that of the input and lags behind it by an angle, ϕ. Some elements actually result in a phase lead rather than a lag, and may also produce an increase in the magnitude. Usually the magnitude and phase of the output depend upon the frequency.

The simplest element in a control loop is the comparator. As shown in Figure 2, this unit inverts the feedback input because of the negative sign.

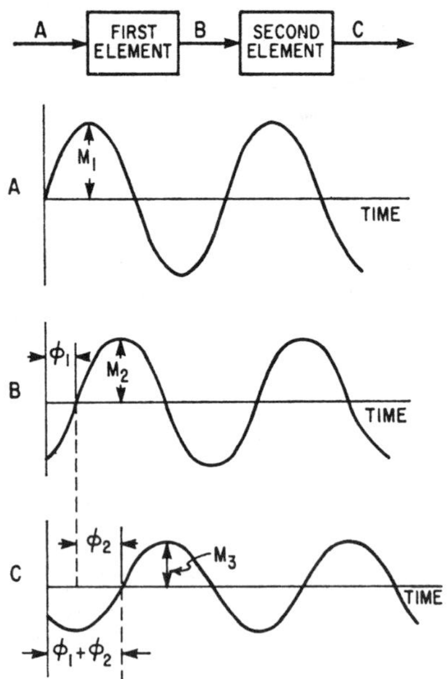

Figure 3. Effect of a series of two linear elements on a sinusoidal input.

The inversion amounts to a phase lag of 180° with no change in magnitude. No change is shown for the set point since it is apparent that if the set point were varied sinusoidally, the output would be an exact replica of the set point variation.

Most control systems consist of a number of elements in series. For simplicity, suppose we consider two elements, as shown in Figure 3. The output of the first element, B, lags behind its input, A, by phase angle, ϕ_1; similarly, C lags behind B by the angle, ϕ_2. Hence, C lags behind A by the sum of the lags, $\phi_1 + \phi_2$. Also, the ratios of the magnitudes for the individual elements, M_2/M_1 and M_3/M_2, when multiplied together yield the overall ratio, M_3/M_1.

II. DEVELOPMENT OF THE BODE STABILITY CRITERION

In Figure 4, a block diagram for a feedback system is shown. The phase lag of 180° for the comparator is indicated together with the several lags, ϕ_1, ϕ_2, ϕ_3 for the process elements, and ϕ_C for the controller. The ratio of the output to the input for each element is also given as r_1, r_2, r_3, and r_C.

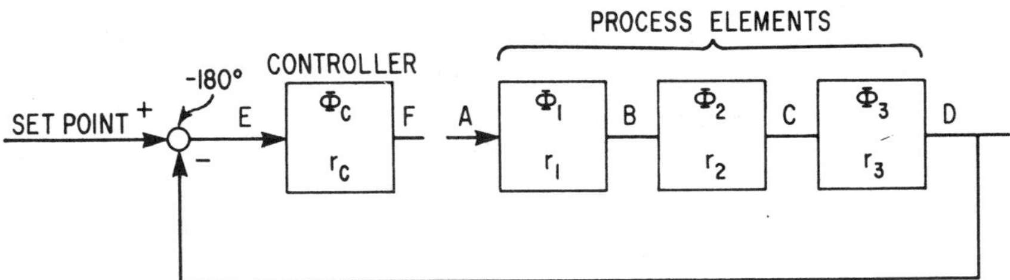

Figure 4.   Block diagram for a feedback system.

Now suppose that the input, A, is sinusoidal with frequency ω, in radians per unit time. The total phase *lag* from A to F is

$$\phi_{A-F} = 180 + \phi_1 + \phi_2 + \phi_3 + \phi_C \tag{1}$$

An interesting situation occurs when the frequency is such that the total lag is 360°. This means that F is *in phase* with A. The magnitude of F is equal to the magnitude of A multiplied by the product of the magnitude ratios:

$$F = Ar_1r_2r_3r_C$$

In particular, if the product, $r_1r_2r_3r_C$, is unity, F not only is in phase with A but also has the same magnitude as A. Therefore, since F is directly connected to A in this feedback system, the oscillation is self-sustaining. If the product is greater than unity, the oscillations will build up in magnitude, but if less than unity, they will die down.

The product, $r_1r_2r_3$, is referred to as the open-loop process gain, and r_C is the *effective* controller gain. From the preceding illustrations, we see that for stability, the product, $r_Cr_1r_2r_3$, must be less than unity at a frequency such that the total phase lag is 180°, excluding the phase lag contributed by the comparator. This is referred to as the Bode stability criterion. Usually one is interested in the maximum controller gain, $r_{C,\max}$, and from the Bode criterion it can be seen that

$$r_{C,\max} = \frac{1}{r_1r_2r_3} \qquad \text{when } \sum \phi = 180° \qquad (2)$$

In summary then, finding the maximum controller gain is a two-step process: (1) the "critical" or "crossover" frequency must be found at which the total phase lag of the elements is 180°, excluding the 180° lag of the comparator; (2) then the gains of each of the elements at this frequency must be determined and substituted in Equation 2. The extension of this example to one involving more or less elements poses no difficulties.

In retrospect, it is interesting to note that the system treats a signal at the critical frequency in the manner of positive feedback. On the other hand, consider the behavior of the system in response to a signal whose frequency is zero—in other words, a steady-state signal. We know that the system is connected up in such a way as to treat steady-state signals in a negative feedback way. Thus, the system treats one particular signal in a positive manner and another in a negative manner.

This discussion may have seemed somewhat artificial with respect to a real system, since it could be argued that the input A will not likely be a pure sine wave, much less at the critical frequency. However, it may be recalled that periodic disturbances can be expressed in terms of a Fourier series containing multiples of the fundamental frequency, and that nonrecurring disturbances can be expressed in terms of a continuous spectrum of frequencies. Thus it is reasonable to speculate that all frequencies will be present in the system. If the critical frequency occurs for only an instant, its magnitude will increase if the controller gain exceeds the stability limit given in Equation 2. Furthermore, it can be seen that it really makes no difference where the disturbance occurs in the loop; it could just as well begin at B or at any other point.

This development of the Bode stability criterion was based on intuitive arguments rather than on mathematically rigorous ones. There are certain types of elements for which it always applies. One class of these are the so-called "minimum phase" elements. These are characterized by transfer functions where the numerator and denominator have no roots in the right half of the complex plane. Most of the transfer functions we have encountered fall in this category. Although the transfer function of a pure time delay is a "nonminimum phase" element with a root in the right half plane, the Bode criterion can still be used for it and its combinations with minimum phase elements.

The Bode criterion usually cannot be applied to systems containing nonminimum phase elements. We have seen that this criterion requires locating the frequency at which the total phase lag in the loop is 180°. However, these elements frequently exhibit a phase lag of 180° at more than one frequency, and this can lead to difficulties and errors in interpreting the criterion.

All distributed parameter processes display nonminimum phase behavior whereas only a few lumped parameter processes do. One notable lumped parameter example is an inherently unstable reactor. More details concerning nonminimum phase elements can be found in References 1i, 2, 3 and 6i. When they are present, a general test of stability should be applied. The Routh criterion is suitable when the transfer functions are in the form of polynomials. Otherwise, the Nyquist criterion can be used. Like the Bode criterion, it is a frequency response technique, but it is derived from the theory of complex variables. Some discussion and illustrations of the Nyquist method can be found in References 4i, 5, 6ii.

III. FREQUENCY RESPONSE OF PROCESS ELEMENTS

A. First- Order

A logical method for finding the frequency response of an element would be to use its transfer function in combination with the transform of a sine wave. For example, consider the case of a first-order element:

$$\frac{Y}{X} = \frac{K}{Ts + 1} \tag{3}$$

If the input, X, is a pure sine wave of amplitude, M, then

$$x(t) = M \sin \omega t$$

and

$$X(s) = \frac{M\omega}{s^2 + \omega^2}$$

Therefore

$$Y(s) = \frac{K}{Ts + 1} \frac{M\omega}{s^2 + \omega^2} \tag{4}$$

This can be inverted to yield

$$y(t) = \frac{MK}{\omega}\left[\frac{T\omega^2}{1 + T^2\omega^2}\,e^{-t/T} + \frac{\omega\,\sin(\omega t + \phi)}{(1 + T^2\omega^2)^{1/2}}\right] \tag{5}$$

where

$$\phi = -\tan^{-1}(\omega T) \tag{6}$$

The first term on the right is the transient, which dies out. The second is the response to the sustained sinusoidal input; this is the term of interest in frequency response analysis.

Although this method can be used, it usually requires the use of inversion tables, and furthermore provides additional information about the transient response that is not required. Fortunately, there is a rather simple way of obtaining the steady-state frequency response from the transfer function alone. This consists of the simple substitution of $j\omega$ for s where j is the imaginary number, $\sqrt{-1}$. An explanation of the theory behind this for nth-order linear systems is given in Reference 4ii. This rule works just as well for a pure delay and for distributed elements. The only restriction is that if the process itself is not inherently stable, then the method is not valid.

As a demonstration, let us apply this method to the first-order transfer function. Substituting $j\omega$ for s, we obtain

$$\frac{Y(\omega)}{X(\omega)} = \frac{K}{j\omega T + 1} \tag{7}$$

This can be rationalized to the following form:

$$\frac{Y(\omega)}{X(\omega)} = \frac{K}{1 + \omega^2 T^2} - \frac{jK\omega T}{1 + \omega^2 T^2} \tag{8}$$

The equivalent representation of this expression in the complex plane is shown in Figure 5. This figure can be interpreted in the following way.

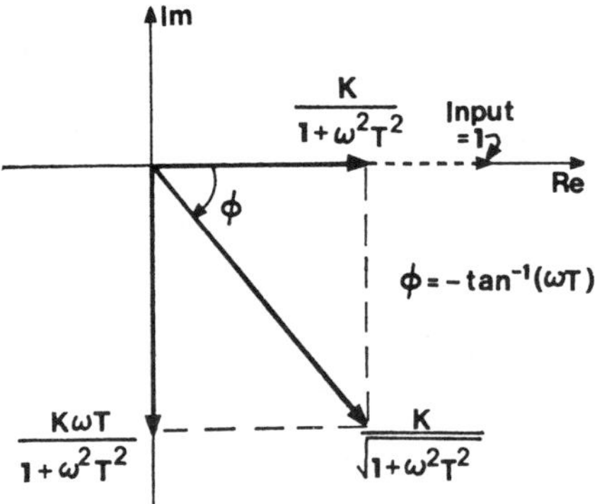

Figure 5. Complex representation of first-order transfer function.

If the input to this first-order process is a pure sine wave of unit magnitude, the output can be represented in the complex plane by two vectors, one real and one imaginary. These two vectors can be added to obtain the resultant vector whose phase angle, ϕ, is negative; in other words, the output lags behind the input. For an input of magnitude, M, the magnitude of the output is $MK/\sqrt{1 + \omega^2 T^2}$ and the phase angle is still ϕ. This confirms the steady-state term of Equation 5. As the frequency increases, the magnitude of the real component decreases more rapidly than that of the imaginary component. Therefore the phase angle of the output asymptotically approaches $-90°$ and its magnitude continuously decreases.

In the development of the stability criterion, the ratio of the amplitude of the output to that of the input was expressed as r. In terms of this first-order example, this ratio is

$$r = \frac{K}{\sqrt{1 + \omega^2 T^2}} \tag{9}$$

This gain consists of the product of a steady-state gain, K, and a frequency-dependent part that is designated by a factor, f:

$$f = \frac{1}{\sqrt{1 + \omega^2 T^2}} \tag{10}$$

A convenient way to represent the frequency response of an element is in the form of a Bode diagram. The amplitude ratio is plotted against frequency on log-log coordinates and the phase angle is plotted on arithmetic coordinates against the frequency on logarithmic coordinates. Frequently, it is expedient to omit the steady-state factor from the amplitude ratio of the plot. Thus, in the case of a first-order element, only the factor, f, would be plotted against frequency. Since this factor approaches unity as the frequency approaches zero, this amplitude ratio is said to be "normalized." Furthermore, a generalized plot for all first-order elements results if the abscissa is made dimensionless using ωT rather than ω itself. Accordingly, a generalized plot for a first-order process is shown in Figure 6.

This diagram has several important features. At low frequencies, the amplitude ratio is nearly unity and the phase angle is close to zero. At a dimensionless frequency of 1.0, the phase angle is $-45°$ and the amplitude ratio is $1/\sqrt{2}$ or 0.707. At high frequencies, the phase angle approaches $-90°$ and the slope of the amplitude ratio curve approaches -1. These features suggest an approximate diagram, as shown in Figure 7, which is sometimes helpful in sketching out the Bode diagram for a system containing several elements. The frequency at which ωT is unity is referred to as the "corner frequency" since this is the point at which the two asymptotes of

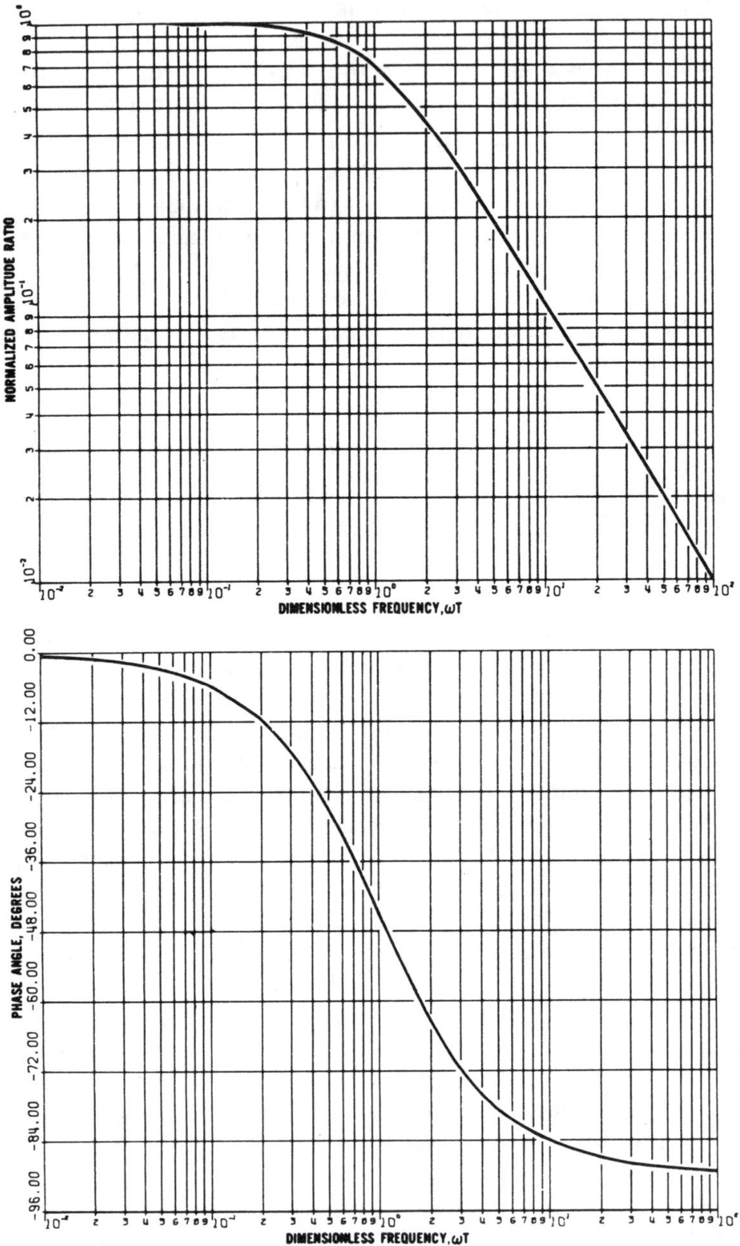

Figure 6. Generalized Bode diagram for a first-order process.

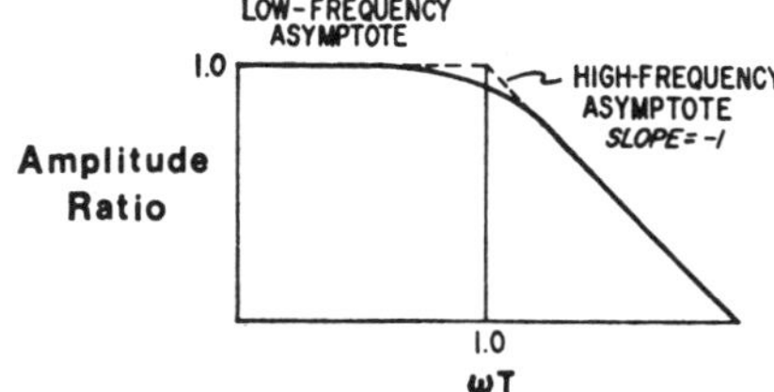

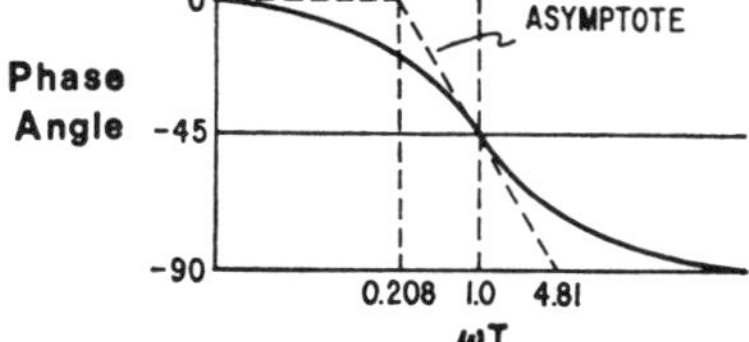

Figure 7. Asymptotic Bode diagram for a first-order process.

the amplitude plot meet. The phase curve requires three lines for approximation. The diagonal line is the tangent at a dimensionless frequency of unity. The phase curve approximation is less accurate than the approximation for the amplitude curve and is not as useful.

The utility of the Bode diagram is particularly apparent for control systems having several elements. Consider, for example, process 3, which was discussed in Section II.B.1.c of Chapter 13. This third-order process was characterized by the time constants of 1, 5, and 10, as shown in the block diagram of Figure 8. The Bode diagram for this process is presented in Figure

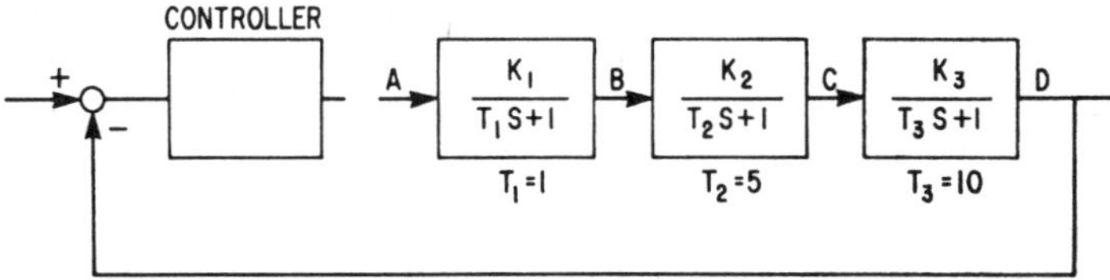

Figure 8. Block diagram for a third-order process.

9. First-order curves are drawn for each element. As mentioned before, the individual steady-state gains are omitted for convenience, leaving only the frequency-dependent factors. Hence, all three amplitude ratio curves are asymptotic to unity at low frequencies.

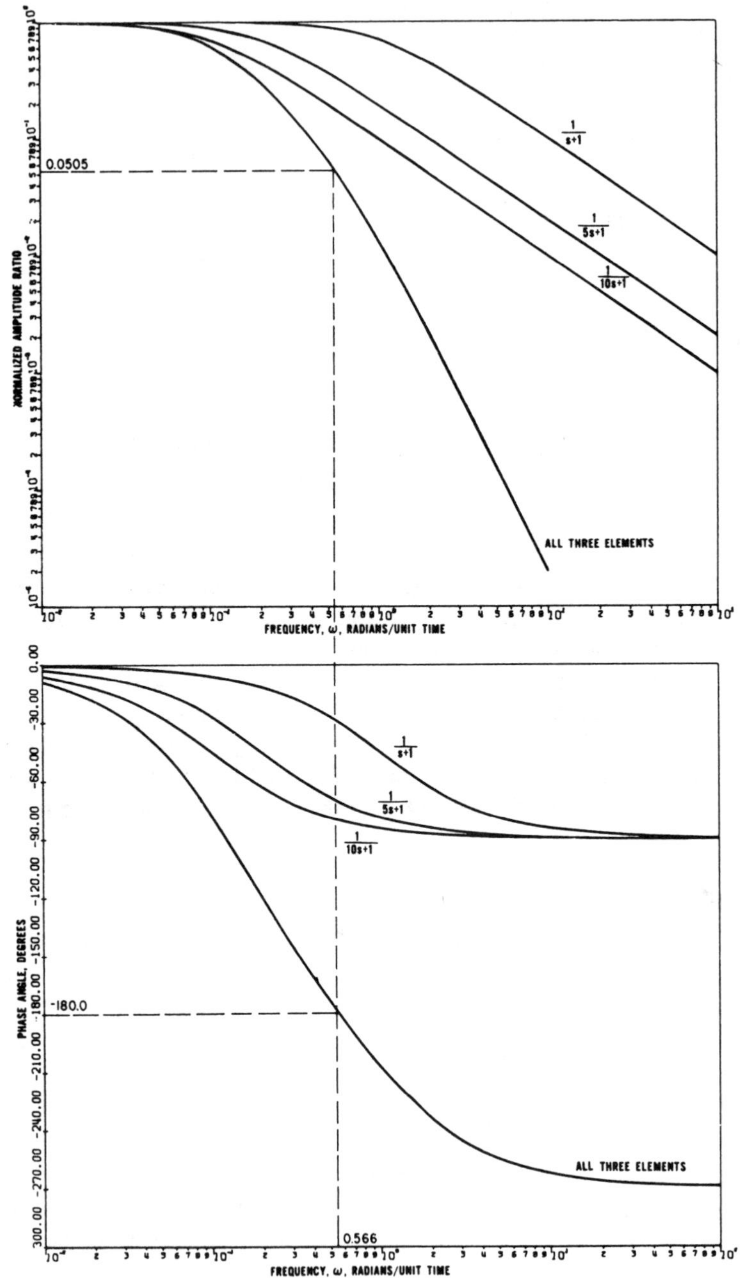

Figure 9. Bode diagram for a third-order process.

The construction of the composite curves for all three elements is particularly simple because of the coordinates of the graphs. Consider the amplitude ratio, D/A:

$$\frac{D}{A} = r_1 r_2 r_3 = \frac{K_1}{\sqrt{1 + \omega^2 T_1{}^2}} \cdot \frac{K_2}{\sqrt{1 + \omega^2 T_2{}^2}} \cdot \frac{K_3}{\sqrt{1 + \omega^2 T_3{}^2}} \tag{11}$$

Thus

$$\frac{D}{A} = K_1 K_2 K_3 f_1 f_2 f_3 \tag{12}$$

where

$$f_1 = \frac{1}{\sqrt{1 + \omega^2 T_1{}^2}}$$

$$f_2 = \frac{1}{\sqrt{1 + \omega^2 T_2{}^2}}$$

$$f_3 = \frac{1}{\sqrt{1 + \omega^2 T_3{}^2}}$$

The f's are plotted on the diagram. Since their product, $f_1 f_2 f_3$, is the magnitude of the composite curve, f_P, either of two methods can be used. The product can be evaluated at a number of abscissa points using the corresponding values of the three curves, or a graphical construction can be employed utilizing the same principle of logarithms as is used in multiplying numbers together on a slide rule. This is illustrated by the sketch in Figure 10. For example, at a particular frequency, ω_1, the distances from the unity line for both f_1 and f_2 are added down from the corresponding point on the f_3-curve to obtain the point for the composite curve, f_P. Furthermore, since the phase lags of the individual elements are additive, a similar graphical construction can be used for the composite phase lag curve.

This third-order process serves as a good example to illustrate the use of the Bode diagram in determining the maximum gain and critical frequency for a system. Referring back to Figure 9, we begin by finding the critical frequency at which the total phase lag is 180°. As indicated, this frequency is 0.566 rad/unit time and the corresponding value of f_P is 0.0505. For stability

$$r_C < \frac{1}{r_1 r_2 r_3} = \frac{1}{K_1 K_2 K_3 f_1 f_2 f_3} = \frac{1}{K_1 K_2 K_3 f_P} = \frac{1}{K_1 K_2 K_3 (0.0505)} \tag{13}$$

We shall assume here that a proportional controller will be used. For an ideal controller of this type, a sine wave input is only magnified by the gain factor, K_C, and this factor is independent of the frequency. Hence, r_C is

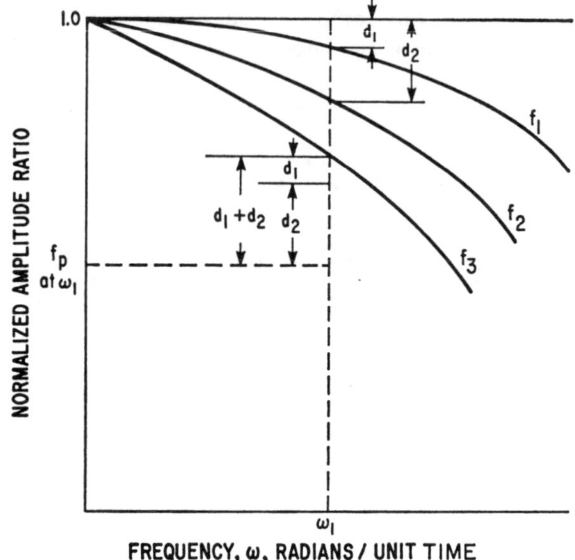

Figure 10.   Graphical construction of composite amplitude ratio curve.

identically K_C. In addition, the controller output is in phase with the input for all frequencies, so its contribution to the phase angle is zero. Therefore, for stability, we see that

$$K_C < \frac{1}{K_1 K_2 K_3 (0.0505)} \tag{14}$$

The overall gain of the loop, K_o, is the product of the individual gains:

$$K_o = K_C K_1 K_2 K_3 \tag{15}$$

Substituting this relation into Equation 14 and rearranging, we find that

$$K_o < 19.8 \qquad \text{for stability}$$

This value checks that found by Equation 29 of Chapter 13.

From this example, it can be seen why there is theoretically no maximum proportional gain for a process characterized by either one or two first-order elements. If there is one of these elements, the maximum phase lag approaches 90°, and if there are two, the maximum total lag approaches 180°. However, in neither case does the phase lag actually reach 180°, so there is no maximum value of the loop gain.

The frequency response of a first-order element served as a good beginning example since so many processes have been modeled by it. We shall now investigate the responses of other types of processes that have been discussed in previous chapters.

B. Pure Capacitance

The transfer function for a pure capacitance has the form

$$\frac{Y}{X} = \frac{K}{Cs} \tag{16}$$

Substituting $j\omega$ for s, we obtain

$$\frac{Y}{X} = \frac{K}{jC\omega} = -\frac{jK}{C\omega} \tag{17}$$

As shown in Figure 11, this function only has an imaginary component in the complex plane. Accordingly, the frequency-dependent factor is

$$f = \frac{1}{C\omega} \tag{18}$$

and the phase lag is a constant:

$$\phi = -90° \tag{19}$$

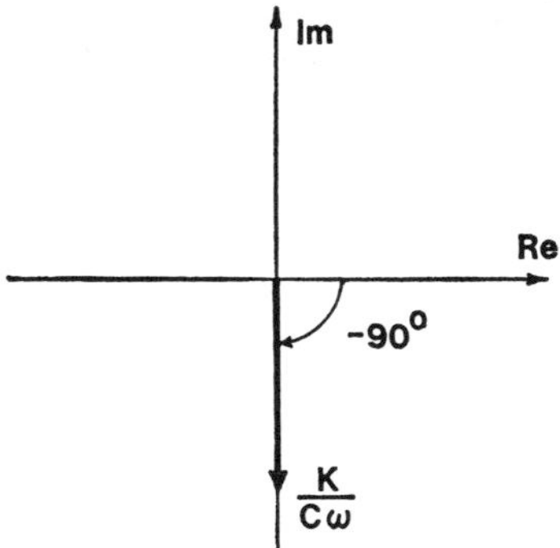

Figure 11. Complex representation of a pure capacitance.

C. Second Order

The standard form of the transfer function for a second-order process is

$$\frac{Y}{X} = \frac{K}{\dfrac{1}{\omega_n^2}s^2 + \dfrac{2\zeta}{\omega_n}s + 1} \tag{20}$$

Upon substituting $j\omega$ for s and rationalizing, the result is

$$\frac{Y}{X} = \frac{K[1 - (\omega/\omega_n)^2]}{[1 - (\omega/\omega_n)^2]^2 + 4\zeta^2(\omega/\omega_n)^2} - j\frac{K(2\zeta\omega/\omega_n)}{[1 - (\omega/\omega_n)^2]^2 + 4\zeta^2(\omega/\omega_n)^2} \tag{21}$$

The frequency-dependent factor of the amplitude ratio is

$$f = \frac{1}{\{[1 - (\omega/\omega_n)^2]^2 + 4\zeta^2(\omega/\omega_n)^2\}^{1/2}} \tag{22}$$

and the phase angle is

$$\phi = -\tan^{-1}\frac{2\zeta\omega/\omega_n}{1 - (\omega/\omega_n)^2} \tag{23}$$

The basic character of the frequency response is only dependent on the damping factor, ζ. If this factor is greater than unity, the response is overdamped and is effectively the combination of two first-order elements in series. The Bode diagram for critical damping and for several overdamped cases is shown in Figure 12. The plot has been generalized by using the dimensionless frequency, ω/ω_n, where ω_n is the natural frequency. The underdamped case is especially interesting since the amplitude ratio can be very high in the region of the natural frequency if the damping factor is low. As indicated by Equation 22, as the damping factor approaches zero, the amplitude ratio approaches infinity at the natural frequency. Bode diagrams for several underdamped cases are shown in Figure 13.

D. Pure Delay

The transfer function for a pure delay is

$$\frac{Y}{X} = e^{-\tau_d s} \tag{24}$$

Substituting $j\omega$ for s, we obtain

$$\frac{Y}{X} = e^{-j\omega\tau_d} \tag{25}$$

This exponential function can be expanded into the sum of a real part and an imaginary part using Euler's equation which is as follows:

$$e^{\pm j\theta} = \cos\theta \pm j\sin\theta \tag{26}$$

Therefore, Equation 25 can be written

$$\frac{Y}{X} = \cos\omega\tau_d - j\sin\omega\tau_d \tag{27}$$

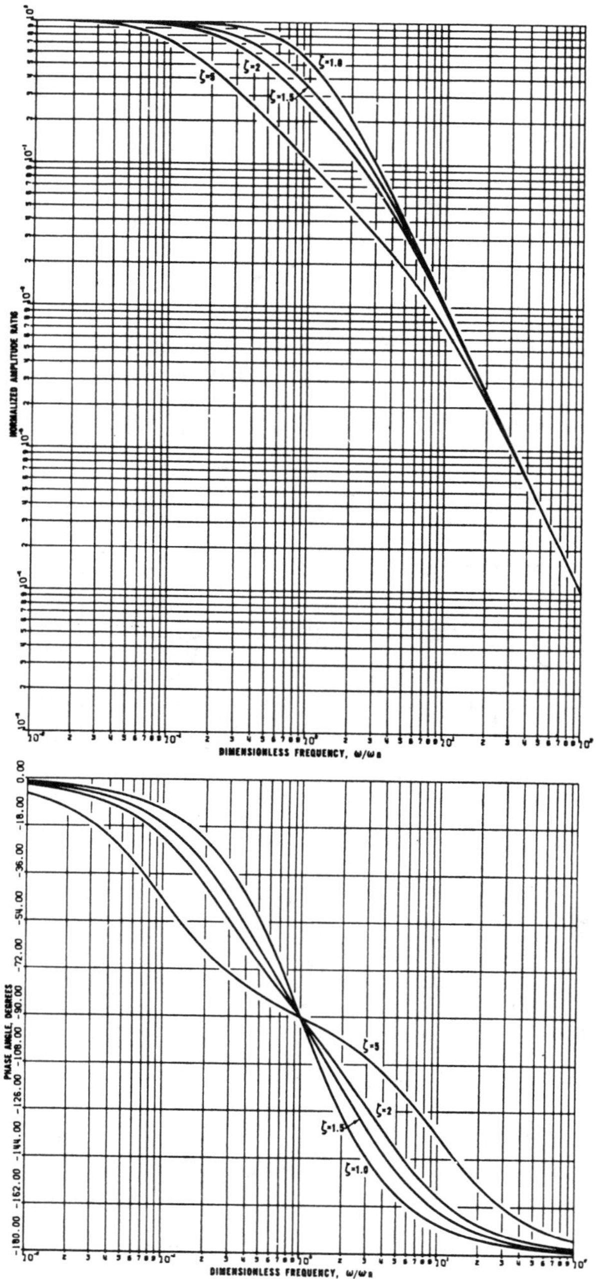

Figure 12. Bode diagrams for critically damped and overdamped second-order processes.

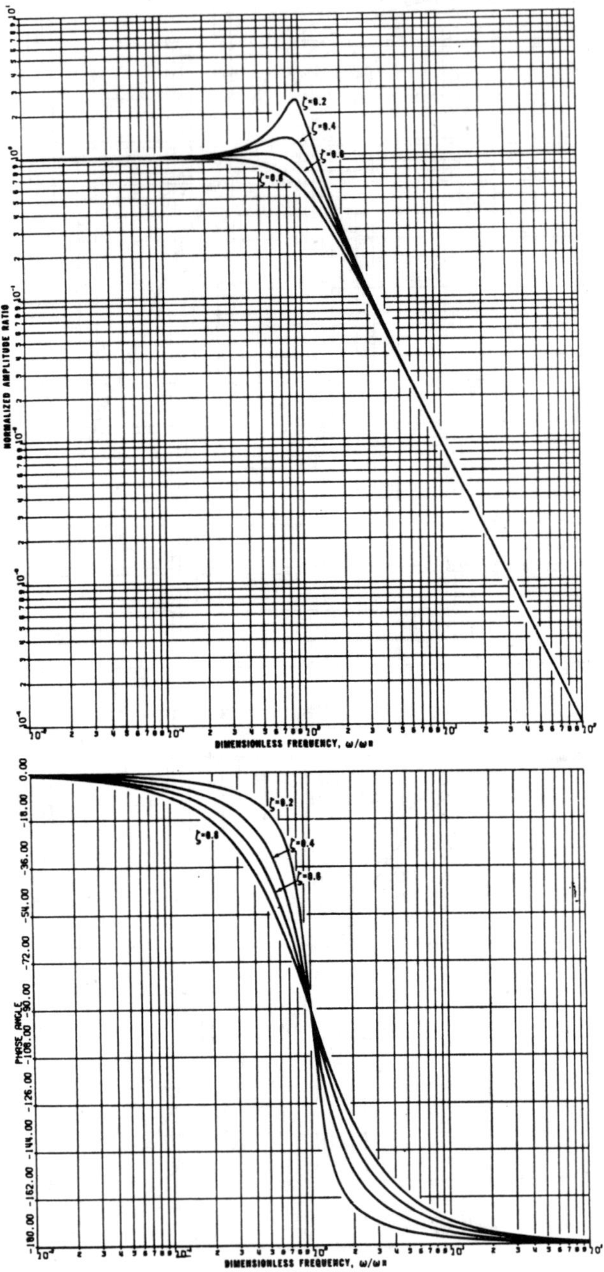

Figure 13. Bode diagrams for underdamped second-order processes.

394

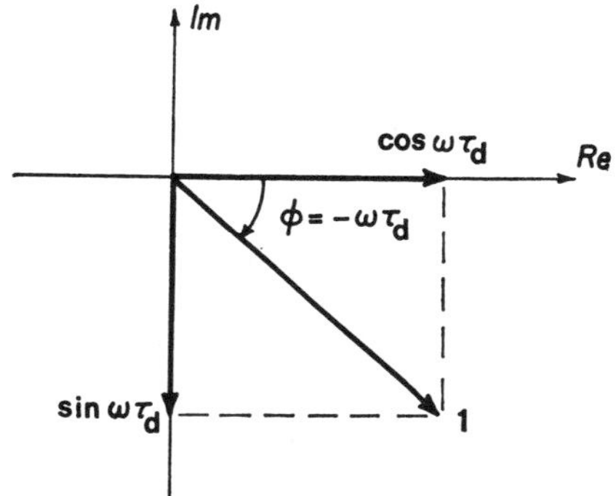

Figure 14. Complex representation of a pure delay.

A complex representation of this function is presented in Figure 14. The phase angle is determined as follows:

$$\phi = \tan^{-1}\left(\frac{-\sin \omega\tau_d}{\cos \omega\tau_d}\right) = -\omega\tau_d \ \text{rad} \tag{28}$$

Radians can be converted to degrees by multiplying by 57.3. The frequency-dependent factor is

$$f = \sqrt{\sin^2 (\omega\tau_d) + \cos^2 (\omega\tau_d)} = 1 \tag{29}$$

The amplitude ratio is always unity regardless of the frequency. The Bode diagram is shown in Figure 15.

Since the phase lag of a pure delay increases without bound as the frequency increases, if there are any pure delays in a process there will be some frequency at which the total lag is 180° with a corresponding upper limit for the gain of the loop. Therefore, even if the process is reasonably well characterized by a first- or second-order element, the additional unlimited phase lag of the pure delay insures a "crossover" of the 180° line.

E. Distributed Processes

Because of the large variety of distributed processes, it is not possible here to present a comprehensive coverage of their frequency response behavior. A few of the simpler cases of Chapter 11 will be examined to indicate the unusual properties of these processes.

As with lumped parameter processes, the frequency response is found by

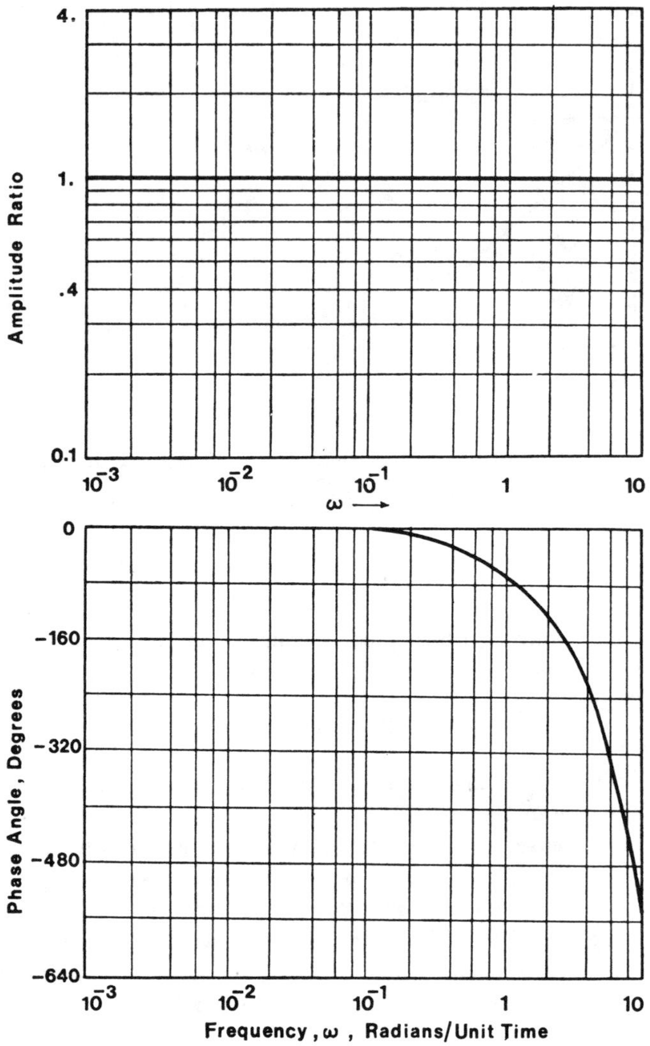

Figure 15.   Bode diagram for a pure delay.

substituting $j\omega$ for s. The calculation of the amplitude and phase lag often leads to quite cumbersome algebraic manipulations. If it is necessary to find the frequency response of a distributed transfer function, the computations are usually best carried out by a digital computer. The complex functions of Fortran are well suited for this purpose and this was the technique used for all of the processes discussed here.

1. Heat Transfer in a Slab

a. SEMIINFINITE SLAB

The transfer function relating the temperature at a distance, L, from the surface to that at the surface is obtained from Equation 12 of Chapter 11:

$$\frac{\overline{T}(L, s)}{\overline{T}(0, s)} = e^{-\sqrt{s/\alpha}\,L} \tag{30}$$

The gain factor, which is the frequency-independent part of the transfer function, is found by letting s approach zero. This limit is unity, and this is the expected result since if the oscillation of the surface temperature is very, very slow, the temperature at distance, L, will essentially keep pace with it. The frequency-dependent amplitude ratio and phase angle are derived in Reference 4iii with the following results:

$$f = e^{-\sqrt{\omega/2\alpha}\,L} \tag{31}$$

$$\phi = -\sqrt{\frac{\omega}{2\alpha}}\,L \ \text{rad} \tag{32}$$

As the frequency increases, the amplitude ratio decreases continuously and the phase lag increases without bound. This phase lag behavior is similar to that found for a pure delay, although not quite the same since the phase lag of a pure delay is proportional to the frequency whereas that for the slab is proportional to the square root of the frequency.

A generalized Bode diagram is obtained by using the dimensionless frequency, $\omega L^2/2\alpha$, and is presented in Figure 16.

b. FINITE SLAB, PERFECTLY INSULATED ON ONE FACE

The transfer function for this case was given in Equation 17 of Chapter 11:

$$\frac{\overline{T}(x, s)}{\overline{T}(0, s)} = \frac{\cosh\,[\sqrt{s/\alpha}\,(L - x)]}{\cosh\,(\sqrt{s/\alpha}\,L)} \tag{33}$$

At the insulated surface where $x = L$, Equation 33 becomes

$$\frac{\overline{T}(L, s)}{\overline{T}(0, s)} = \frac{1}{\cosh\,(\sqrt{s/\alpha}\,L)} \tag{34}$$

The amplitude ratio and phase angle are derived in Reference 1ii and the results are:

$$f = [\cosh^2\,(\sqrt{\omega/2\alpha}\,L) - \sin^2\,(\sqrt{\omega/2\alpha}\,L)]^{-1/2} \tag{35}$$

$$\phi = -\tan^{-1}\,[\tan\,(\sqrt{\omega/2\alpha}\,L) \cdot \tanh\,(\sqrt{\omega/2\alpha}\,L)] \tag{36}$$

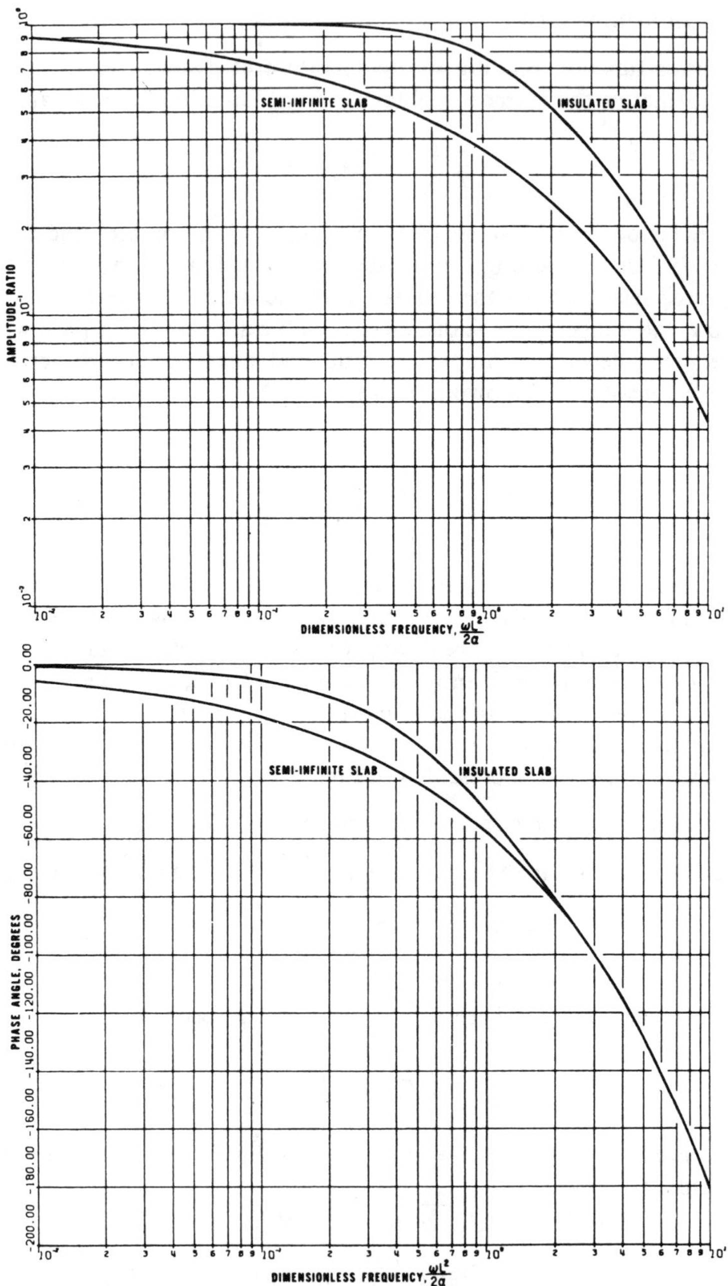

Figure 16. Bode diagrams for heat transfer in a slab.

398

The Bode diagram is plotted in Figure 16 for comparison with that for a semiinfinite slab. The phase curves are essentially coincident at dimensionless frequencies greater than about 2.5 because the hyperbolic tangent is approximately unity for values of the argument greater than 2.5.

Comparing Figures 15 and 16, we see that the phase lags for the slab cases increase with frequency in much the same way as that for a pure delay. Furthermore, the amplitude ratios for slabs fall off with increasing frequency in a way similar to that for a first-order process. This suggests approximating the transfer function for slabs by a combination of those for a pure delay and a first-order element:

$$\frac{\overline{T}(L,\,s)}{\overline{T}(0,\,s)} \cong \frac{e^{-\tau_d s}}{Ts\,+\,1} \tag{37}$$

An illustration of this for the case of an insulated slab is presented in Reference 8i. The time constant, T, was set equal to $L^2/2\alpha$ and the delay, τ_d, was set equal to $L^2/10\alpha$ where α is the thermal diffusivity. The approximation for both the amplitude and phase curves is very good out to a phase lag of about $180°$.

2. Pneumatic Transmission Line Terminated by a Small Bellows

Section II.B.1 of Chapter 11 noted that when a transmission line is terminated by a small bellows, it can usually be assumed that the transfer function relating the pressure in the bellows to that at the sending end, $P(L,\,s)/P(0,\,s)$, is identical in form with that in Equation 33 of this chapter for heat transfer in a slab. Therefore, the Bode diagram in Figure 16 for an insulated slab can be used.

The transmission line problem reduces to the heat transfer problem because the impedance of the bellows is much much larger than that of the line and the inertance of the air is negligible. A quantitative justification of these assumptions based upon considerations of frequency response is given in Reference 7.

IV. FREQUENCY RESPONSE OF CONTROLLERS

In this section we shall be mainly concerned with the frequency response of controllers whose transfer functions are of the "ideal" forms. The modifications imposed by nonideality will be discussed in a general way later on.

A. Proportional Controller

As mentioned earlier in the chapter, a proportional controller only magnifies the input by a constant factor. Hence, its amplitude ratio is a constant and it introduces no phase shift.

B. Proportional-Integral Controller

The transfer function for a proportional-integral controller is

$$G_C = K_C\left(1 + \frac{1}{T_R s}\right) \tag{38}$$

Substituting $j\omega$ for s, we obtain

$$G_C = K_C\left(1 - j\frac{1}{\omega T_R}\right) \tag{39}$$

The frequency-dependent part can be represented in the complex plane as shown in Figure 17. The orientation of the vectors is the same as that in Figure 5 for a first-order element. The resultant vector is always in the fourth

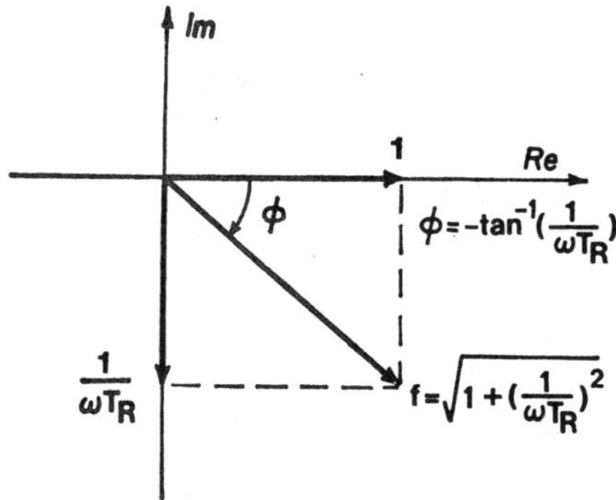

Figure 17. Complex representation of a proportional-integral controller.

quadrant, but it begins with a phase lag of 90° at zero frequency and asymptotically approaches 0° as the frequency increases. The frequency-dependent factor and phase angle are:

$$f = \sqrt{1 + \frac{1}{\omega^2 T_R^2}} \tag{40}$$

$$\phi = -\tan^{-1}\left(\frac{1}{\omega T_R}\right) \tag{41}$$

Bode diagrams are shown in Figure 18 for reset times of 0.5, 1.0, and 2.0. The gain factor, K_C, is not included in the amplitude ratio since it is not frequency dependent. For a given frequency, the effect of reducing the reset time is to introduce further lag and to increase the effective gain. The plot shown for a reset time of unity can be interpreted as a generalized plot if the abscissa is construed to be the dimensionless frequency, ωT_R.

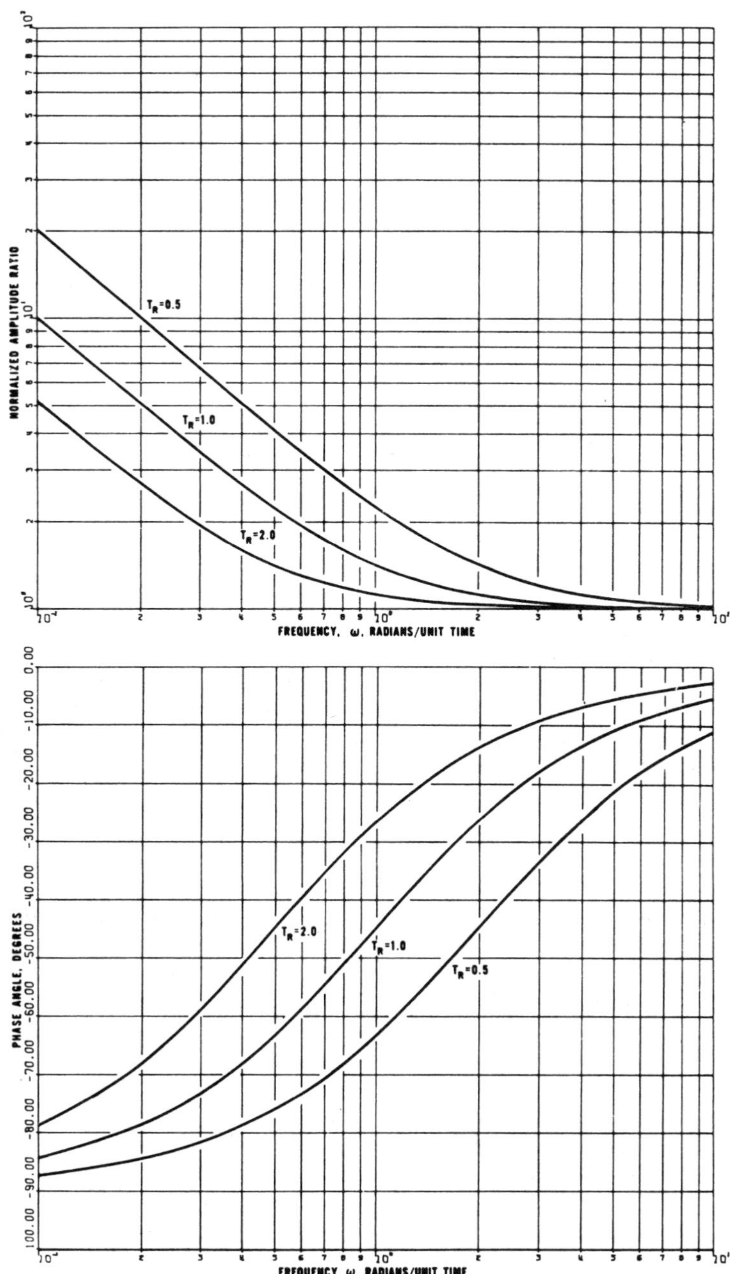

Figure 18. Bode diagrams for an ideal proportional-integral controller.

C. Proportional-Derivative Controller

The transfer function for proportional-derivative action is

$$G_C = K_C(1 + T_D s) \qquad (42)$$

Substitution of $j\omega$ for s yields the representation given in Figure 19 for the frequency-dependent part. Unlike all process elements and controller actions previously discussed, this one results in phase *lead*. The lead is zero

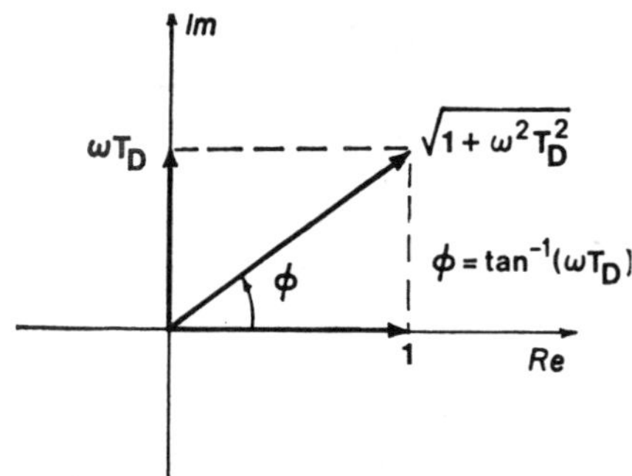

Figure 19. Complex representation of a proportional-derivative controller.

at zero frequency, but increases asymptotically to 90° at high frequencies. The frequency-dependent factor and phase angle are given by the equations:

$$f = \sqrt{1 + \omega^2 T_D^{\,2}} \qquad (43)$$

$$\phi = \tan^{-1}(\omega T_D) \qquad (44)$$

Bode diagrams are shown in Figure 20 for derivative times of 0.5, 1.0, and 2.0. For a fixed value of frequency, as the derivative time is increased, the effective gain and phase lead are increased. The diagram for a derivative time of 1.0 can be interpreted as a generalized plot if the abscissa is considered to be ωT_D.

D. Proportional-Integral-Derivative Controller

The transfer function for an ideal three-mode controller is

$$G_C = K_C\left(1 + \frac{1}{T_R s} + T_D s\right) \qquad (45)$$

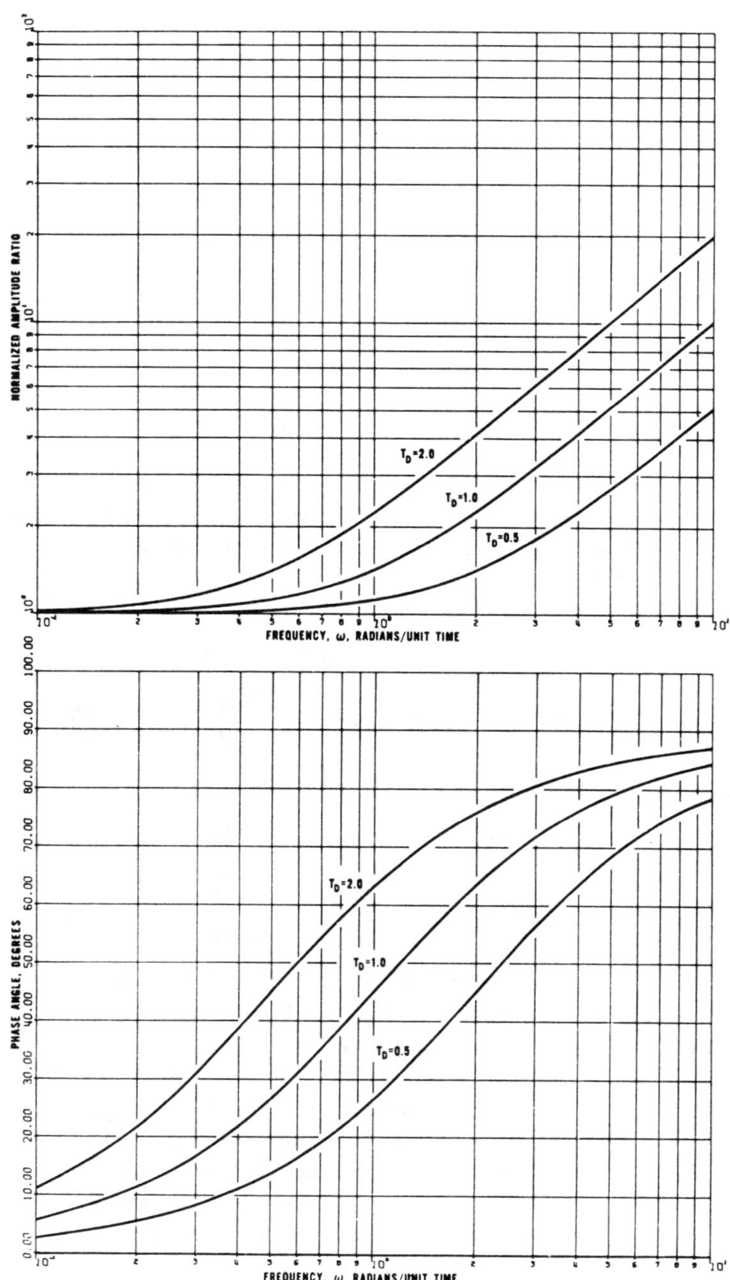

Figure 20. Bode diagrams for an ideal proportional-derivative controller.

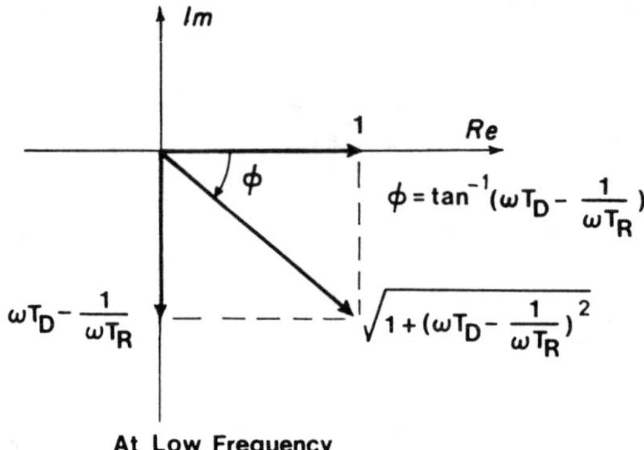

Figure 21. Complex representation of a proportional-integral-derivative controller.

A complex plane representation is presented in Figure 21, which applies at low frequencies. At low frequencies, the imaginary component is negative but at high frequencies, it becomes positive. Hence, the phase angle lags at low frequencies and leads at high. The frequency-dependent factor and phase angle are as follows:

$$f = \sqrt{\left(\omega T_D - \frac{1}{\omega T_R}\right)^2 + 1} \tag{46}$$

$$\phi = \tan^{-1}\left(\omega T_D - \frac{1}{\omega T_R}\right) \tag{47}$$

Bode diagrams are presented in Figure 22 for the following cases:

Case	T_R	T_D
1	4.0	0.25
2	1.0	1.0
3	0.25	4.0

The gain, K_C, has not been included. In case 1, where the derivative time is much less than the reset time, the shape of the amplitude curve at low frequencies is essentially the same as that of a proportional-integral controller, as shown in Figure 18. At high frequencies, the shape of the amplitude curve is similar to that of a proportional-derivative controller, as shown in Figure 20. This suggests that the two actions have very little effect upon one another. To verify this, Bode diagrams are presented in Figure 23 for a proportional-integral controller with a reset time of 4, and for a proportional-derivative controller with a derivative time of 0.25. The curves for the three-mode controller of case 1 are also presented. By inspection, it can be seen that the

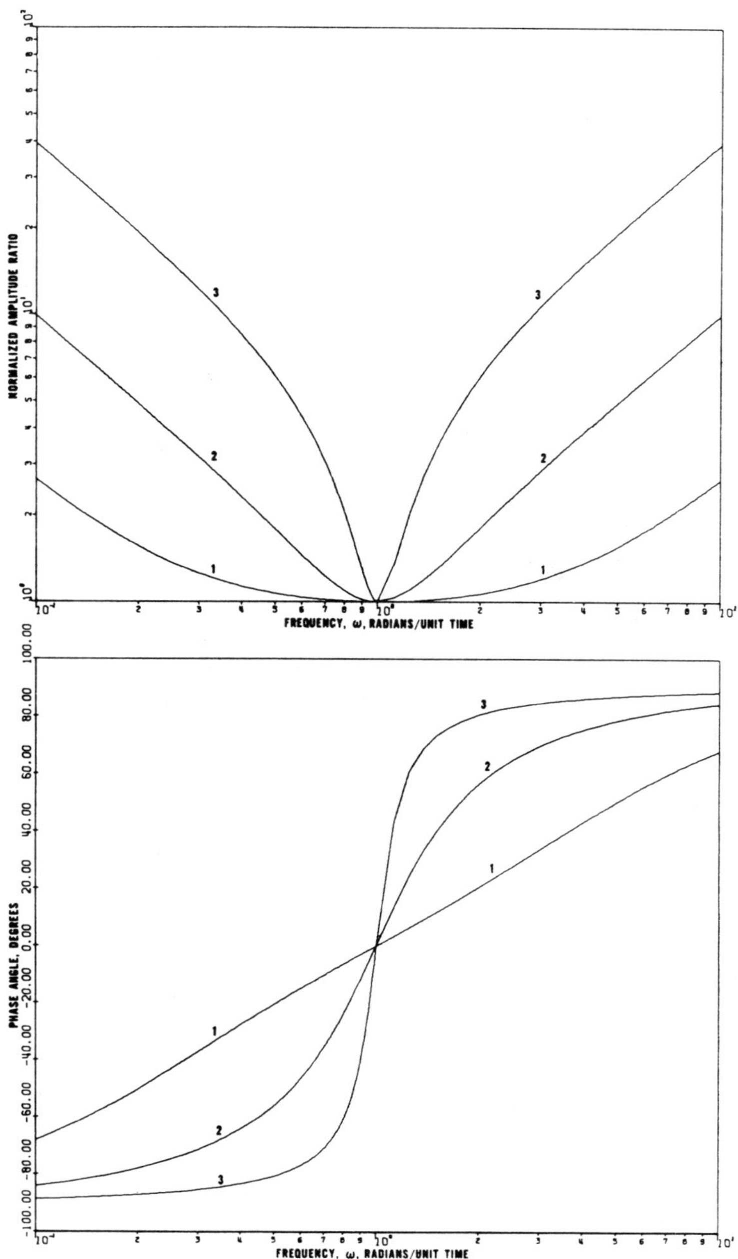

Figure 22.   Bode diagrams for a proportional-integral-derivative controller.

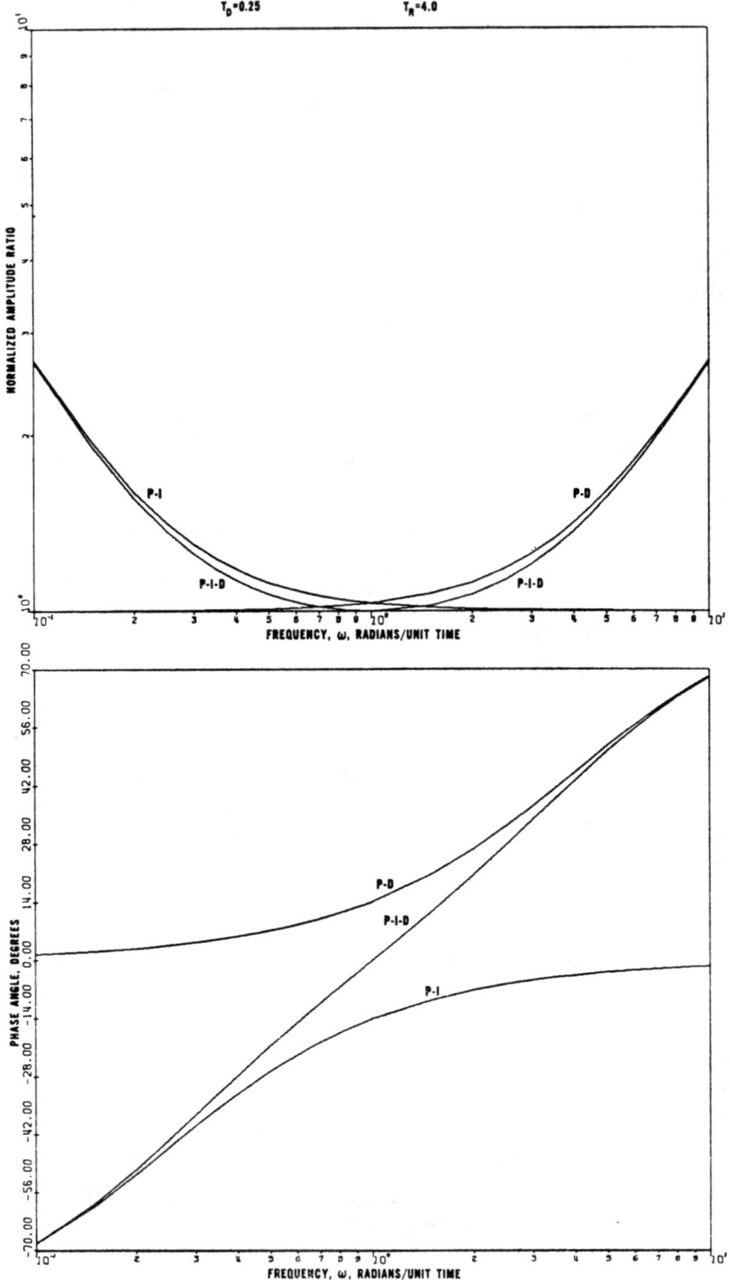

Figure 23. Bode diagrams to show that P-I-D $\cong$ P-I + P-D when $T_D \ll T_R$.

406

summation of the proportional-integral and proportional-derivative curves closely approximates the behavior of the three-mode controller.

Cases 2 and 3 of Figure 22 show that as the derivative time increases with respect to the reset time, the behavior can no longer be described in terms of the addition of proportional-integral and proportional-derivative actions. In all cases, the amplitude curves are symmetric about the frequency, $\omega = (1/T_D T_R)^{1/2}$, and the phase angle is zero at this frequency. When the derivative time exceeds the integral time, the effective gain is greater than unity over most of the frequency range.

E. Nonideal Behavior of Controllers

The frequency response of a real controller can deviate from that predicted from the ideal form of its transfer function for several reasons. One reason is that the true transfer function of an actual controller only approximates that of its ideal counterpart. For example, as the frequency approaches zero, the gain of an ideal proportional-integral controller approaches infinity and the phase angle approaches $-90°$. In an actual controller of this type, as the frequency approaches zero, the gain approaches a finite value and the phase angle approaches zero. As the frequency increases from zero, the phase angle passes through a minimum that is considerably less negative than the value of $-90°$ predicted from the ideal transfer function. Proportional-derivative action displays similar limited behavior. For an ideal controller, as the frequency increases, the phase lead approaches $90°$ and the gain approaches infinity. For an actual controller of this type, the maximum phase lead usually falls in the range between $40°$ and $60°$. At high frequencies, the gain levels off rather than approaching infinity. Furthermore, at very high frequencies, the gain falls off and the phase angle becomes negative.

A second kind of nonideality arises with three-mode controllers. It was mentioned in Section VI.A of Chapter 6 that real controllers are usually constructed so that the derivative and integral actions interact. This means, for example, that when the nominal value of the derivative setting is changed, both the derivative and integral times are effectively changed. A comprehensive discussion of the nonideal behavior of controllers can be found in Reference 8ii.

In the discussion of system design that follows, ideal controllers will be assumed and effective values of reset and derivative times will be used. This is not a serious limitation, since the deviations of the frequency response of real controllers usually occur at very high and very low frequencies beyond the range of interest in most Bode analyses. Furthermore, with respect to the interactions of dial settings, these relationships would be known so that the effective values of reset and derivative time could be calculated.

V. FREQUENCY RESPONSE IN SYSTEM DESIGN

As illustrated in Section III.A of this chapter, if the maximum gain under proportional action is to be calculated, then only the frequency response of the process elements need be considered. Usually, however, one is concerned with the effects of integral and derivative actions on the controllability of a process. Therefore, it is necessary to consider the frequency responses of both the process and the controller.

A. An Approximate Index of Controllability

In designing a complete control system, a set of optimum controller settings might be established on the basis of a study of the transient response of the system to various load disturbances or to a change in set point. The criterion function for the optimization could be the integral of the absolute value of the error or the integral of the square of the error, to name just two possibilities. Such studies are ordinarily not made and are rarely justifiable.

Harriott (8iii) suggests that an index for assessing various controller actions and settings is the product of the maximum overall gain and the critical frequency. Basically, the idea behind this is that the magnitude of the error tends to decrease as the overall gain increases, providing, of course, that the gain does not exceed the stability limit of the system. The critical frequency is important since, in many cases when a system is tuned, it displays slightly oscillatory behavior at somewhat less than the critical frequency, and the magnitude of the error becomes negligible after a few cycles. This can be seen in Figures 12 and 14 of Chapter 13. Therefore, in comparing two systems, the one having the higher critical frequency will tend to reduce the error to a negligible value faster than the one with the lower critical frequency. Harriott concludes that the integral of the absolute error is roughly inversely proportional to the index.

Although this method is easily applied, it is relatively approximate and can lead to serious errors. In particular, it should not be used if the dynamics of the measuring element is significant in either of the two systems being compared. Also, the method overemphasizes the significance of the controller gain when disturbances mainly occur near the end of the series of dynamic elements characterizing the process.

B. Gain Margin and Phase Margin

If the Bode diagram for a process can be drawn, then there are two commonly used methods for selecting the controller gain. One method makes the choice on the basis of a "gain margin" and the other on the basis of a "phase margin." Each of these methods usually results in a value of gain that is

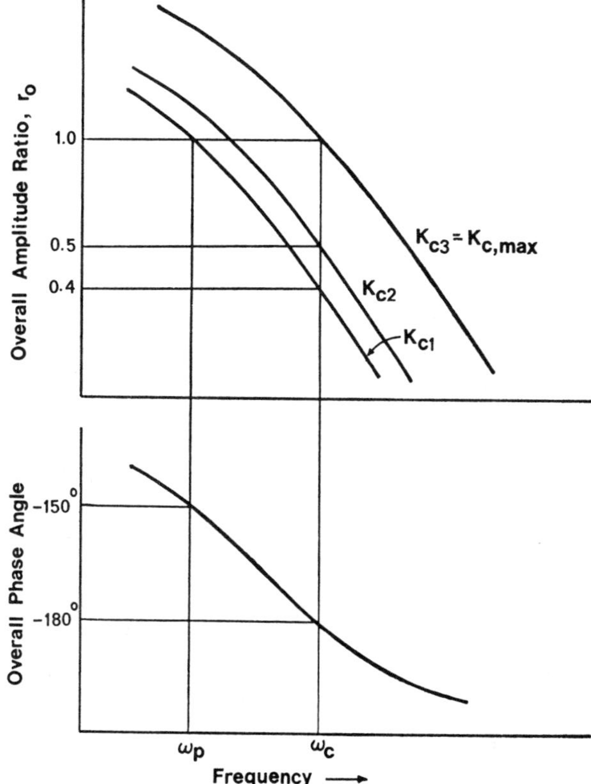

Figure 24.   Bode diagrams to illustrate gain margin and phase margin.

sufficiently less than the maximum gain to insure some margin of safety.

To understand these two methods, refer to the sketch of a Bode diagram shown in Figure 24. The amplitude ratio here is the overall ratio for the controller plus the process, which is assumed here to consist of two elements. Thus:

$$r_o = r_c r_1 r_2 = K_c f_c K_1 f_1 K_2 f_2$$

Three curves are shown for the amplitude ratio plot, one for each of three values of the controller gain. The critical frequency has been located in accordance with the Bode criterion. Thus, if the controller gain is set at K_{c3}, the overall gain will be 1.0 and the system will be on the verge of instability, oscillating at ω_c. If the gain is set at K_{c2}, the overall gain will be 0.5 at ω_c, which is one half of the maximum gain; in other words, the gain margin is $1/0.5$ or 2.0. If, instead, the gain is K_{c1}, the gain margin will be $1/0.4$ or 2.5.

The other method for selecting the controller gain is based on the concept of a phase margin. The idea here is to define a safety margin for stability in terms of the difference between 180° and the phase angle at which the overall gain is 1.0. Suppose, for example, that a phase margin of 30° is selected. Then the frequency, ω_p, is located where the phase angle is 180° − 30°, or 150°. Reading upward at this frequency, it can be seen that the amplitude curve for a gain of K_{C1} intersects the horizontal line at an overall gain of 1.0. Therefore, a controller gain of K_{C1} is said to provide a phase margin of 30°.

The gain constants such as K_C, K_1, and K_2 have been omitted from the Bode diagrams drawn in this chapter. These gains, of course, only shift the amplitude curves up or down on the diagram, so it is possible to use these methods of gain and phase margin without redrawing the curves to include them. To illustrate this, let us reconsider process 2, which was discussed in Section II.B.1.b of Chapter 13. This second-order process had time constants of 5.0 and 10.0. The Bode diagram for this process is shown in Figure 25. The diagram for a proportional-integral controller is also shown, for which a reset time of 2.0 has been arbitrarily selected. The process gain, K_P, and the controller gain, K_C, have been omitted from the amplitude curve.

Suppose we wish to find the controller gain for a gain margin of 2.0. Reading across the phase-angle line at a value of $-180°$, the critical frequency is found to be 0.23; then reading directly up to the amplitude curves, the factors are $f_P = 0.272$, $f_C = 2.45$, and their product for the process plus controller is 0.666. At the maximum controller gain, the overall gain is given by

$$r_o = 1 = K_{C,\mathrm{max}} K_P f_P f_C$$

Substituting in the numerical values, we find

$$K_{C,\mathrm{max}} = \frac{1}{K_P(0.272)(2.45)} = \frac{1.5}{K_P}$$

Therefore, for a gain margin of 2.0, the controller gain should be set at one-half of the maximum, which is $0.75/K_P$.

Figure 25 can also be used to illustrate the use of phase margin in setting the controller gain. Let us assume that a phase margin of 30° is desired. Therefore, a line is drawn horizontally across at a phase angle of $-150°$. It just happens that the phase lag of the process plus controller attains this value at the lowest frequency in the diagram, which is 0.1 rad/unit time. Reading upward, we see that the product, $f_P f_C$, has a value of 3.22 at this frequency. According to the phase-margin criterion, the controller gain is set such that

$$K_C K_P f_C f_P = 1.0$$

$$K_C = \frac{1}{K_P(3.22)} = \frac{0.31}{K_P}$$

Now compare the value found by this criterion with that obtained

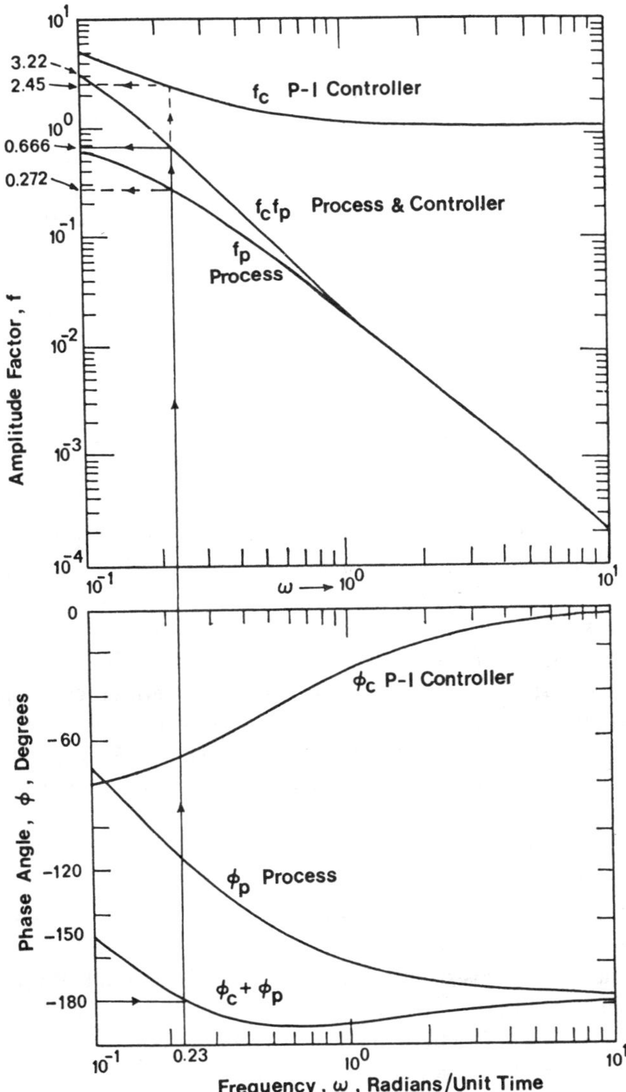

Figure 25. Bode diagrams for process 2 and proportional-integral controller illustrating gain and phase margins. $T_1 = 5.0$; $T_2 = 10.0$; $T_R = 2.0$

from the gain-margin criterion. The phase-margin method results in a controller gain that is about 40% of that obtained using the gain-margin criterion, and therefore it is substantially more conservative *in this example*. It is conceivable that the two methods could result in the same value of controller gain, although this would certainly be a coincidence. Depending upon the shape of the amplitude and phase curves, it could happen that a phase margin of 30° might result in only a very small gain margin, or a gain margin of 2.0 might result in a very small phase margin. Therefore, it is common to specify both the gain and phase margins, and the controller gain is set to meet the more conservative of the two. A phase margin of 30° is a fairly common specification and a gain margin of 1.7 is often used (4iv).

There are two reasons for using margins as large as these. If the gain or phase margin is too small, then the transient responses to various disturbances are likely to be relatively oscillatory and a long time will be required for these to damp out. Furthermore, shifts in operating conditions may increase the process gain so that the gain margin is less than unity, and therefore the system will be unstable.

As an extension of this illustration, let us consider process 2 again, but with a reset time of 6.43, which was the value used in Chapter 13 when this process was under proportional-integral control. It will be recalled that this value was obtained by the reaction curve method.

Bode diagrams for the process, controller, and the process plus the controller are carried in Figure 26. Unlike the previous case, there is no crossover of the 180° line. The reason for this can be seen by comparing Figures 25 and 26. In Figure 26, where the reset time is 6.43, the reset action contributes a relatively large phase lag at low frequencies where the process contributes relatively little lag; then at high frequencies, the controller lag is small even though the process lag is large. The result is that the phase trends of the process and controller are essentially compensatory. However, in Figure 25 for a reset time of 2.0, the compensation is less effective, and the total phase lag crosses over the 180° line.

Since there is no crossover in Figure 26, there is no value of the controller gain for which the system will be unstable, and therefore the maximum gain is infinite. The reaction curve method gave a suggested gain of 9.33 for proportional-integral control so the gain margin is infinite, as it is for all finite values of the gain. However, it can be shown that a gain of 9.33 corresponds to a phase margin of 18.6°; this is the subject of Problem 8 at the end of this chapter.

C. Effects of Integral and Derivative Action

In many systems offset is undesirable, and therefore integral action will be used. The two illustrations in the previous section have demonstrated that

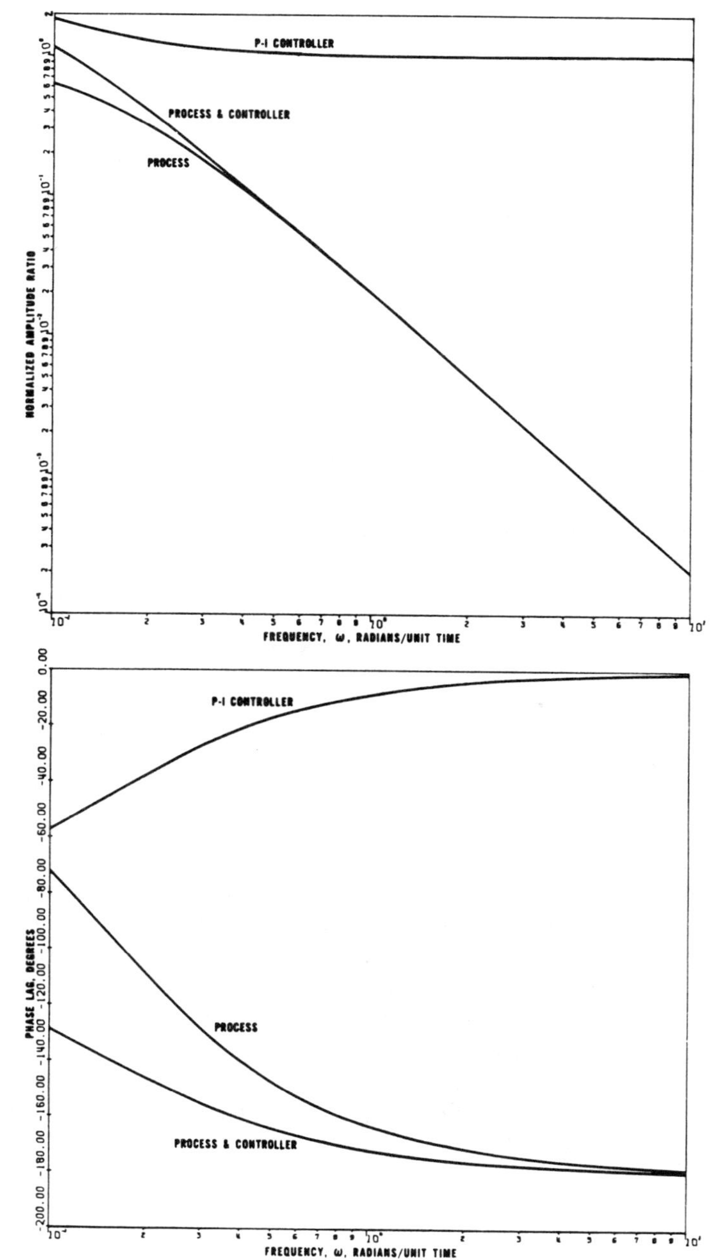

Figure 26. Bode diagrams for process 2 and proportional-integral controller without crossover. $T_1 = 5.0$; $T_2 = 10.0$; $T_R = 6.43$

413

the phase lag from integral action may not have very much effect upon the system if the reset time is large, as it was in Figure 26. However, a large reset time may mean that a relatively long time will be required for the offset to disappear. If faster removal of offset is desired, then a shorter reset time will be necessary, with more tendency for oscillatory behavior; this means that a lower proportional gain will probably be necessary.

Since, as shown in Figure 20, derivative action contributes phase lead, this suggests that the addition of this action to the proportional-integral action will compensate for some of the phase lag of the integral action. Then either a lower reset time or a higher controller gain can be used. To illustrate this, let us consider process 3 of Section II.B.1.c of Chapter 13. This third-order process was described by the following transfer function:

$$G_P = \frac{K_P}{(10s + 1)(5s + 1)(s + 1)}$$

The Bode diagram for the process is shown in Figure 27. The process gain, K_P, has been excluded from the amplitude ratio. At the highest frequency, the phase angle is close to the asymptotic value of $-270°$.

Now consider the effect of using a proportional-integral controller with a reset time of 5.0. The transfer function of the controller is

$$G_C = \frac{K_C(5s + 1)}{5s}$$

Therefore, for the combination of the process plus the controller, the resulting transfer function is

$$G = G_C G_P = \frac{K_C K_P}{5s(10s + 1)(s + 1)}$$

The reset time was chosen so as to cancel the factors, $5s + 1$, in the numerator and denominator. Actually though, the choice is not an unreasonable one, since the reaction curve method gave a reset time of 9.53 for a proportional-integral controller. The resulting Bode diagram for the combination of the process and the proportional-integral controller is also shown in Figure 27 with the gains, K_C and K_P, excluded. The integral action adds significant phase lag to the process, causing the crossover to shift from a frequency of about 0.57 to 0.32. At the same time, the amplitude ratio is increased from about 0.051 to 0.18. Thus, with proportional-integral action, the maximum loop gain is 1/0.18 or 5.4 and for a gain margin of 2, the overall gain would be set at 2.7.

When derivative action is added to the previous proportional-integral controller, the effect is quite significant. The Bode diagram for the process

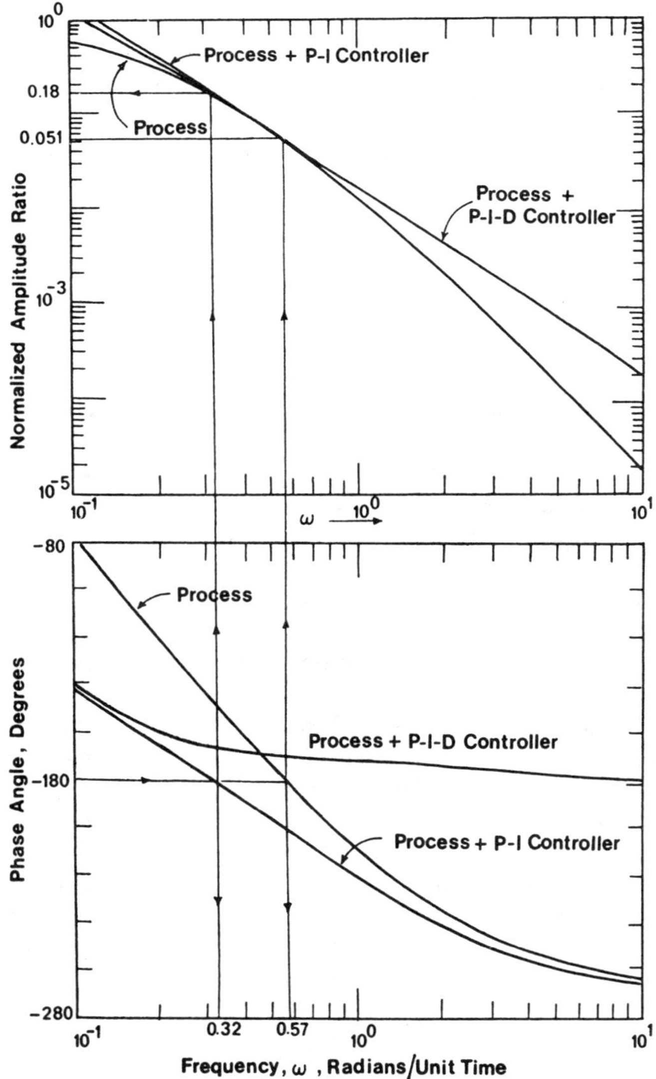

Figure 27. Bode diagram for process 3 and process 3 plus P-I or P-I-D controllers.

plus a three-mode controller is also shown in Figure 27. The derivative time is 1.0 and the reset time, 5.0. The reset time has been kept the same as in the proportional-integral case to facilitate comparisons. Furthermore, the value of 5.0 is close to the setting of 5.72 obtained in Chapter 13 for this process under three-mode control. The derivative time of 1.0 is also realistic since the reaction curve method gave a setting of 1.43 for three-mode control. The phase lead of the derivative action has the effect of making the phase curve asymptotic to the 180° line. Therefore, the maximum gain is theoretically infinite and the gain margin is infinite for all finite gains. On the other hand, if the criterion of a 30° phase margin is employed, the resulting overall gain is about 1.3.

These illustrations are, of course, oversimplified. Only two or three time constants have been taken into account. Real processes always contain some delays and more elements than those used here. This results in some finite maximum gain, although it may be quite high. The illustrations do, nonetheless, illustrate the effects of the controller actions.

These examples have illustrated the effects of integral and derivative actions in controlling particular kinds of processes. One was characterized by two time constants, and the other by three. In particular, we saw that the third-order process greatly benefited by the addition of derivative action. Let us now consider the addition of these actions to a system in a somewhat broader sense.

Integral action is normally used to avoid offset whereas derivative action is usually employed to improve transient response and perhaps to allow the usage of a smaller reset time. The Bode diagram for the open-loop system can be very helpful in indicating whether a relatively large or small reset time should be used, and also in revealing what kinds of systems will benefit from derivative action. Basically, the answers to these questions lie in the slopes of the phase and amplitude curves in the region of crossover.

Consider first the addition of reset action to proportional. As we have seen, this action adds phase lag. Therefore, if the slope of the phase curve of the process is quite flat in the region of crossover, the addition of reset action will shift the crossover point to the left substantially. This will usually reduce the maximum loop gain, since the gain curve rises as the frequency decreases. In this case, to avoid too large a shift in frequency and the corresponding decrease in maximum gain, a relatively large reset time should be used. On the other hand, if the slope of the process phase curve is steep at crossover, a small reset time can be employed.

Some compromise will usually be involved. In a general way, as the reset time is reduced, the time required for the removal of offset is also reduced. However, the corresponding reduction in the critical frequency and in the maximum gain of the system means that the response will tend to be more sluggish.

Turning now to derivative action, if the slope of the phase curve of the process is relatively flat at crossover, then the phase lead contributed by the controller will shift the crossover frequency to the right substantially. If at the same time the amplitude ratio is falling fairly rapidly, then the new maximum gain will be increased significantly; this, in combination with the higher critical frequency, will result in improved response. This was the behavior displayed by the third-order process discussed above and resulted from the fact that the time constants describing it were of widely differing magnitudes.

As with reset action, some compromise is usually required in setting the derivative time. In the first place, the maximum phase lead of a controller with limited action is in the neighborhood of 60°. Secondly as the derivative action is increased, the frequency-dependent gain of the controller action itself increases, as shown in Figure 20. Thus, the falling amplitude ratio of the process itself may be counteracted by the rising ratio of the controller. The overall result then may be a reduction in the maximum system gain, in spite of the increase in crossover frequency.

A number of theoretical studies have been made to aid in the prediction of optimum controller settings for various types of processes. A summary of these can be found in Reference 8iv. One of the parameters in the correlations is the "ultimate period, P_u." This is the period at which the process would oscillate if it were on the verge of instability under proportional control; this period is equal to $2\pi/\omega_c$, where ω_c is the critical frequency. In some of these studies, the optimum reset time has been found to fall in the range between one-third and three times the ultimate period. There have been comparatively few studies regarding optimum derivative times. As we shall see shortly, one rule of thumb is to set the derivative time at one-eighth of the ultimate period.

VI. CONTROLLER TUNING BY THE CONTINUOUS CYCLING METHOD

Normally, the frequency response of a process is not known. It may be possible to estimate some approximate controller settings in advance of installing the controller, but usually some tuning procedures must be carried out once the system is in operation. The reaction curve method for tuning a controller was described in Chapter 13. An alternate method of Ziegler and Nichols is the continuous cycling method (9).

A. Description of the Method

In contrast with the reaction curve method, the continuous cycling method is a closed-loop procedure. Only proportional action is involved. Therefore, if the controller has integral action, the reset time is set at infinity, or if calibrated in terms of reset rate, the rate is set at zero repeats per minute. The derivative time is set at zero if the controller includes this action. The procedure is implemented by first bringing the system to its design condition. Then with the controller in the automatic or closed-loop mode, small changes in set point are made, gradually increasing the proportional gain until the system continuously oscillates. In effect, this directly establishes the crossover frequency of the process and the corresponding maximum controller gain. The Ziegler-Nichols recommended settings are as follows.

Proportional

$$K_C = 0.5K_{C,\text{max}} \tag{48}$$

Proportional-Integral

$$K_C = 0.45K_{C,\text{max}} \tag{49}$$

$$T_R = \frac{P_u}{1.2} \tag{50}$$

Proportional-Integral-Derivative

$$K_C = 0.6K_{C,\text{max}} \tag{51}$$

$$T_R = \frac{P_u}{2.0} \tag{52}$$

$$T_D = \frac{P_u}{8} \tag{53}$$

Note that $K_{C,\text{max}}$ in each of the preceding cases is that found in the proportional-only test.

For proportional action alone, the gain is set at half of the maximum value, so the gain margin is two. With the addition of integral action, the additional phase lag added by the controller reduces the crossover frequency and the maximum gain. To compensate for this reduction in gain, the recommendation is that the controller gain be set 10% below the value for proportional action alone. When derivative action is added, the phase lead contributed by this increases the crossover frequency. Since this usually increases the maximum gain, the recommendation is that the controller gain be increased above the value for proportional action alone.

B. Comparison of Settings by the Reaction Curve and Continuous Cycling Methods

The continuous cycling method has its foundations in frequency response theory, whereas the reaction curve method is based on transient response theory. Although it is theoretically possible to calculate the frequency response of a process from its transient response, the relationship between the two is not simple. Therefore it would not be expected that the two tuning methods would result in identical settings.

To cite an extreme example, consider process 2, which has time constants of 5 and 10. Theoretically, the system cannot be made unstable under proportional action. Therefore, there is no finite maximum gain and the ultimate period is zero, so the continuous cycling method suggests an infinite controller gain and zero reset and derivative times. On the other hand, referring back to Section II.B.1.b of Chapter 13, we find that the reaction curve method gave finite values of gain, reset time, and derivative time.

For process 3, with time constants of 1, 5, and 10, it was shown in Section III.A. of this chapter that the maximum gain is 19.8. The critical frequency is 0.566 and this corresponds to an ultimate period of 11.1 time units. The recommended settings of the continuous cycling method are summarized in the following table, together with those obtained from the reaction curve method. In all cases, the reaction curve method results in a lower recommended gain than the continuous cycling method, but the reset and derivative times by the two methods are essentially the same.

	Recommended Settings	
Controller Action	Continuous-Cycling Method	Reaction-Curve Method
Proportional	$K = 9.9$	$K = 7.08$
Proportional-integral	$K = 8.91$ $T_R = 9.26$	$K = 6.37$ $T_R = 9.53$
Proportional-integral-derivative	$K = 11.9$ $T_R = 5.55$ $T_D = 1.39$	$K = 8.50$ $T_R = 5.72$ $T_D = 1.43$

It was noted in Section III.C. of Chapter 13 that the reaction curve method was developed to result in underdamped response with about a one-quarter decay ratio. This was also the criterion upon which the continuous cycling method was based. Since such "tight" tuning is not always desirable, the continuous cycling method is not universally applicable to all processes.

SUMMARY

This chapter has been a brief introduction to frequency response as it relates to control theory. It is a powerful tool in the analysis of systems. Although the frequency response of a process is seldom known, it may be possible to sketch out a rough approximation of it, and this may be sufficient for the purposes of selecting appropriate controller actions and approximate settings.

The introductory paragraphs of this chapter noted that the frequency response of a closed-loop feedback system can be used to predict its transient response. Some aspects of this are covered in Reference 8v. Frequency response techniques have been widely employed for comparing the dynamics of a process against those of various postulated models. For example, this type of testing has been used for stirred-tank heat exchangers and reactors to show that even though they are basically distributed processes, they can be satisfactorily described by lumped parameter models (10, 11).

Probably one of the most practical applications of frequency response theory arises in the design of dampers to remove pressure and flow pulsations in systems. Some aspects of this subject are covered in Reference 12. The use of frequency response in determining the transmission characteristics of pneumatic transmission lines is treated in Reference 7.

REFERENCES

1. C. R. Webb, *Automatic Control*, McGraw-Hill Book Co., Inc., New York, 1964, (i) pp. 189–190, (ii) pp. 186–188.
2. H. Chestnut and R. W. Mayer, *Servomechanisms and Regulating System Design*, Vol. 1, Second Edition, John Wiley and Sons, Inc., New York, 1959, pp. 160, 323.
3. J-C. Gille, M. J. Pelegrin, and P. Decaulne, *Feedback Control Systems*, McGraw-Hill Book Co., Inc., New York, 1959, pp. 154–157.
4. D. R. Coughanowr and L. B. Koppel, *Process Systems Analysis and Control*, McGraw-Hill Book Co., Inc., New York, 1965, (i) pp. 235–236, 268–276, (ii) pp. 213, 214, (iii) p. 348, (iv) p. 237.
5. J. L. Bower and P. M. Schultheiss, *Introduction to the Design of Servomechanisms*, John Wiley and Sons, Inc., New York, 1958, Ch. 5.
6. Y. Takahashi, M. J. Rabins, and D. M. Auslander, *Control and Dynamic Systems*, Addison-Wesley Publishing Co., Reading, Mass., 1970, (i) pp. 366–367, 383, (ii) pp. 372–382.
7. G. H. Farrington, *Fundamentals of Automatic Control*, John Wiley and Sons, Inc., New York, 1951, pp. 138–145.

8. P. Harriott, *Process Control*, McGraw-Hill Book Co., Inc., New York, 1964, (i) pp. 106–107, (ii) pp. 123–129, (iii) pp. 100–104, (iv) Ch. 9, (v) Ch. 7.
9. J. G. Ziegler and N. B. Nichols, "Optimum Settings for Automatic Controllers," *Trans. ASME*, **64** (1942), 759.
10. T. W. Weber and P. Harriott, "Dynamics of Heat Removal From an Agitated Tank," *Ind. Eng. Chem. Fund.*, **4** (1965), 155.
11. R. J. Fanning and C. M. Sliepcevich, *A.I.Ch.E. Journal*, **5** (1959), 240.
12. P. S. Buckley, *Techniques of Process Control*, John Wiley and Sons, Inc., New York, 1964, Ch. 16.

PROBLEMS

PROBLEMS RELATING TO SECTIONS I TO IV

Problem 1.   In Example 2 in Section IV.B.9 of Chapter 8, it was shown that the time constant and gain factor for a tank vary with operating conditions. Assume that the dynamics of the process can be characterized by this first-order process plus an effective delay of 0.636 min.

(a) Find the critical frequency and maximum value for the product of $K_C K_V$ when the flow rate is at the normal value of 55 ft^3/min.
(b) Suppose that the entering flow rate falls to 35 ft^3/min. Find the new critical frequency and the maximum product.
(c) Repeat the calculations for a flow rate of 75 ft^3/min.
(d) If it is desired to keep the controller gain and loop gain constant over this range of flow rates from 35 to 75 ft^3/min, what should be the effective characteristic of the valve? How do you reconcile the results here with those of part *a* of the example?
(e) Looking back at this example as a whole, why did the results come out the way they did?

$$\begin{aligned}
\textit{Answers:}\ \ &\text{(a)}\ \omega_c = 2.5\ \text{rad/min} \qquad (K_C K_V)_{\max} = 196\ \text{ft}^3/\text{min/ft}\\
&\text{(b)}\ \omega_c = 2.49\ \text{rad/min} \qquad (K_C K_V)_{\max} = 195\ \text{ft}^3/\text{min/ft}\\
&\text{(c)}\ \omega_c = 2.51\ \text{rad/min} \qquad (K_C K_V)_{\max} = 197.5\ \text{ft}^3/\text{min/ft}\\
&\text{(d)}\ \text{Linear}
\end{aligned}$$

Problem 2.   A process has the following transfer function:

$$G_P = \frac{s}{Ts + 1}$$

(a) Draw a vectorial representation in the complex plane similar to those done in Section III of this chapter.

(b) Find an expression for the amplitude ratio and one for the phase angle.
(c) At very low frequencies, what is the behavior of the amplitude ratio and that of the phase angle?
(d) At very high frequencies, what is the behavior of the amplitude ratio and that of the phase angle?

Problem 3.   A pure delay of duration, D, is under pure integral control with integral time, T_i.

(a) Find an expression for the minimum integral time.
(b) If the delay is under proportional-integral control, find an expression for the critical frequency, ω_c.

$$Answers: \text{(a) } T_{i,min} = \frac{2D}{\pi}$$

$$\text{(b) } \tan^{-1}\left(\frac{1}{\omega_c T_R}\right) + \omega_c D = \pi$$

PROBLEMS RELATING TO SECTION V (FREQUENCY RESPONSE IN SYSTEM DESIGN)

Problem 4.   A brine solution is being blended with water in a 1000-gal tank. The entering water rate is 95 gal/min and the brine rate is 5 gal/min. The mixing in the tank is not instantaneous; thus, there is an "effective mixing time" that is characterized by a pure time delay of 15 sec. The concentration of the leaving solution is continuously measured by a device whose effective delay is 10 sec.

(a) Without doing any detailed calculations, estimate the phase lag contributed by the tank at crossover. Based on this estimate, calculate the crossover frequency and ultimate period.
(b) Check the estimate made in part *a* by exact calculations.
(c) What is the maximum loop gain?
(d) If the sampling delay and mixing delay were each reduced by a factor of two, approximately what would be the improvement in the integral of the error?
(e) Generalize your findings—that is, approximately how does the integral of the error vary with the delay?

$$Answers: \text{(a) } \omega_c = 3.77 \text{ rad/min}$$
$$P_u = 1.67 \text{ min}$$
$$\text{(c) } 37.7$$
$$\text{(d) A factor of 4}$$
$$\text{(e) Inversely as the square of the delay}$$

Problem 5. For a process characterized by three first-order elements, the maximum gain and crossover frequency are given by the expressions:

$$K_{max} = (1 + R_2 + R_3)\left(1 + \frac{1}{R_2} + \frac{1}{R_3}\right) - 1$$

$$\omega_c = \frac{1}{T_1}\sqrt{\frac{1 + K_{max}}{R_2 + R_3 + R_2 R_3}}$$

where

$$R_2 = \frac{T_2}{T_1}$$

and

$$R_3 = \frac{T_3}{T_1}$$

Suppose the time constants of a system are 0.1, 2, and 10. The time constants of 2 and 10 might be the effective time constants of the process. For example, in a stirred-tank reactor, the larger of the two would be closely related to the reactor and the smaller might be related to the cooling jacket. The smallest time constant, 0.1, might be associated with the temperature measuring element. Consider the individual effects of 20% changes in each of these time constants upon the performance of the control system. As a basis for comparison, use the index of controllability described in Section V.A of this chapter. In which direction should each of these time constants be changed to improve the performance? As a practical matter, which ones might really be altered at the discretion of the control engineer?

Problem 6. An exothermic reaction is carried out in a continuous-flow, stirred-tank reactor whose temperature is controlled by manipulating the flow of cooling water through a coil. The temperature of the tank contents is measured by a bulb in a well. The parameters and operating conditions are as follows:

Coil holdup	400 lb water
Tank holdup	3000 lb
Heat capacity of tank fluid	1 Btu/lb, °F
Area for heat transfer	50 ft^2
Overall coefficient of heat transfer	100 Btu/hr, ft^2, °F
Tank feed rate	5000 lb/hr
Coil feed rate	15,000 lb/hr
Bulb-well effective time constants	30 and 5 sec
Control valve time constant	4 sec

The heat of reaction is essentially independent of temperature.

(a) Assume that the transfer function given in Equation 21 of Chapter 12 applies. What is the maximum overall gain and ultimate period?

(b) Assess each of the following suggestions for improving the controllability of this process using the index of controllability described in Section V.A. Therefore in each case it will be necessary to find the maximum gain and the crossover frequency.

(1) Use a cooling coil having half the cross-sectional area for flow of the previous one, and one-fourth the area for heat transfer. Assume that the controlling resistance to heat transfer is the tank-side film resistance so that the overall coefficient is essentially unchanged.

(2) Remove the bulb from the well and protect it with a corrosion-resistant plastic coating. The bulb will then behave as a first-order element with a time constant of 10 sec.

(3) Increase the speed of the agitator so as to double the overall coefficient of heat transfer between the tank contents and the coolant. The outside film coefficient between the tank contents and the wall of the well is also doubled, resulting in new effective time constants for the well-bulb combination of 15 and 5 sec.

(c) If you were specifying the controller, would you recommend incorporating reset and/or derivative action?

$$Answer: \text{ (a) } K_{o,\max} = 48.3$$
$$P_u = 264 \text{ sec}$$

Problem 7.   For a process consisting of three first-order elements having time constants of 1, 5, and 10, the reaction curve method gives an overall gain value of 7.08 for a proportional-only system.

(a) What phase margin does this correspond to?
(b) What gain margin does this correspond to?

$$Answers: \text{ (a) } 29.4°$$
$$\text{(b) } 2.8$$

Problem 8.   In Section II.B.1.b of Chapter 13, the transient response of process 2 was illustrated for a proportional-integral controller with a gain of 9.33. This gain was obtained using the reaction curve method. Show that this gain results in a phase margin of 18.6°.

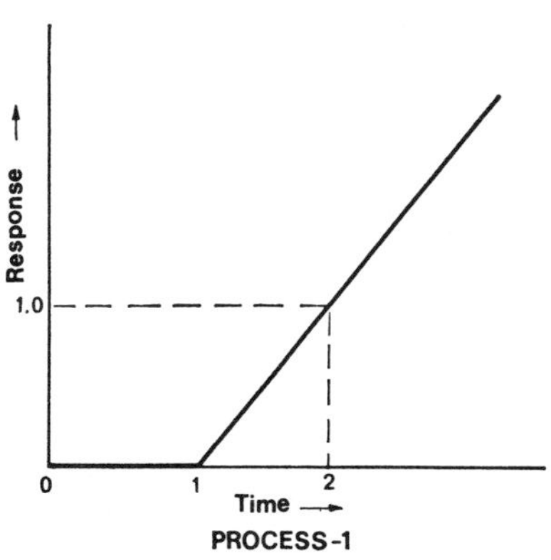

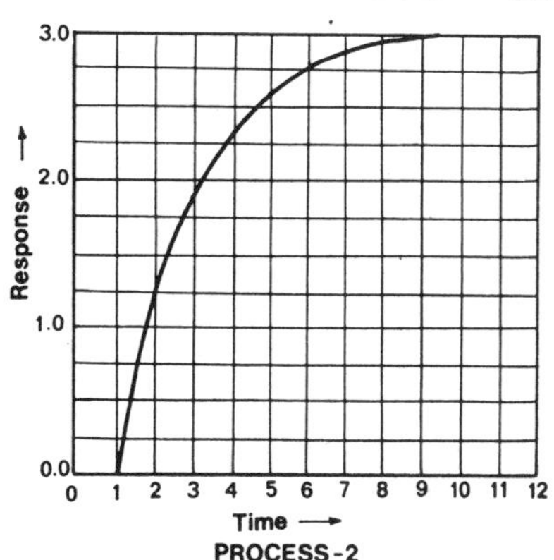

Figure 28.   Process reaction curves for Problem 9.

Problem 9.   Process reaction curves are shown in Figure 28 for two processes. The purpose of this problem is to determine which one of these two processes will benefit most from the addition of derivative action to proportional action, based on the index of controllability discussed in Section V.A of this chapter. As a basis for calculations, assume that the derivative action will be adjusted so as to contribute 30° of phase lead at crossover.

Problem 10.   Consider a process consisting solely of a pure delay under proportional control. Using frequency response considerations, would the addition of derivative action appear to be beneficial? As a means of reaching a conclusion, carry out the following calculations.

(a)   Find the maximum gain and crossover frequency under proportional action alone. Then assume that when the derivative action is added to the proportional action the phase lead contributed by the derivative action at the new crossover frequency is 30°. Compare the proportional-only and proportional-derivative cases using the index of controllability given in Section V.A.

(b)   Repeat part *a* but use the derivative time given by Equation 53 of this chapter. What is the resulting phase lead contributed by the derivative action at crossover now?

(c)   What is your conclusion regarding the value to be gained by adding derivative action to proportional in controlling processes having large pure delays?

Problem 11.   Frequency response data are presented below for two processes:

	Process 1		Process 2	
Frequency, *rad/unit time*	*Normalized* *Amplitude Ratio*	*Phase Lag,* *degrees*	*Normalized* *Amplitude Ratio*	*Phase Lag,* *degrees*
0.01	1.2	30	1.0	8
0.02	1.28	62	1.05	20
0.03	1.30	95	1.1	32
0.04	1.25	140	1.08	40
0.06	1.1	218	1.02	63
0.08	0.82	293	0.94	85
0.1	0.56	360	0.82	106
0.15	0.15	480	0.53	146
0.2	0.02	600	0.35	180
0.5			0.041	252
0.8			0.01	275

(a) Find the ultimate period and maximum gain for each process.

(b) For each process, if derivative action is added to proportional, what derivative time will result in 30° of lead from the derivative action at the new crossover frequency? What are the resulting crossover frequency and maximum gain?

(c) Repeat part *b* but with the derivative action contributing 60° of lead.

(d) Using the index of controllability given in Section V.A, evaluate the results of the parts *a* to *c*. What conclusions can be drawn regarding the addition of derivative action to these two processes? What general conclusions can be drawn about the suitability of derivative action to processes?

PROBLEMS RELATING TO SECTION VI (CONTROLLER TUNING BY THE CONTINUOUS CYCLING METHOD)

Problem 12.   A temperature control system has process time constants of 20 min and 5 min. The control valve and the thermometer bulb both have time constants of 10 sec. A 1-psi change in controller output changes the controlled flow 25 gpm from the normal value of 200 gpm. The process temperature is 175°F for 200 gpm and 174°F for 210 gpm. The temperature transmitter has a span of 80°F.

(a) If proportional action is to be used, what should the proportional band be set at according to the continuous cycling method of tuning? What sensitivity in psi/inch does this correspond to?

(b) Approximate the system by three time constants and use the equation given in Problem 5 to find the maximum gain. Compare your result with that found in part *a*. Does the approximation check well? If so, give an explanation of why it did.

(c) Do you think that the control of this process would benefit from derivative action? Explain briefly in a few sentences.

Answers: (a) 0.92%
330 psi/in.

Problem 13. One of the simplest two-parameter characterizations of a process is that of a first-order element plus a time delay. Consider the following three cases.

(a) The time delay is much much greater than the time constant.
(b) The time delay is exactly equal to the time constant.
(c) The time delay is much much less than the time constant.

For each of these, find the recommended settings for proportional control and for proportional-integral control, first using the reaction curve method and then using the continuous cycling method. Tabulate the results and summarize your conclusions.

Problem 14. One of the simplest two-parameter characterizations of a process is that of a first-order element plus a time delay. Suppose that for a particular process, the time constant, T, is much much larger than the time delay, D (5000 times larger, for example).

(a) Sketch the process reaction curve and find the recommended settings of the reaction curve method, first for proportional, and then for proportional-integral control. Express your findings in terms of T and D. Is the recommended gain close to unity?

(b) Make a sketch of the Bode diagram for this process, showing both the amplitude and phase curves. Of particular interest are the shapes of these curves in the vicinity of the crossover frequency.

(c) Find the recommended settings of the continuous cycling method, first for proportional, and then for proportional-integral control. Make sensible approximations in this calculation.

(d) Tabulate your findings from parts *a* and *c*. Which method is more conservative?

(e) Would you expect derivative action to be beneficial in controlling this process? Explain.

Answers:

(a) Recommended settings are given in the answer for part *d*. The recommended gain is not close to unity.

(d)

Method	Proportional Recommended K	Proportional-Integral Recommended K	T_R
Reaction curve	T/D	$0.9T/D$	$D/0.3$
Continuous cycling	$0.785T/D$	$0.707T/D$	$D/0.3$

Subject Index